AF878492

Springer Tracts in Civil Engineering

Springer Tracts in Civil Engineering (STCE) publishes the latest developments in Civil Engineering—quickly, informally and in top quality. The series scope includes monographs, professional books, graduate textbooks and edited volumes, as well as outstanding Ph.D. theses. Its goal is to cover all the main branches of civil engineering, both theoretical and applied, including:

Construction and Structural Mechanics
Building Materials
Concrete, Steel and Timber Structures
Geotechnical Engineering
Earthquake Engineering
Coastal Engineering
Hydraulics, Hydrology and Water Resources Engineering
Environmental Engineering and Sustainability
Structural Health and Monitoring
Surveying and Geographical Information Systems
Heating, Ventilation and Air Conditioning (HVAC)
Transportation and Traffic
Risk Analysis
Safety and Security
To submit a proposal or request further information, please contact: Pierpaolo Riva at Pierpaolo.Riva@springer.com, or Li Shen at Li.Shen@springer.com

More information about this series at http://www.springer.com/series/15088

Masa Noguchi
Editor

ZEMCH: Toward the Delivery of Zero Energy Mass Custom Homes

Springer

Editor
Masa Noguchi
Melbourne School of Design
The University of Melbourne
Melbourne
Australia

ISSN 2366-259X ISSN 2366-2603 (electronic)
Springer Tracts in Civil Engineering
ISBN 978-3-319-31965-0 ISBN 978-3-319-31967-4 (eBook)
DOI 10.1007/978-3-319-31967-4

Library of Congress Control Number: 2016940823

© Springer International Publishing Switzerland 2016
This work is subject to copyright. All rights are reserved by the Publisher, whether the whole or part of the material is concerned, specifically the rights of translation, reprinting, reuse of illustrations, recitation, broadcasting, reproduction on microfilms or in any other physical way, and transmission or information storage and retrieval, electronic adaptation, computer software, or by similar or dissimilar methodology now known or hereafter developed.
The use of general descriptive names, registered names, trademarks, service marks, etc. in this publication does not imply, even in the absence of a specific statement, that such names are exempt from the relevant protective laws and regulations and therefore free for general use.
The publisher, the authors and the editors are safe to assume that the advice and information in this book are believed to be true and accurate at the date of publication. Neither the publisher nor the authors or the editors give a warranty, express or implied, with respect to the material contained herein or for any errors or omissions that may have been made.

Printed on acid-free paper

This Springer imprint is published by Springer Nature
The registered company is Springer International Publishing AG Switzerland

Preface

Housing needs and demands are dynamic having been changed in the course of time. The shifts in socio-demographics highlight the emergence of non-traditional households and influence housing configurations and performances. Homes need to be *customisable* in order to raise the level of social sustainability in accommodating users' individual requirements, desires and expectations. In the light of fuel poverty issues arising in various countries in which the drastic hike of energy cost is a serious concern, the notion of housing affordability has been extended to encompass both initial and operating costs. In view of economic sustainability at macro and micro levels, homes need to be *affordable*. Global warming accelerated with excessive carbon dioxide (CO_2) emission is becoming conspicuous. Generally, a house consumes a significant amount of energy before and after occupancy, and the associated CO_2 emission is contributing partially to the climate change. Towards securing environmental sustainability, housing needs to be net carbon neutral (or *zero energy*) in consideration of CO_2 emission derived from the overall energy use.

In response to market needs and demands for social, economic and environmental sustainability of housing in developed and developing countries, the *zero energy mass custom home* (ZEMCH) integrated lean design and construction concept was envisaged and discussed globally. Towards the ZEMCH delivery, an emerging notion of *mass customisation* was scrutinised initially. It emerged in the same year as the general concept of *sustainable development* was widespread in 1987. The oxymoron was recognised eventually as a means to lessen housing design and construction costs whilst achieving the customisability through economies of scope rather than economies of scale.

In order to crystallise a wide spectrum of hopes and fears around the design, production and marketing approaches to the ZEMCH delivery in global contexts, *ZEMCH Network* was established in 2010. Today, the R&D collaboration network consists of over 500 partners from nearly 40 countries and the enrolment is constantly on the rise. Originally, the ZEMCH Network was formed by a group of academics, who participated in the 2010 industry-academia knowledge transfer technical visits to production and sales facilities of low to zero energy mass customised housing manufacturers in Japan. The technical tour, later called *ZEMCH Mission to Japan*, dates back to the 2006 operation and it celebrates the 10th anniversary in 2016.

This book is the collective knowledge gained through global ZEMCH R&D activities, consisting of 12 chapters that packaged essential engineering design, construction, and commercialisation techniques and strategies applicable to the ZEMCH delivery: Chapter 1 identifies the general notion of sustainable development as a start to consider how ZEMCH practice can meet global housing market needs and demands; Chap. 2 revisits mass housing developments with the aim to articulate the necessity for enhancement of production efficiency without sacrificing design customisability; Chap. 3 introduces prefabrication seen as a means to standardise housing products and processes for construction efficiency; Chap. 4 summarises the origin of mass customisation and its application to the delivery of quality affordable homes; Chap. 5 crystallises a new notion of mass personalisation applicable to affordable housing developments; Chap. 6 summarises inclusive design techniques that aim to accommodate housing users' changing needs and demands over their lifetime; Chap. 7 clarifies how energy is used in housing before and after occupancy; Chap. 8 recaps various passive design approaches being applied to lessening energy demands of housing; Chap. 9 showcases a variety of active systems available for supplementing energy use in housing with mechanical innovations; Chap. 10 introduces the global movement and practice of zero energy homes; Chap. 11 demonstrates the significance of building energy and environmental performance simulation; and Chap. 12 unveils Japan's successful business operation essential for the ZEMCH delivery.

ZEMCH movement came to life and emerged from the grass roots needs and demands of housing in developed and developing countries. First, I would like to express my sincere gratitude to all authors, who contributed to the development of this book, which is indeed the first of this kind. Their expertise and experience streamlined ZEMCH technical knowledge and made the multidisciplinary contexts comprehensive to a wide range of audience. Second, I am thankful to all chapter leaders of this book, Sara Wilkinson, Kheira Anissa Tabet Aoul, Arman Hashemi, Victor Bunster, Karim Hadjri, Haşim Altan, Jun-Tae Kim and Laura Aelenei, for not only their first authorship, but also their dedication to the team management and leadership that helped secure the quality outcomes.

Lastly, I would like to thank all ZEMCH Network's global partners and regional experts centre directors for their generous, constructive support and cooperation. So far, ZEMCH Network established four regional centres in Brazil, Italy, UAE and UK, being responsible for the organisation and operation of annual industry-academia knowledge transfer events—i.e. ZEMCH Mission to Japan, ZEMCH International Conference and ZEMCH Design Workshop. The success is rooted in the network partners' individual efforts and collaborative actions.

ZEMCH R&D activities have been thriven for continuous improvement of the built environments in developed and developing countries, budding out the global movement for people and society.

Melbourne, Australia Masa Noguchi

Contents

Chapter 1
Sustainable Development

Sara Wilkinson, Mona Hajibandeh and Hilde Remoy

Abstract The continuous increase in the number and size of urban regions across the world pose great challenges for sustainable development. Given the connection between energy use, greenhouse gas emissions and climate change and the reality that the built environment emits around half of total emissions, the construction industry has considerable potential to reduce emissions and a key role in mitigating global warming. Other local challenges include for example loss of species and habitats, social degradation of neighborhoods, and an overall erosion of sustainability. Urbanisation patterns and the lifestyles of urban dwellers also affect the planet on wider scales and contribute to shaping bio-physical processes on planetary scales and affect how humans mentally connect with the Biosphere. However there is evidence our current understanding of the concept of sustainability, and thus sustainable development, is fragmented and unclear. There are a plethora of terms used to cover sustainable buildings, such as ecological, green, Gaian, zero energy, eco-friendly and environmental; all of which come in, and out, of fashion over time; *do they mean the same thing or are they different?* Furthermore, do the stakeholders within the built environment demonstrate a clear understanding of the concept of sustainability or; are they muddled and confused? The consequence of unclear thinking and a lack of understanding is that ultimately the built environment stakeholders are unlikely to deliver 'sustainability' efficiently or even at all, with the broader and more onerous consequences for society as a whole. In addition what are

S. Wilkinson (✉)
The University of Technology Sydney, Sydney, Australia
e-mail: Sara.Wilkinson@uts.edu.au

M. Hajibandeh
Erasmus University Rotterdam, Rotterdam, The Netherlands
e-mail: Monahaji@gmail.com

M. Hajibandeh
Tehran University of Art, Tehran, Iran

H. Remoy
Delft University of Technology, Delft, The Netherlands
e-mail: h.t.remoy@tudelft.nl

© Springer International Publishing Switzerland 2016

M. Noguchi (ed.), *ZEMCH: Toward the Delivery of Zero Energy Mass Custom Homes*, Springer Tracts in Civil Engineering,
DOI 10.1007/978-3-319-31967-4_1

the implications for education and should academics be broadening the debate? This lecture examines the environmental, economic, social, political and philosophical thinking underpinning the concept of sustainable development and shows how different perspectives reflect very different ways of thinking about sustainability and sustainable development. It aims to create a better understanding while offering creative solutions.

1.1 Introduction

The continuous increase in population and, in the number and size of urban regions across the world, pose great challenges for sustainable development especially in the built environment. Given the connection between energy use, greenhouse gas emissions and climate change and the reality that the built environment emits around half of total emissions, the construction industry has considerable potential to reduce emissions and a key role in mitigating global warming (UNEP 2007). Other local challenges include for example loss of species and habitats, social degradation of neighbourhoods, and an overall erosion of sustainability. At the time of writing, 'sustainability' and 'sustainable development' is the preferred term and typically embraces economic, environmental and social considerations identified by Elkington (1997); though the term sustainability was first defined a decade earlier in the Brundtland Report (WCED 1987).

Urbanisation patterns, as well as the lifestyles of urban dwellers, affect the planet on wider scales and contribute to shaping bio-physical processes on planetary scales and affect how humans mentally connect with the Biosphere. However there is evidence our current understanding of the concept of sustainability, and thus sustainable development, is fragmented and unclear.

There are a plethora of terms used to cover sustainable buildings, such as ecological, green, Gaian, zero energy, eco-friendly and environmental; all of which come in, and out, of fashion over time; *do they mean the same thing or are they different?* Furthermore, do the stakeholders within the built environment demonstrate a clear understanding of the concept of sustainability or; are they muddled and confused? The consequence of unclear thinking and a lack of understanding is that ultimately the built environment stakeholders are unlikely to deliver 'sustainability' efficiently or even at all, with the broader and more onerous consequences for society as a whole. In addition what are the implications for education and should academics be broadening the debate?

This chapter examines the environmental, economic, social, political and philosophical thinking underpinning the concept of sustainable development and shows how different perspectives reflect very different ways of thinking about sustainability and sustainable development. It aims to create a better understanding while offering creative solutions. The chapter starts with a discussion of the history of sustainable development and how it is defined. The Brundtland Report (WCED 1987) is described and gives context to subsequent evolution in thinking about sustainable development as posited by Elkington (1997) in the Triple Bottom Line concept. The chapter moves on to explain the different levels of sustainability and the attributes and perspectives that inform the different viewpoints. A quiz is provided for readers to self-assess their own conceptual understanding of sustainability in Sect. 1.6. Section 1.3 describes sustainable urbanism and notion of the compact city, before the issues of high tech versus low tech approaches are described. The challenges of measuring sustainability form the content of Sect. 1.5 of the chapter as illustrate the risks and benefits of a focus on absolute measurement paradigms in the built environment. Penultimately, there is a discussion that endeavours to summarise what sustainable development in the context of this book. The aim of the chapter is to set the context for the following chapters and to appraise readers of the essentials of, and a framework for, the conceptual understanding of sustainable development and sustainability.

1.2 An Introduction to and History of Sustainable Development

The theoretical framework for sustainable development evolved between 1972 and 1992 through a series of international conferences and initiatives. The United Nations (UN) Conference on the Human Environment, held in Stockholm in 1972, was the first major international gathering to discuss sustainability at the global scale. The conference created considerable momentum, and a series of recommendations led to the establishment of the UN Environment Programme (UNEP) as well as the creation of numerous national environmental protection agencies at the national level. The recommendations from Stockholm were further elaborated in the 1980 World Conservation Strategy which was a collaboration between the International Union for the Conservation of Nature, the World Wildlife Fund (WWF) and UNEP—which aimed to advance sustainable development by identifying priority conservation issues and key policy options (Drexhage and Murphy 2010).

The term, sustainable development, was promoted in Our Common Future, a report published by the World Commission on Environment and Development (WCED) in 1987. It is also known as the Brundtland report, after Gro Harlem Brundtland the chair of the commission. Our Common Future defined sustainable development as: "development which meets the needs of the present without compromising the ability of future generations to meet their own needs" (WCED 1987: 43). This definition introduced two important principles; those of inter-generational and intra-generational equity—that is consideration of the generations to come and also those less fortunate and already here. Acceptance of the report by the United

Nations (UN) General Assembly gave the term political credence; and in 1992 leaders set out the principles of sustainable development at the UN Conference on Environment and Development (UNCED) in Rio de Janeiro, Brazil, which is also known as the Rio Summit or the Earth Summit. The Earth Summit adopted the Rio Declaration on Environment and Development and Agenda21, a global plan of action for sustainable development (Kates et al. 2005).

Three seminal instruments of environmental governance were established at the Rio Summit: the UN Framework Convention on Climate Change (UNFCCC), the Convention on Biological Diversity (CBD), and the non-legally binding Statement of Forest Principles. Following a recommendation in Agenda21, the UN General Assembly officially created the Commission on Sustainable Development (CSD) in 1992.

Since that time a number of important international conferences on sustainable development have been held—including the 1997 Earth Summit+5 in New York and the 2002 World Summit on Sustainable Development (WSSD) in Johannesburg. These meetings were primarily reviews of progress; and reported that a number of positive results had been achieved, but implementation efforts largely had been unsuccessful at the national and international level. The negotiations at the WSSD in 2002 demonstrated a major shift in the perception of sustainable development—away from environmental issues toward social and economic development. This shift, which was driven by the needs of the developing countries and strongly influenced by the Millennium Development Goals (MDGs),[1] is but one example of how sustainable development has been pulled in various directions over its 25-year plus history. Defining and implementing sustainable development has had to deal with the tensions between the so-called three pillars of sustainability (economic, environmental and social), and the prevailing political and economic influences at different points in time (Drexhage and Murphy 2010).

At, and since the Rio Summit, sustainable development has found its most prominent hook, at least in terms of media and political attention, around the issue of climate change. Responses to address climate change, in terms of both mitigation and adaptation, are linked to sustainable development. The Fourth Assessment Report of the Intergovernmental Panel on Climate Change (IPCC 2007; Chap. 12.1.1) pointed out the iterative relationship between climate change and sustainable development, and that the two can be mutually reinforcing.

[1]The UN Millennium Declaration was adopted in 2000 and committed countries to reach eight Millennium Development Goals by 2015. The eight goals included: halving extreme poverty, halting the spread of HIV/AIDS, providing universal primary education, eliminating gender disparity in education, reducing the under-five mortality rate, reducing the maternal mortality rate and achieving universal access to reproductive health, developing a global partnership (to address the needs of the poorest countries, to further an open non-discriminatory trade system, and to deal with developing country debt); and ensuring environmental sustainably (by integrating sustainable development into country policies and programs, reducing biodiversity loss, improving access to safe drinking water and sanitation, and improving the lives of slum dwellers) (UN 2010).

1.3 Definitions of Sustainable Development

Since the Brundtland Commission first defined sustainable development in 1987 (WCED 1987), dozens, if not hundreds, of scholars and practitioners have articulated and promoted their own alternative definition; yet a clear, fixed, and immutable meaning remains elusive. Cook and Golton (1994) and Wilkinson (2012) have claimed it is a contested concept, that it is means all things to all men. That is, depending on ones worldview, various interpretations are possible. Further more these interpretations have varying degrees of action and consequently, sustainability (Wilkinson 2012; Washington 2015).

Despite this ambiguity and openness to interpretation, sustainable development has evolved a core set of guiding principles and values, based on the Brundtland Commission's standard definition to meet the needs, now and in the future, for human, economic, and social development within the restraints of the life support systems of the planet. Further, the connotations of both of the phrase's root words, "sustainable" and "development" are generally quite positive for most people, and their combination imbues this concept with the notion that sustainability is a worthwhile value and goal—a powerful feature in diverse and conflicted social contexts.

Sustainable development requires the participation of diverse stakeholders who hold diverse perspectives, with the ideal of reconciling different and sometimes opposing values and goals toward a new synthesis and subsequent coordination of mutual action to achieve multiple values simultaneously and even synergistically. As real-world experience has shown, however, achieving agreement on sustainability values, goals, and actions is often difficult and painful work, as different stakeholder values are forced to the surface, compared and contrasted, criticised and debated. Sometimes individual stakeholders find the process too difficult or too threatening to their own values and either reject the process entirely to pursue their own narrow goals or critique it ideologically, without engaging in the hard work of negotiation and compromise (Washington 2015). Critique is nonetheless a vital part of the conscious evolution of sustainable development—a concept that, in the end, represents diverse local to global efforts to imagine and enact a positive vision of a world in which basic human needs are met without destroying or irrevocably degrading the natural systems on which we all depend (Kates et al. 2005).

1.4 The Brundtland Report

In 1984, the UN General Assembly created the World Commission on Environment and Development, an independent committee of twenty-two members, headed by Gro Harlem Brundtland, the Prime Minister of Norway. Designed to examine global environment and development to the year 2000 and beyond, the Commission sought to reassess critical problems, to formulate realistic proposals for solving them, and to raise the level of understanding and commitment to the issues of environment and development, on the basis of what was called "Sustainable Development". In 1987 the Commission published the report titled Our Common Future. This report was instrumental in preparing the UN Conference on Environment and Development held in Rio de Janeiro in 1992 (Schubert and Lang 2005).

Our Common Future gave impetus to global, regional and national environmental policies, and took a very technocentric approach rather than presenting a negative report about the destruction of finite natural resources, it offered an agenda advocating economic growth based on policies that do not harm, and they advocated can even enhance, the environment. Ecocentrics of course take a different perspective with regards to economic growth and sustainable development (Wilkinson 2012). The Commission recognised, through the definition adopted, that economy and ecology had a symbiotic relationship, and human progress through development should not bankrupt resources for future generations.

The paradox of the Brundtland Report is that it supports ecological sustainability and capitalist development at the same time. The Brundtland Report stated that "The satisfaction of human needs and aspirations is the major objective of development." (WCED 1987). However the development that economic growth spawns increases the profits of those in control of that development, and the inequitable distribution of wealth increases (Washington 2015). Profit is the mechanism by which economics makes decisions and the satisfaction of human needs and aspirations is not.

Brundtland (WCED 1987) concluded that the cause of environmental destruction is lack of economic growth. It did this because it identified poverty as the only cause of environmental destruction and identified economic growth as the only solution to poverty. The obvious conclusion was that growth in the developing world was necessary to reduce poverty to help the environment. Furthermore all developing countries needed high export growth for rapid development (The Brundtland Report 1996). The conclusion drawn was, to stop environmental destruction, make developing countries increasingly dependent on foreign trade, make them increasingly westernised, and encourage corporate growth (Kates et al. 2005). The report's focus on economic growth obscures the solution to environmental destruction and social injustices. Brundtland should promote sustainable use, and not sustainable development; an activity is sustainable if it can be continued indefinitely. A sustainable use is an activity that will not affect its resource base to the point, where that resource can no longer be used for that activity.

1.5 Elkington and the Triple Bottom Line Concept

John Elkington wrote 'Cannibals With Forks: The Triple Bottom Line of 21st century Business in 1997 nearly two decades ago. In this book he posited that everyone benefits if a paradigm of sustainable business is adopted. Significantly Elkington (1997) broadened the concept from environmental quality to embrace

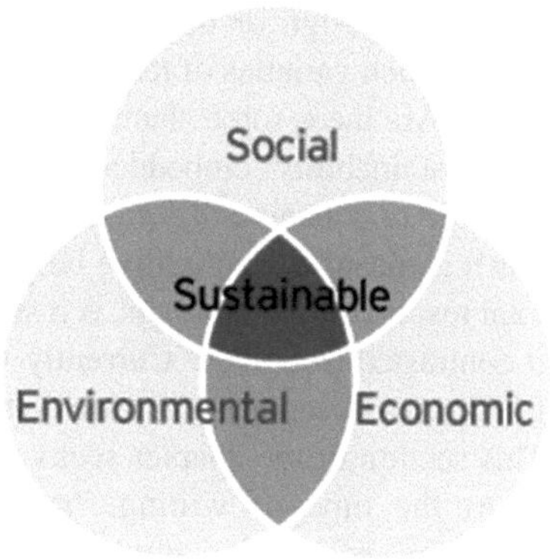

Fig. 1.1 Elkington's triple bottom line

social justice and economic prosperity, the so-called three pillars of sustainability or the triple bottom line (Fig. 1.1). The notion posited by Elkington was that when the three pillars align, true sustainability is achieved.

The three bottom lines are connected, interdependent and partly in conflict. In order to deliver the triple bottom line, businesses are required to think and act in seven dimensions or 7D. These dimensions are markets, values, transparency, life cycle technology, partnerships, time-perspective and corporate governance. Elkington (1997) argues that although capitalism and sustainability do not make easy bedfellows, he believes capitalism can be part of the solution. Evidence to date of effective environmental action does not particularly bear out his belief. This, we shall see in Sect. 2.3, is a technocentric perspective. Elkington (1997) proposes ways in which to think and act in 7D in the book. He sees opportunity for entrepreneurial businesses to get ahead of others by adopting the three pillars. To some extent many of his predictions regarding accountability of performance through corporate social responsibility (CSR) and global reporting of sustainability has come to pass (Wilkinson et al. 2004), though at the same time the environmental footprint of businesses has grown, as has social and economic inequality (Washington 2015). The major contribution of the book was a framework is presented in which to discuss and act on sustainability issues. The message is positive and Elkington (1997) believes humankind has the capacity to avert global disaster, without changing our prevailing economic paradigm, it is a position that not everyone concurs with (Washington 2015).

1.6 The Spectrum of Sustainability

There is a plethora of terms used to encompass the concept of sustainability, especially within the built environment. Green, green, greener, ecological, natural, sustainable, environmentally sensitive, zero energy, Gaian, and environmentally conscious design or building are some of the terms adopted by stakeholders and actors. Such varieties of terms beg the questions; do concepts overlap or are they the same? Are there some shared aspects between concepts and if so, what are they? Is the 'sustainability' embedded within some concepts questionable? Furthermore is it possible to conceive of a sustainable building in an absolute or a relative form? By this it is meant can a building be genuinely sustainable when considering the earth's total resources (absolute) or, is it simply more sustainable than a building to which is it contrasted (relative)? Currently with building rating systems such as BREEAM, LEED and Green Star they are conceived as being sustainable in an absolute sense. This section of the chapter seeks to elucidate some of these questions.

At the time of writing, 'sustainability' is the preferred term and typically embraces economic, environmental and social considerations identified by Elkington (1997); though the term sustainability was first defined a decade earlier in the Brundtland Report (WCED 1987). Additionally the concept is informed by political and philosophical thought, and all aspects were taken into account within a literature review. The literature revealed that there are distinct characteristics and sub groups within the concept of sustainability which needed to be de-constructed and ordered to clarify shared characteristics and those which were separate.

1.6.1 The Concept of Ecocentrism

A key division between the groups is ecocentrism and anthropocentrism (Pepper 1984; Dobson 1990; Brown 1995). In summary, an ecocentric worldview perceives ecosystems as part of an integrated environmental system with organisms, biological communities and ecosystems creating the mantle of life surrounding the planet. Ecocentrism is advocated by an environmental movement known as Deep Ecology (Naess 1990; Brown 1995) and is grounded in seeking the common good of the human and non-human world (Purser and Montuori 1995). Within the ecocentric worldview, three groups exist; transpersonal ecologists, deep ecologists and moderate ecologists. The transpersonal group subscribe to the most extreme views whilst the moderate ecologists hold the least extreme views and ideals within ecocentrism. In some instances views are shared but to differing degrees, whereas on other issues some groups subscribe to a view whilst others do not. The list of key views are summarised for all groups in Table 1.1.

In addition ecocentrics are radically egalitarian where entities such as animals, humans, rivers, seas and lakes are all believed to have equal and intrinsic value. Ecocentrics' argue that only when this worldview is adopted will we substitute environmentally destructive policies for more benign policies. Paradoxically in asking humankind to take responsibility for whole of the ecosphere eccentrics' are expressing anthropocentrism. Furthermore in reality, the egalitarian ecocentric world would collapse into nihilism if no distinctions of value are made where for

Table 1.1 Ecocentric and anthropocentric standpoints (Wilkinson 2012)

Stand-point	Trans-personal ecology	Deep ecology	Moderate ecology	Accommodating environmentalism	Cornucopian environmentalism
Belief system	Religious level of belief	Bio-ethics and intrinsic value	Primary value of ecosystems	Intra and inter-generational equity	Support for traditional ethical reasoning
				Instrumental value in nature	Rights of humans
		Accepts 'carrying capacity' of earth argument	Accepts 'carrying capacity' of earth argument		
	Emotional and irrational			Rational and pro science	
		Lacks faith in technology			Faith in science and technology
Population	Population cull	Reduce population	Zero population growth	Silent	
Resource consumption		Extreme preservationist	Resource preservation	Resource conservationist	Resource exploitative
World view	Ecocentric			Anthropocentric	
	Lacks faith in technology				Faith in technology
Waste	Reuse, repair and then recycle			Recycle	
Economic	Capitalism is not sustainable Rejects consumerism	Heavily regulated economy Capitalism is not sustainable Do not favour overseas trade 'eco'nomics. Rejects consumerism Little overseas trade.	Zero economic growth. Capitalism is not sustainable Do not favour overseas trade 'eco'nomics. Rejects consumerism. Little overseas trade	Managed growth. Capitalism is sustainable. Consumerism is acceptable Overseas trade is acceptable.	Maximise growth Capitalism is sustainable Substitution theory prevails Laissez faire economics Green consumerism is accepted Promotes consumerism Promote foreign trade/agreements
Energy	Preservationist	Preservationist	Conservationist	Conservationist	Nuclear is acceptable, conserve and increase consumption
	Very strong sustainability		Strong sustain-ability	Weak sustainability	Very weak sustainability

example the value of a child in a ghetto tenement is equal to that of a family of rats (Brown 1995).

Taken to the extreme, there is a concern that ecocentrism lends itself to an ideology of domination, where eco police enforce eco policy (Dobson 1990). Whilst some reduction in mankind's interference with the ecosphere is desirable, it is argued that some forms of ecocentrism would lead to the rejection of human rights in favour of the ecosphere, for example propositions of a human population cull advocated by the transpersonal ecology group (Naess 1990). Within social and political systems, ecocentrics tend to dislike centralised systems and materialism (Cook and Golton 1994) and this is a stance, which puts them at odds with current prevailing paradigms.

1.6.2 The Concept of Anthropocentrism and Technocentrism

The dominant world view has been anthropocentric, where mankind is perceived to have a dominant role, only humans possess intrinsic value, and are the rightful masters of 'nature' as well as being the origin and source of all values (Cook and Golton 1994). As such, anthropocentrism is a very different worldview to ecocentrism (Brown 1995). It is contended that to deliver sufficient sustainability to avert overwhelming levels of climate change, it is necessary "to persuade civil society to make a break from the anthropocentric perspective where the environment affects and benefits humans" (Salinger 2010). Within the anthropocentric paradigm resources are extracted without replenishment and non-reusable materials such as plastics and nuclear waste accumulate. Some argue that anthropocentrism is based in the positivist, objective-thinking characteristics in our scientific, mechanistic and technological world view which emerged from the 17th century onwards (Brown 1995). Anthropocentrism is held by ecocentrics to be the root cause of the ecological crisis (Cook and Golton 1994). Anthropocentrics believe that mankind is able to provide a technological fix to the environmental problems and another term for this approach is technocentric (Cook and Golton 1994).

However it is too simplistic to see a clear divide between ecocentrism and anthropocentrism, as in real life the boundaries are blurred and the issues are complex (Pepper 1984). One issue between an ecocentric worldview as opposed to an anthropocentric one is; where does the line between fair use and abuse lie (Purser and Montuori 1996)? Or where does economic development become exploitative? Furthermore authors such as Pearce (1993) and Pepper (1984) perceived further sub groups or categories within ecocentrism and anthropocentrism.

1.6.3 *The Concepts of Accommodating and Cornucopian Environmentalism*

Within anthropocentrism those on the left, known as 'accommodating environmentalists' tend to be gradual reformers believing in careful economic and environmental management but without radical change to social economic and political structures (Cook and Golton 1994). Those on the right are known as 'cornucopian environmentalists' believe in unfettered economic growth and humankind's right to utilise the worlds resources as they see fit. Within the ecocentric camp there is a divide between those on the right; 'deep ecologists', who put a greater emphasis on the limits to growth or carrying capacity of the earth, and those on the left 'moderate ecologists' who believe in decentralised political and social institutions. Deep ecologists believe in compulsory restraints on human population growth and on resource consumption.

Economically anthropocentrics belong to the neo classical school, believing that economic growth is possible, they reject intervention in the economy to tax or incentivise sustainability measures. There is evidence that this stance is beginning to change and evolve in capitalist economies with an increase in the scope of environmental legislation. For example, in 2010 the disclosure of energy consumption in commercial buildings in Australia became mandatory (Warren and Huston 2011) and in the UK similar legislation, known as Energy Performance Certificates (EPCs), was mandated in 2007 (DirectGov 2012). A more contentious legislation in Australia is the introduction of a carbon pricing mechanism which was commence in July 2012, the notion of 'taxing' carbon pollution met with significant resistance in the Australian parliament during 2011. There was concern about the potential impact on the economy and the amount of the carbon price compared to other countries; it remains to be seen how the policy is accepted by the electorate in the forthcoming 2013 election. To date the Australian government has largely offset the potential negative political and economic impacts of the pricing mechanism with generous government assistance to households. It is hard to say whether there is a temporary or permanent shift in the neo classical economic philosophy adopted by cornucopian environmentalists towards an economic outlook more attuned to accommodating environmentalism. In Australia, it is possible that this carbon pricing policy has resulted from a coalition government of the Labour and the Green party and represents the compromise Labour were prepared to make for political leadership. What is a concern is that within the built environment improved economic performance through a perceived increase in capital value is the main argument used to persuade property owners and investors to adopt sustainability (Eichholtz et al. 2009; Fuerst and McAllister 2011; Newell 2008).

In summary a spectrum of ideas and values exists within the concept of sustainability which goes from dark green to light green, or as some have suggested to grey; implying that the pursuit of weak sustainability does not deliver sustainable outcomes (Söderbaum 2011; Cooper 1994). The range of standpoints identified in the literature is expressed in Table 1.1. Five distinct groups were identifiable with two classified as anthropocentric (accommodating and cornucopian environmentalism) and three being ecocentric (transpersonal, deep and moderate ecology).

In Table 1.1 it is apparent that the most radical group, the 'transpersonal ecologists' are so embroiled in 'eco-sophical' arguments and debate that they are unable to form a coherent group who are capable of action (Dobson 1990). The 'deep ecologists' and 'moderate ecologists' share a number of beliefs but also have distinct and separate positions on some issues, for example, both groups believe capitalism is not sustainable. However 'deep ecologists' believe in bio-ethics and in the intrinsic value of nature whereas 'moderate ecologists' believe in the primary value of ecosystems; a less extreme view. A similar situation prevails for the anthropocentrics, the environmentalists. The two anthropocentric groups share views on the value of science and rational thought. They differ on the 'rights of humans' which dominate for the 'cornucopian environmentalists' however for the 'accommodating environmentalists' there is instrumental value in nature. Another way of presenting these beliefs and standpoints figuratively is shown in Fig. 1.2 as the spectrum of sustainability concepts.

Figure 1.2 illustrates the disconnect between transpersonal ecology and anthropocentrism/environmentalism. Elsewhere there is some overlap between the groups in their value systems and beliefs. There is a broader divide between ecocentrism and anthropocentrism where one is considered to deliver strong sustainability and the other weak sustainability. The question is: is weak and very weak

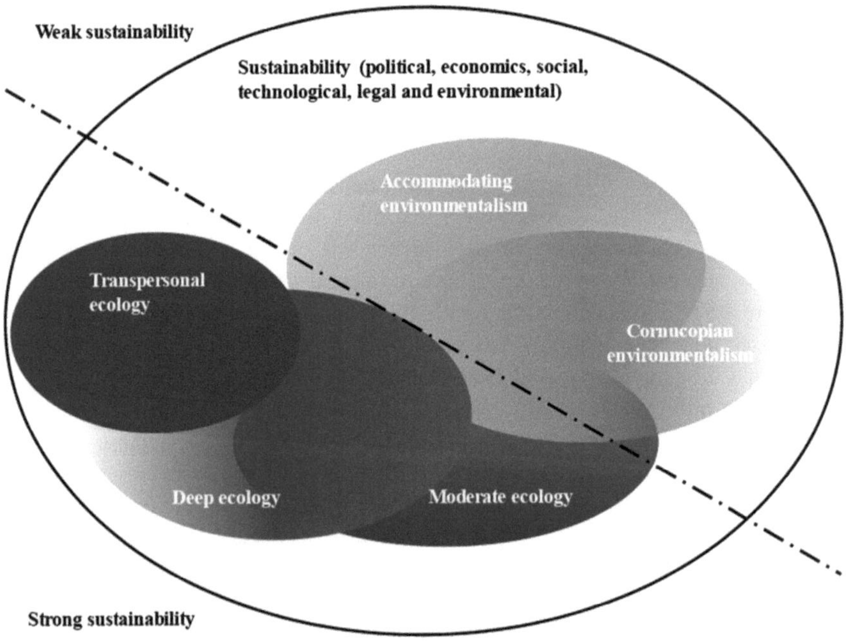

Fig. 1.2 The spectrum of sustainability concepts (Wilkinson 2013)

sustainability going to deliver sufficient changes for the generations to come and those already here? Brown (1995) asserts this level of sustainability will fall short of what is needed.

1.7 The Relationship of Built Environment to the Conceptual Model

The built environment is responsible for significant environmental impacts. Buildings use resources during construction with the extraction of resources, energy and water resources are used in the transport and manufacturing of construction materials and components. Considerable amounts of waste are also created at this stage. During the building's operational phase energy resources are used in lighting, heating and cooling and water is used in building services. Occupant or building user health is also impacted by the materials used during construction. At the end of the building lifecycle, unless materials are re-used or recycled, they will be transported to landfill where the resources are lost in perpetuity.

Within the built environment, construction companies are a sub group who have an impact on the sustainability of the buildings that they construct, design and sometimes operate and in this regard their conceptual understanding of sustainability is very important. It has become a current practice for many organisations, and not just construction companies, to adopt 'corporate social responsibility' or CSR as a means of organising, structuring, managing and reporting their environmental impact (Wilkinson et al. 2004). As a requirement of CSR companies provide information about their sustainability targets, policies and strategies, usually on their websites. This information is deemed to be an accurate and unbiased account of their respective stance and attitudes towards sustainability.

Figure 1.3 adapts Fig. 1.2 and shows construction companies as a sub set of built environments. Other actors within the built environment include designers, building users, owners and policy makers however their conceptual understanding, though important and collectively significant, is outside the scope of this study.

There is variance in the conceptual understanding of sustainability; and that it is, as Cook and Golton noted; 'an essentially contestable concept' (Cook and Golton 1994). An essentially contestable concept was coined by Gallie (1956) and exists where individuals On a positive note there are encouraging signs that some of the positions advocated by accommodating environmentalism are gaining traction and this reflects a shift from the findings of Cook and Golton's study of the UK construction sector in the early 1990s (Cook and Golton 1994). Given that Brown (1995) has stated that weak sustainability will not, in itself, provide the solutions to the problem humankind is facing; there are significant issues that we need to address as an industry, as individuals and as a community. For all our green building rating tools and schemes we are, with our current level of understanding of sustainability, in imminent danger that we will hit the targets but miss the point.

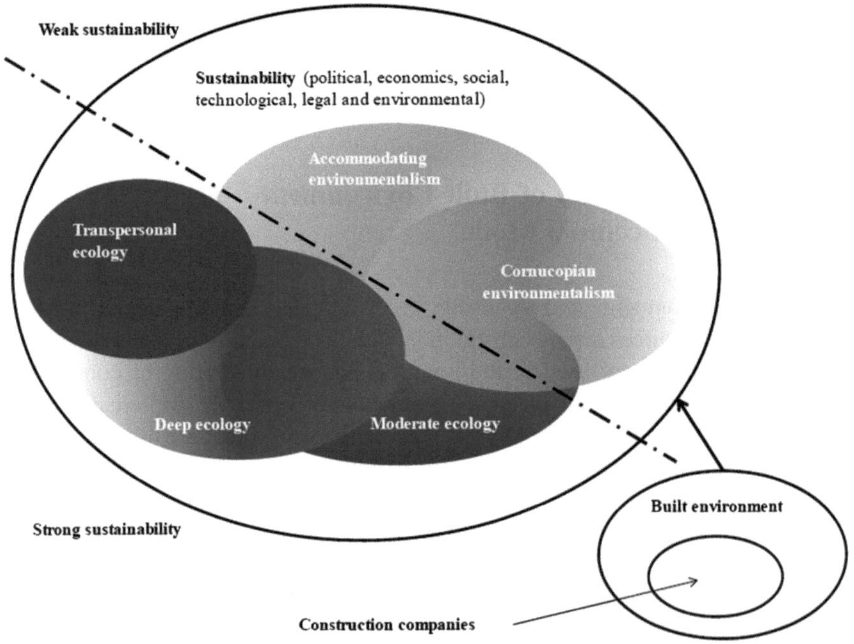

Fig. 1.3 The relationship of built environment and construction companies to the spectrum of sustainability concepts (Wilkinson 2012)

1.8 Sustainable Development and Urban Settlements

Urbanisation is a complex dynamic process playing out over multiple scales of space and time (Alberti et al. 2003). It is both a social phenomenon and physical transformation of landscapes that is now clearly at the forefront of defining humanity's relationship with the biosphere. Urban landscapes represent probably the most complex mosaic of land cover and multiple land uses of any landscape and as such provide important large-scale probing experiments of the effects of global change on ecosystems (e.g. global warming and increased nitrogen deposition). Urbanisation and urban landscapes have recently been identified by the Millennium Ecosystem Assessment as research areas where significant knowledge gaps exist (Millennium Ecosystem Assessment 2005).

The ideal of the good city is for all times. Planners and designers have always played a central role in the creation and development of vital and liveable cities. Depending on the spirit of the times, this sometimes meant the promotion of zoned, single-use urban forms; at other times, it meant the pursuit of a compact, spatially mixed and intensively used city (Williams et al. 2000).

Sustainable Urbanism, as a defined term, is the application of sustainability and resilient principles to the design, planning, and administration/operation of cities.

There are a range of organizations promoting and researching sustainable urbanism practices including governmental agencies, non-governmental organizations, professional associations, and professional enterprises around the world. Related to sustainable urbanism is the Ecocity movement (also known as Ecological Urbanism) which specifically is looking to make cities based on ecological principles, and the Resilient Cities movement addresses depleting resources by creating distributed local resources to replace global supply chain in case of major disruption. Green urbanism is another common term for sustainable urbanism. Sustainability can be defined as the practice of maintaining processes of productivity indefinitely—natural or human made—by replacing resources used with resources of equal or greater value without degrading or endangering natural biotic systems (Hendrix 2014; Lynn et al. 2014).

Now more than ever, cities are hot spots responsible for threatened global ecological boundaries. Climate change impacts and global environmental change are challenges for urban dwellers, planners, and managers. To develop opportunities for the sustainable development of cities, researchers from multiple disciplines are studying the feedback, dynamics, and behaviour of urban systems in the face of change (Chelleri and Olazabal 2012).

Although, highest per capita car ownership rates are still in the USA, Canada, Australia and Western Europe. Along with these trends are increases in fuel consumption and emissions. Given these indicators, the extent to which changes in urban form, facilitated through spatial planning, can have an impact on sustainable transport is rightly questioned (Williams et al. 2000).

There is a plethora of terms used to encompass the concept of sustainability;

1.8.1 New Urbanism

New Urbanism is an urban design movement, which arose in the USA in the early 1980s, promoting walkable, mixed-use neighbourhoods and transit-oriented development, seeking to end suburban sprawl and promote community. Characteristics include narrow streets, wide sidewalks and higher densities, qualities which we can all find in the European cities (Boeing et al. 2014).

1.8.2 Smart Growth

This is a transportation and urban planning theory that promotes growth in compact walkable city and town centres to avoid sprawl. Smart growth promotes compact, walkable, transit-oriented, bicycle-friendly environments, with local schools and mixed-use development with a range of housing choices (Boeing et al. 2014). Smart growth is used mostly in North America, whereas in the UK and Europe the term ‘compact city’ or ‘urban intensification’ are used to describe similar notions.

The concept of the compact city has influenced government planning policies in the UK and several European countries. The ecological city approach seems complementary to the other two approaches in terms of their respective areas of strengths and weakness (Jepson and Edwards 2010).

1.8.3 Green Urbanism

Green Urbanism is a conceptual model for zero-emission and zero-waste urban design, which arose in the 1990s, promoting compact energy-efficient urban development, seeking to transform and re-engineer existing city districts and regenerate the post-industrial city centre. It promotes the development of socially and environmentally sustainable city districts. It probably contains the most similar ideas with sustainable urbanism. They both put emphasis on urban design with nature, as well as shaping better communities and lifestyles. However, the principles of green urbanism are based on the triple-zero frameworks: zero fossil-fuel energy use, zero waste, and zero emissions. Sustainable Urbanism, on the other hand, is more focused on designing communities that are walkable and transit-served so that people will prefer to meet their daily needs on foot (Cerveroa and Sullivana 2011).

There are two contradictory, yet predominant, theories related to strategies for designing and planning of sustainable city form. They are the "compact city" idea from the European context and "urban sprawl" theory from Australia and North America. These two theories consider two main factors of land use and density patterns in cities' form. It has been concluded and then argued that the necessity of "mixed use" function which should be combined with "density" measures simultaneously to create a new balanced model which can be termed "compact mixed use city form" as the alternative for achieving urban sustainability (Masnavi 2011). Both these ideal-types of urban development have been encouraged because of their alleged social, cultural and economic advantages. Yet the compact city model is credited with being less detrimental to the environment (Williams et al. 2000).

1.9 Theory of the Compact City

Compact, liveable urban neighbourhoods attract more people and business. Compactness, or density, plays an important role in sustainable urban development because it supports reductions in per-capita resource use and benefits public transit developments. The density of new development across the U.S. averages roughly two dwelling unit per acre, which is too low to support efficient transit and walk-to destinations. Such low-density development is a characteristic of urban sprawl, which is the major cause of high dependence on private automobiles, inefficient infrastructure, increased obesity, loss of farmlands and natural habitats, pollution, and so on (Frumkin 2002). For these reasons, sustainable urbanism requires minimum development densities roughly four times higher than two dwelling units per acre.

Overall, compact development generates fewer pollutants to the natural world. Research has shown that low-density development can exacerbate non-point source pollutant loadings by consuming absorbent open space and increasing impervious surface area relative to compact development. While increasing densities regionally can better protect water resources at a regional level, higher-density development can create more impervious cover, which increases water quality problems in nearby or adjacent water bodies (Richards et al. 2011).

Increasing neighbourhood population density also supports improved public transit service. Concentrating development density in and around transit stops and corridors maximizes people's willingness to walk and thus reduces car ownership and use. Sustainable urbanism seeks to integrate infrastructure design increase with density, because a concentrated mixed-use development required less per capita infrastructure usage compared to detached single-family housing (Farr 2008).

1.10 Theory of the Urban Dispersal

In Europe and North America there is a growing concern about the development of urban form, especially deconcentration of urban land use in the form of urban sprawl. This has unintentional consequences such as city centre decline, increased reliance on the use of the private car, and the loss of open space. While governments try to regulate the development of urban form, there are no easy solutions. However, policies such as 'new urbanism' and 'smart growth' in North America, and 'compact city' and 'multifunctional land use' policies in Europe, though difficult to implement, have the potential to curb urban sprawl and the further growth in car use (Dieleman and Wegener 2004).

Most of studies concluded that sprawl has both positive and negative effects. The most complete and rigorous North American studies concluded that overall, sprawl is more costly than compact development for both operating and capital costs (Burchell et al. 1992, 2000). The greatest savings gained from growth controls were in land consumed and infrastructure built especially water, sewer, and road facilities.

Urban sprawl or suburban sprawl describes the expansion of human populations away from central urban areas into low-density, mono-functional and usually car-dependent communities. In addition to describing a particular form of urbanisation, the term also relates to the social and environmental consequences associated with this development (Neuman 2005).

The term urban sprawl is highly politicised, and almost always has negative connotations. It is criticised for causing environmental degradation, and intensifying segregation and undermining the vitality of existing urban areas and attacked on aesthetic grounds. Due to the pejorative meaning of the term, few openly support urban sprawl as such. The term has become a rallying cry for managing urban growth (James et al. 2013), have summarised the various definitions of urban sprawl in the planning literature to create a working definition of the concept as: '… unplanned, uncontrolled, and uncoordinated single use development that does not provide for a functional mix of uses and/or is not functionally related to surrounding land uses and which variously appears as low-density, ribbon or strip, scattered, leapfrog, or isolated development' (Daneshpour and Shakibamanesh 2011).

1.11 Urban Resilience

The continuous increase in the number and size of urban regions across the world, and the simultaneous shrinking of cities in some regions, pose great challenges for sustainable development. Urbanization patterns and the lifestyles of urban dwellers also affect the planet on wider scales in time and space. They contribute to shape bio-physical processes on planetary scales, and affect how humans around the world mentally connect with the Biosphere. However, through their local to global linkages, cities can play a key role in the quest to continuously and increasingly support sustainable development (Chelleri and Olazabal 2012).

Based on the definition of Holling (2001), Alberti et al. (2003) have defined urban resilience as the degree to which cities are able to tolerate alteration before reorganising around a new set of structures and processes. They assert that urban resilience can be measured by how well a city can simultaneously balance ecosystem and human functions. When most people think of urban resilience, it is generally in the context of response to impacts (e.g. hazard or disaster recovery), however what we learn from our understanding of resilience in regional social-ecological systems is a society that is flexible and able to adjust in the face of uncertainty and surprise is also able to capitalise on positive opportunities the future may bring (Barnett 2001).

Assuming that the process towards urban sustainability includes a combination of encompassing transitions (i.e. transformations) as a response to natural resource scarcity and climate change (adapted from Kemp and van Lente 2011), the theory of resilience is a promising framework (Chelleri and Olazabal 2012). Certainly, it supports the conceptualisation and development of tools to plan and manage urban sustainability transitions, providing a long-term perspective based on the three key concepts of learning, adaptation and transformation (Walker 2004; Folk et al. 2010). Resilience plays a crucial role in achieving sustainability (Brand and Jax 2007) and Promoting resilience means changing the nature of decision-making to recognize the benefits of self-sufficiency and new forms of governance which focus more on social equity, learning and the capacity to adapt (UN 2012).

In summary, the goal of urban sustainability poses great challenges to urban planners and policymakers; the cross-scale nature of urban interactions places these challenges at the centre of global scale solutions. As discussed here, urban transformation can be positioned around three challenges: reducing resource consumption, integrating social and environmental criteria alongside economic interests in decision-making and mitigating the impacts of and adapting to climate change. To help cities in this process of transformation a significant aim of urban sustainability must be to reduce potential vulnerabilities, especially to climate change, and notably in low-income countries where rapid urbanisation brings significant threats. Moving from vulnerability to resilience is the key to this process.

1.12 High Tech Versus Low Tech Paradigms and the Conundrum of Measurement

The Brundtland Commission defined sustainable development as 'development that meets the needs of the present without compromising the ability of future generations to meet their own needs' (WCED 1987). From this proposition sustainability goals have been defined and, to a limited extent achieved. This has lead to greenhouse gas emission restrictions and targets for the reduction of energy and material use. Worldwide, the building industry is responsible for 25 % of the road-traffic, 35 % of the waste produced and 40 % of the energy consumption and CO_2 emissions (UNEP 2009).

Where sustainable development is perceived as development without growth, this philosophy lead to ideas about extended lifespan (De Jonge 1990), cradle-to-cradle developments that consider recycling or upcycling of second hand building materials (McDonough and Braungart 2002) and the circular economy (MacArthur 2013) that reduces waste production and energy use in construction. Viewed on this basis, sustainable development of the built environment is understood as durable; focusing on a long life span, sustainable; focusing on energy use, and adaptable; focusing on loose fit and space use. Furthermore sustainable development demands balancing these three criterion.

1.12.1 High-Tech

Expanded functional lifespan or multiple functional lifespans contribute to sustainable development, although these contributions are not as easily recognised and accounted for as the more technical, energy saving solutions. In comparison buildings with climate facades, maximised thermal insulation and reduced CO_2 emissions are seen as sustainable, and these aspects are recognised in sustainability rating tools, such as LEED, Green Star and BREEAM (DGBC 2009; Eichholtz et al. 2008). However, the sustainability measures that are adopted, are not necessarily related to the structural parts of the building, and do not say much about the quality of the building. Moreover, sustainability-rating tools are based on different rating systems and the highest achievable levels of the rating tools differ (Reed et al. 2009).

The highest level of BREEAM is the rating 'Outstanding', which has a far higher level of sustainability compared to the highest rating in Green Star (6 Stars) or LEED (Platinum) (Reed et al. 2009). None of the existing rating tools have well incorporated criteria to measure refurbishment only BREEAM has a refurbishment tool, called BREEAM In Use. This is surprising, given that only 1–2 % is typically added to the total existing building stock each year (Reed et al. 2009). Furthermore, the required performance of buildings such as offices, for example, which includes increasing comfort and emission reducing building characteristics, changes over time. Measures reducing the energy use and emissions from office buildings should therefore be seen as part of an adaptable building concept (Kendall 1999). Several researchers propose the use of a life cycle costing approach (LCA) as these are more complete than the sustainability rating tools. As an example, the model of Eco-costs/Value Ratio (EVR) allows comparing new construction to renovation and maintenance (De Jonge 2005). This type of LCA model could be used in the initial phase of new developments, to calculate the possible benefits of buildings with future adaptation and transformation possibilities, compared to traditional standardised buildings.

1.12.2 Low-Tech

Low-tech sustainability is an umbrella concept covering a range of sustainability measures, all focusing on the quality of the building, for example use value, adaptability, extended building lifespan, material use and relationship with the surroundings. Examples of low-tech measures are buildings that work with the climate and site, specific (micro) climatic design solutions, natural materials rather than man made materials, solutions for passive houses, adobe walls to provide high thermal mass, and so on. However, in medium to high-rise buildings and high-density city centres these solutions are not always viable. A solution that is viable on all scale levels is extended building lifespan. The remainder of this section will focus on this solution. Extended building life span is a low-tech proposal to improve the sustainability of the built environment by increasing the durability of the buildings, considering the buildings technical, functional and economic lifespan (De Jonge 1990). Adaptation contributes to prolonging the building lifespan. In housing, if a building is adapted, an extended lifespan of 30 years following the adaptation is expected (Douglas 2006). In commercial property, offices respond much faster to both market and user preferences, and so the extended lifespan expectation of each adaptation will be lower.

Buildings with a long lifespan are sustainable as such; they contribute to lowering the construction industry's waste production, reducing construction related traffic, and reducing energy use for construction (Lichtenberg 2005). Depending on a buildings lifespan, the building materials are more or less important to the total environmental load of the building. Buildings constructed for a long lifespan should be built with focus on sustainable energy solutions, the type of building materials is less important, whilst if a building is built to last for less than 20 years, it should be built focusing on a sustainable use of building materials and recycling or re-use of the materials after demounting or deconstructing it. In the case of a building with an expected lifespan of 20 years, attention should be paid to both energy consumption and building materials (Van den Dobbelsteen 2004). In calculating the Life Cycle Costs of adaptation versus demolition and new construction, adaptation was found to be more favourable (De Jonge 2005). Increasing a buildings functional lifespan through adaptation is possible by extending the original functional lifespan, or by enhancing the buildings potential to transform or change use. However extending the building lifespan requires investors who are willing to invest and real estate investors typically consider investment periods of 10–15 years, which is relatively short. Investors need market evidence of higher exit yields to compensate for higher initial costs. Public investors, like the Government Buildings Agency, are parties that play a natural role as forerunners, and are currently studying the possibilities of investing in longer life spans (Geraedts and Remøy 2015).

1.13 Adaptability—Prolonging Building Lifespan

Adaptable buildings are designed with a view to later adaptations and can accommodate several functional life spans, thereby avoiding functional obsolescence. Adaptable buildings are designed for longer lifespans, and hence are sustainable by definition (Remøy et al. 2011; Remøy 2010). Habraken (1972) was one of the first researchers to study adaptability, though originally for another purpose: he discussed housing as an act that is not completed by the developer or the builder, but instead offers a system of supports and finishing elements, in which the inhabitants choose the finishing elements. Both Le Corbusier and the Japanese Metabolism inspired Habraken's work. Le Corbusier, in many of his housing projects, described the apartment as an infill in a larger structure, i.e. in plan Obus he suggested a structure consisting of floors supported by columns, and stairs, electricity and sewerage added to the structure. Within this structure the inhabitants were free to build their own home on a "building lot" (Frampton 1992). Though inspired by Le Corbusier's ideas, Habraken did not agree with Le Corbusier's proposed structure. While Le Corbusier saw the structural elements floors and columns as necessary, Habraken had a structure in mind that was closer connected to architecture, including the dwellings outer walls.

Habraken's idea of adaptability was based on the idea that a dwelling is a product of its inhabitants. The ideas were a reaction to the movement "New Objectivity" that opted for slum clearance and massive new construction, and defined structures that accommodate coincidence. Seeing mass-production of housing as inevitable, he sought to offer freedom of choice within the frames of mass-construction.

Duffy (1990), in considering mainly office building use, was interested in the buildings capacity to adapt to the changing requirements of its user, resulting in a robust building. He defined buildings as systems with several subsystems or layers and recognised shell, services and scenery, whereas scenery is everything that can be altered without influencing the functioning of the services or the shell. The services are including electricity, sewerage and ventilation, and servicing elements like elevators, while the shell includes both the building's facade and its construction. His way of defining adaptability is based on refurbishments of office buildings and which layers may be altered in order to renew the working environment without influencing the technical functioning of the office building itself. The distinction between different layers is related to the dissimilar lifespans of the layers. Though buildings in Europe and North America have a life expectancy of 50–70 years, the other elements or layers have far shorter lifespans: The servicing systems have a lifespan of 15–20 years, the spatial divisions and fittings 5–10 years, and the workstation equipment and furniture less than 5 years (Nutt 1988).

Duffy's approach was adapted by Brand who categorises the different parts of a building in six layers; site, structure, skin, services, space plan and stuff (Brand 1994), and by Leupen (2006). Leupen recognises five layers; structure, skin, scenery, services and access. A high independency of the layers makes adaptations possible. According to Leupen (2006), buildings that consist of several layers are sustainable and are more likely to be adapted than buildings where the different layers are dependent on each other.

1.14 Cherished Buildings

The most important function of a building design is its quality and ability to generate admiration or "love", because beloved buildings are often adapted and reused and, in this way, are sustainable (Van Kasteren 2002). This quality is difficult to describe but it is evident for instance in the old Venetian "palazzo" or Amsterdam warehouses. The architecture of the solid is "absolute"; the form of the building is defined architecturally and not following function (Leupen et al. 2005). These types of urban buildings are designed and built for a technical lifespan of 200 years, considering that several functional alterations and technical upgrades can be performed during the buildings life. The investment perspective is that of a normal development; for a housing corporation approximately 30 years, so that after 30 years the buildings may be technically or functionally upgraded.

1.15 IFD: Industrial, Flexible and Demountable

On the other hand, the idea behind Industrial, Flexible and Demountable (IFD) is industrial manufacturing of the building parts (building becomes assembly of building parts), flexible buildings in use (adapting becomes replacing), and demountable buildings (demolishing becomes disassembly). Industrial mass production aims at optimising the use of labour, materials and machines by industrial manufacturing methods and by processing all building material in factories. On the building site, the building components are assembled. The method claims flexibility in the freedom of spatial flexibility for the first user, easy replacement of components to fit possible new uses, flexibility in the compartmenting of the building and the possibility for adding or subtracting square metres according to the wishes of the user. At the end of the buildings life cycle, it can be disassembled and the parts may be re-used if they are not worn out (Bouwmeester 2001, 2003; Groenendijk et al. 2000).

An example of an IFD-building, designed and built according to the ideas of IFD, is the XX office. The XX office building designed by XX architects was built in Delft in the Netherlands in 1998 (Fig. 1.4). The technical lifespan of the building was set to 20 years to match the economic lifespan of the building. After 20 years, the building can be dismantled and the building parts recycled (Fig. 1.5). All the joints in the building are demountable and ensures the disassembly (Post 1998). A problem with the IFD approach is that it is both systemic and requires specific settings in order to work. Also, the idea of the recyclable building is questioned by new environmental research focusing on a cradle-to-cradle perspective of materials, claiming that recycling of building materials (because these themselves are always composite) is still a down-cycling of the material since large amounts of energy is needed for the material to be transported, re-manufactured and reused (McDonough and Braungart 2002). According to the theory of the circular economy, sustainable development means to keep products, components and materials at their highest utility and value (MacArthur 2013), which means that reuse of building components is a 'second best' if buildings cannot be reused entirely.

Fig. 1.4 Office, Delft Netherlands

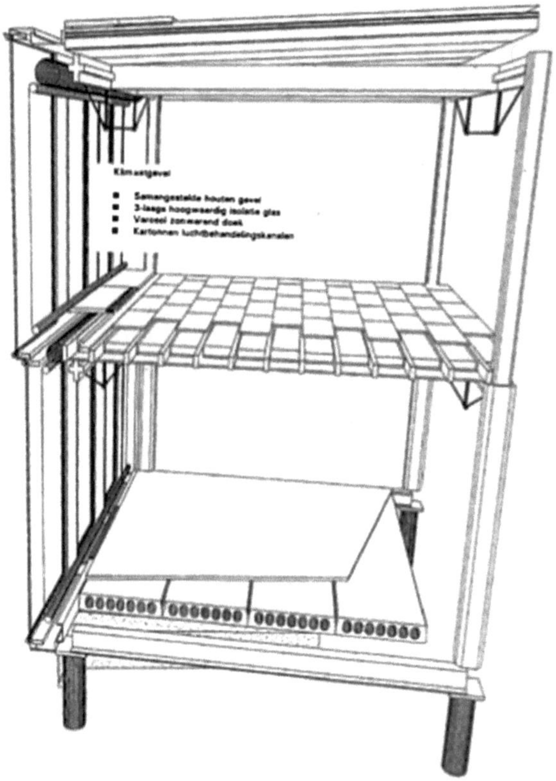

Fig. 1.5 Section office, Delft Netherlands

1.16 The Discrepancy Between Green Building Rating Tools and Life-Cycle Thinking

Current rating tools are focused on new construction, recurrent building operations and maintenance (Conejos et al. 2015). Worldwide, more than 600 tools are used to measure or evaluate the social, environmental and economic dimensions of sustainability. The most used rating tools are LEED, Green Star and BREEAM (Reed et al. 2009; DGBC 2009; Eichholtz et al. 2008). These three tools provide a broad assessment of the environmental impact of a building. Each tool is based on a checklist, and points are awarded based on answers to the questions in the checklist (Jansz 2012). The tools are used to rate a building, and henceforth to market the building according to its sustainability label. The tools have been developed in different geographical, climatical and cultural contexts, and hence focus more or less on different aspects. As an example, Green Star is developed in Australia and focuses more than the other tools on water usage and less on energy use than LEED, developed in the US, and BREEAM, developed in Northern Europe. Next to these large, holistic tools, a range of smaller instruments exist that focus on one single aspect of sustainability. For example, energy is seen as an important aspect so sustainability in the Dutch context, and so the Dutch government developed the EPC (Energy Performance Coefficient) in 1995, followed by the energy label in 2008. Whereas the EPC is a tool to certify construction plans, the energy label is developed as a design tool, and gives recommendations about how to improve the energy label of a design (Jansz 2012).

An international comparison of rating tools (Reed et al. 2009) has shown some of the challenges of measuring sustainability. The assessment criteria used in the different tools are different. Comparing 15 assessment criteria, the study showed that all criteria were used in several tools, but no tool used all criteria. The assessment methods and weighting of the criteria are also different, hence buildings that get the top rating by one tool, score average with another tool. Moreover, the differences between tools is unclear. A common basis and agreement about criteria, assessment methods and weighting of criteria is needed.

The building's life span plays a minor role in current methodologies, and the standard life span applied in the models is not supported by hard data. In 2015, BREEAM is the only rating tool that has an application to assess buildings in use. None of the existing rating tools appraises adaptive reuse performance (Conejos et al. 2015). Several authors (De Jonge 2005; Langston 2011) called for the use of lifecycle assessment models. However, commonly used life cycle assessment models apply a standard life span expectation, which means that the building's qualities or shortcomings are not taken into account. Haapio and Viitaniemi (2008) emphasise that the effect of the assumed life span on the results of environmental assessments should be thoroughly analysed. The first explorations of estimating the lifespan of buildings and including this estimated service life (ESL) in an environmental assessment showed a profound influence (Jansz 2012; Van den Dobbelsteen and Van der Linden 2005). To be able to fully rate the sustainability of new, adapted and existing buildings, a life cycle assessment tools is needed that fully incorporate the lifespan of buildings (ESL), including the material use and the energy use. A rating tool needs to be developed that includes durability, sustainability and adaptability. Others question over–reliance on the tools in some markets where a slavish adoption of tools can be seen to drive design and result in attainment of targets; i.e. a rating level of 5 stars to give a market edge and premium to value, but the stakeholders here may have missed the point of sustainability (Wilkinson et al. 2015).

1.17 Conclusions

This chapter has introduced the concept of sustainable development explaining the Brundtland definition posited in 1987. It has been shown that numerous variations of this definition have appeared in the years since, cherry picking different aspects to suit the needs of the author(s). Some definitions emphasise economic aspects over environmental and social aspects. The concept was expanded through

Elkington triple bottom line concept in the 1990s. Section 1.2 the different types and levels of sustainability were introduced and discussed. Sustainability can have ecocentric or technocentric perspectives, which influence the social, political and economics beliefs of individuals. In summary the concept of sustainability shares Gallie's (1956) attributes of a 'contested concept'; that it is in essence, something that means 'all things to all men'. It is a notion that varies from person to person, market to market, location to location, land use to land use, country to country, and of course over time. Given this, the best we are able to claim is that one building is, or may be, relatively speaking more sustainable than another; it is impossible to speak of sustainable property development in an absolute sense. In this case, we face the situation attributed to Emmanuel Kant, the philosopher in the 1750s, that often 'it is necessary to make decisions on information sufficient for action but insufficient to satisfy the intellect'.

The philosophical, political, economic and social constructs underpinning our definitions of sustainability are broad and complex. In some respects our current application of sustainability in the built environment draws on a narrow and limited understanding and inclusion of selected and therefore limited aspects of sustainability. This narrow perspective has to be broadened over time as we are in danger, as a sector of 'hitting the targets but missing the point'. The point being that our planetary resources is finite and our consumption levels and pollution levels continue to increase. The chapter explained briefly the characteristics of sustainable urban development from the urban scale to the building scale, noting that factors such as geographical location, land use type impact on the breadth and depth of sustainability delivered. A discussion of high-tech versus low-tech approaches to sustainability highlighted the characteristics of each paradigm. Finally the challenges of measuring sustainability were introduced in the examination of green building rating tools. Further chapters will draw out in detail the factors, including sustainability factors that each stakeholder considers in the process.

References

Alberti MJMM, Shulenberger E, Bradley G, Ryan C, Zumbrunnen C (2003) Integration humans into ecology: opportunities and challenges for studying urban ecosystems. BioScience 53:169–179

Barnett J (2001) The meaning of environmental security. Ecological politics and policy in the new security era. Zed Books Ltd., New York

Boeing G, Church D, Hubbard H, Mickens J, Rudis L (2014) LEED-ND and livability revisited. Berkeley Plann J 27(1)

Bouwmeester H (2001) Demonstratieprojecten IFD-bouwen 2000 een omslag in bouwen een omslag in denken. Stuurgroep Experimenten Volkshuisvesting, Rotterdam

Bouwmeester H (2003) Demonstratieprojecten IFD-bouwen op de drempel van 2002 een doorbraak 23 voorbeelden van industrieel, flexibel en demontabel bouwen. Stuurgroep Experimenten Volkshuisvesting, Rotterdam

Brand S (1994) How buildings learn; what happens after they're built. Viking, New York

Brand FS, Jax K (2007) Focusing the meaning (s) of resilience: resilience as a descriptive concept and a boundary object. Ecol Soc 12(1):23

Brown CS (1995) Anthropocentrism and ecocentrism: the quest for a new worldview. Midwest Q 36(2):191
The Brundtland Report (1996) State of the world 1996: A worldwatch institute report on progress toward a sustainable society. Earthscan, 1996
Burchell RW, Anglin R, Beaton WP, Brail RK, Danielson MN, Foxley SJ, You JK (1992) Impact assessment of the New Jersey interim state development and redevelopment plan
Burchell RW, Lowenstein G, Dolphin WR (2000) The costs of sprawl. National Academy Press, Washington, DC
Conejos S, Langston C, Smith J (2015) Enhancing sustainability through designing for adaptive reuse from the outset; a comparison of adaptstar and adaptive reuse potential (ARP) models. Facilities 33(9/10):531–552
Cook S, Golton B (1994) Sustainable development concepts and practice in the built environment. A UK perspective. In: CIB TG 16, sustainable construction. Tampa Florida, USA, 6–9 Nov 1994
Cooper I (1994) Sustainability—broadening the perspectives, cities and sustainability, science and engineering research council, pp 9–13
Chelleri L, Olazabal M (2012) Multidisciplinary perspectives on urban resilience. Basque Center for Climate Change, Spain
Daneshpour A, Shakibamanesh A (2011) Compact city; dose it create an obligatory context for urban sustainability? Int J Architectural Eng Urban Plann 21(2):110–118
De Jonge H (1990) The philosophy and practice of maintenance and modernisation. In: Proceedings property maintenance management and modernisation, Singapore
De Jonge T (2005) Cost effectiveness of sustainable housing investments. DUP Science, Delft
DGBC (2009) Dutch Green Building Council. http://www.dgbc.nl. Accessed 01 Apr 2015
Dieleman F, Wegener M (2004) Compact city and urban sprawl. Built Environ 30(4):308–323
DirectGov (2012) http://www.direct.gov.uk/en/HomeAndCommunity/BuyingAndSellingYourHome/Energyperformancecertificates/DG_177026. Accessed 01 Mar 2012
Dobson A (1990) Green political thought: an introduction. Routledge, London
Douglas J (2006) Building adaptation. Butterworth-Heinemann, Oxford
Drexhage J, Murphy D (2010) Sustainable development: from Brundtland to Rio 2012. United Nations Headquarters, New York, International Institute for Sustainable Development (IISD), pp 21–26
Duffy F (1990) Measuring building performance. Facilities 8(5):17–21
Eichholtz P, Kok N, Quigley JM (2009) Doing well by doing good? An analysis of the financial; performance pf green office buildings in the USA. RICS, London
Elkington J (1997) Cannibals with forks: the triple bottom line of 21st century business. Capstone, Oxford
Eichholtz PM, Kok N, Otten R (2008) Executive compensation in UK property companies. J real estate Financ Econ 36(4):405–426
Farr D (2008) Sustainable urbanism. Wiley, Oxford
Folk C et al (2010) Resilience thinking: integrating resilience, adaptability and transformability. Ecol Sci 15(4)
Frampton K (1992) Modern architecture; a critical history, 3rd edn. Thames and Hudson, London
Frumkin H (2002) Urban sprawl and public health. Assoc Schools Public Health 117:201–217
Fuerst F, McAllister P (2011) Green noise or green value? Measuring the effects of environmental certification on office values. Real Estate Economics 39(1):45–69
Gallie WE (1955–1956) Essentially contested concepts. Proc Aristotelian Soc 56(1956):187–198
Geraedts RP, Remøy H (2015) Buildings with future value—a method to assess the adaptive capacity of buildings. Open House International
Groenendijk P, Vollaard P, Vos H, van Braam Morris M (2000) Industrieel, flexibel en demontabel bouwen demonstratieprojecten IFD-bouwen 1999: zien is geloven. Stuurgroep Experimenten Volkshuisvesting, Rotterdam
Haapio A, Viitaniemi P (2008) A critical review of building environmental assessment tools. Environ Impact Assess Rev 28(7):469–482

Habraken NJ (1972) Supports: an alternative to mass housing. Architectural Press, London
Holling CS (2001) Understanding the complexity of economic, ecological, and social systems. Ecosystems 4(5):390–405
IPCC (2007) Climate change the physical science basis. In: AGU fall meeting abstracts, Vol 1. p 01
James P et al (2013) Managing metropolises by negotiating mega-urban growth. In: Topfer HM (ed) Institutional and social innovation for sustainable urban development. Routledge, London
Jansz S (2012) The effect of the estimated service life on the sustainability of vacancy strategies. TU Delft, Delft
Jepson EJ Jr, Edwards MM (2010) How possible is sustainable urban development? An analysis of planners' perceptions about new urbanism, smart growth and the ecological city. Plann Practice Res 25(4):417–437
Kates RW, Parris TM, Leiserowitz AA (2005) What is sustainable development? Goals, indicators, values and practice. Environ Sci Policy Sustain Dev 47(3):8–21
Kemp R, van Lente H (2011) The dual challenge of sustainability transitions. Environ Innovation Societal Transitions 1(1):121–124
Kendall S (1999) Open building: an approach to sustainable architecture. J Urban Technol 6 (3):1–16
Langston C (2011) Estimating the useful life of buildings. Paper presented at the Australian University Building Educators Association, Gold Coast
Leupen B (2006) Frame and generic space. 010 Publishers, Rotterdam
Leupen B et al (2005) Time-based architecture. Uitgeverij 010, Rotterdam
Lichtenberg J (2005) Slimbouwen. Aenas, Boxtel
Lynn R et al (2014) Communicating sustainability for the green economy. Sharpe, New York
MacArthur E (2013) Towards the circular economy. Vol 2, opportunities for the consumer goods sector
McDonough W, Braungart M (2002) Cradle to cradle remaking the way we make things by William McDonough and Michael Braungart. North Point Press, New York
Masnavi MR (2011) Sustainable urban forms; planning and design strategies. the compact city, urban sprawl, and mixed use development in theory and practice. LAP-Lambert Academic Publishing
Melvin K Hendrix (2014) Sustainable backyard polyculture: designing for ecological resiliency smashwords eBook edition, ISBN 9781310285547, License Notes, Available via https://www.smashwords.com
Millennium Ecosystem Assessment (2005) Ecosystems and human well-being: synthesis world resources institute. Island Press, Washington DC
Montuori A, Purser RE (1995) Deconstructing the lone genius myth: Toward a contextual view of creativity. J Hum Psychol 35(3):69–112
Naess A (1990) Ecology, community and lifestyle: outline of an ecosophy. Cambridge University Press, Cambridge
Neuman M (2005) The Compact City Fallacy. J Plann Educ Res 25:11–26
Newell G (2008) The strategic significance of environmental sustainability by Australian-listed property trusts. J Property Investment Financ 26(6):522–540
Nutt B (1988) The strategic design of buildings. Long Range Plan 21(4):130–140
Pearce D (1993) Blueprint 3—measuring sustainable development. Earthscan Publication Limited, London
Pepper D (1984) The roots of modern environmentalism, Routledge, London
Purser RE, Montuori A (1996) Ecocentrism is in the eye of the beholder. Academy of management. Acad Manag Rev 21(3):611
Post J (1998) The xx building. Retrieved 23 Feb 2009, from www.xxarchitecten.nl
Reed R, Bilos A, Wilkinson S et al (2009) International comparison of sustainable rating tools. J Sustain Real Estate 1(1):1–22
Remøy H (2010) Out of office—a study on the cause of office vacancy and transformation as a means to cope and prevent. IOS Press, Amsterdam
Remøy H, De Jong P, Schenk W (2011) Adaptable office buildings. Property Manag 29(5):443–453

Richards L, Anderson G, Santore MK (2011) Protecting water resources with higher density developments. US Environmental Protection Agency
Cerveroa R, Sullivana C (2011) Green TODs: marrying transit-oriented development and green urbanism. J Sustain Dev World Ecol 18(3):210–218
Salinger J (2010) The climate journey over three decades: from childhood to maturity, innocence to knowing, from anthropocentrism to ecocentrism. Clim Change 100:49–57
Söderbaum P (2011) Sustainability economics as a contested concept. Ecol Econ 70:1019–1020
Schubert A, Láng I (2005) The literature aftermath of the Brundtland report 'Our Common Future'. A scientometric study based on citations in science and social science journals. Environ Dev Sustain 7(1):1–8
United Nations (2010) The millennium development goals report. United Nations, New York
UN (2012) Resilient people, resilient planet: a future worth choosing. United Nations, New York
United Nations Environment Program (UNEP) (2007) United nations environment programme annual report: 2006. UNEP, Nairobi
United Nations Environment Program (UNEP) (2009) Buildings and climate change. Summary for decision-makers. Retrieved from http://www.unep.org/sbci/pdfs/SBCI-BCCSummary.pdf. Accessed 1 Apr 2015
Van den Dobbelsteen AAJF, Van der Linden A (2005) Managing the time factor in sustainability—a model for the impact of the building lifespan on environmental performance. In: Yang J (ed) Smart and sustainable built environments. Blackwell Publishing, Oxford, pp 223–233
Van den Dobbelsteen AAJF (2004) The sustainable office an exploration of the potential for factor 20 environmental improvement of office accommodation. TU Delft, Delft
Van Kasteren J (2002) Buildings that last: guidelines for strategic thinking. NAI Uitgevers, Rotterdam
Walker BE (2004) Resilience adaptability and transformability in social-ecological systems. Ecol Soc 9(2)
Warren CMJ, Huston S (2011) Promoting energy efficiency in public sector commercial buildings in Australia. Paper presented at the conference of proceedings of RICS construction and property. COBRA: construction and building research conference, Salford, United Kingdom, 12–13 Sep 2011
Washington H (2015) Demystifying sustainability: Towards real solutions. Routledge
Williams et al (2000) Achieving sustainable urban form. E and F SPON, London
Wilkinson SJ (2012) Conceptual understanding of sustainability in Australian Construction firms. In: CIB Montreal Canada June 2012
Wilkinson SJ (2013) Conceptual understanding of sustainability in Australian property organisations. J Property Manag 31(3)
Wilkinson SJ, Pinder J, Franks A (2004) Conceptual understanding of corporate social responsibility in the UK construction and property sectors. In: CIB Toronto, Canada 2004
Wilkinson SJ, Sayce S, Christensen P (2015) Developing property sustainably. Routledge, Oxon, UK
World Commission on Environment and Development WCED (1987) Our common future. Oxford University Press, Oxford

Chapter 2
Mass Housing: Challenges, Contemporary Paradigms and Future Potentials

Kheira Anissa Tabet Aoul, Khaled Galal Ahmed and Sahera Bleibleh

Abstract In a world of increasing global population and constant transformation, where climate change and limited natural resources have become undeniable facts, there is a growing demand for housing that efficiently and adequately responds to the needs of the masses, while limiting environmental impacts. Given this setting, the objective of this introductory chapter is multi-fold and intends to carry a holistic review of mass housing paradigms, trends, challenges and production in order to uncover its future potentials. Henceforth, the generic challenges and the complex and inter-related driving forces shaping the mass housing production are first underlined. Following, mass housing is explored through its conceptual and sometimes shifting paradigms; from social egalitarian state, welfare state, to the market driven context and the participatory concepts. In each pattern, the various policies and processes are discussed, while representative case studies from different contexts are presented and critically reviewed to uncover both their achievements and drawbacks. This chapter ends with an exploration of the anticipated approaches for mass housing in both developed and developing countries. This encompasses the predicted environmental, economic and social sustainability as well as the technical challenges and the expected, or rather recommended, scenarios for confronting such challenges. In this regard, the perceived future mass housing approaches, decision-making processes, forms and technical opportunities are briefly discussed.

K.A. Tabet Aoul (✉) · K.G. Ahmed · S. Bleibleh
United Arab Emirates University, Al Ain, UAE
e-mail: Kheira.anissa@uaeu.ac.ae

K.G. Ahmed
e-mail: kgahmed@uaeu.ac.ae

S. Bleibleh
e-mail: sahera.bleibleh@uaeu.ac.ae

© Springer International Publishing Switzerland 2016

M. Noguchi (ed.), *ZEMCH: Toward the Delivery of Zero Energy Mass Custom Homes*, Springer Tracts in Civil Engineering,
DOI 10.1007/978-3-319-31967-4_2

2.1 Introduction

Housing has been universally recognised and accepted as one of the most elementary human need. The United Nations International Standards under its Universal Declaration of Human Rights enshrined in its laws the right to an adequate standard of living, including adequate housing (Article 25.1 1948). The right to an adequate standard of living involves, that everybody shall enjoy the essential sustenance rights, which translates into adequate food, clothing, housing and the necessary conditions of care when needed. The universally accepted fundamental human right for an adequate housing that is safe, secure, sanitary, accessible and affordable has been acknowledged by most nations' governments (Hohmann 2013) and translates into policies and housing programs initiatives. The "adequate" housing provision is to a large extent determined by the prevailing social, economic, political, cultural, climatic, ecological and other conditions in any given context. Access to adequate housing for all, however, remains for many nations the most urgent, complex and highly contested challenge. Countries that have struggled to seriously resolve their housing issues have often adopted and implemented the concept and principle of mass housing. Effective mass housing delivery involves many actors and has often called for the establishment of national housing policies to guide development in the sector. Therefore, mass housing as a concept could be approached from different angles, ranging from contextual responses, policies, processes, funding scenarios, technical construction solutions to the adequacy of meeting users' needs. This chapter aims to first introduce the concept of mass housing, through a review of the determining factors in the genesis and development of mass housing solutions, then explores its traditional generic and current challenges, in the context of its shifting conceptual paradigms, to finally lead to prospective developmental opportunities.

2.2 Mass Housing Definition, Ideology and Concepts

The basic and most recurrent definition of housing is the act of providing shelter or lodging. Similarly, mass housing may be defined as the production of a large number of housing units in a given context. This however, will be a reductionist definition for a concept that embodies complex multidimensional considerations. There is an agreement across different cultural and political contexts that mass housing is based on two fundamental concepts: first, it is founded on a standardised design and industrial building techniques. Second, it relies on the belief of an egalitarian living conditions as a social goal, provided by state authorities (Urban 2012). Mass housing is often described as a match between architecture and social policy. In general terms though, mass housing may be best designated as housing provided on a large scale and in multiple units exclusively by government or in collaboration with the private sector for purposes of public acquisition either on owner-occupier or rental basis (Ibem and Amole 2010). The purpose of mass housing is to provide decent housing at reasonably reduced costs, to households unable to afford the heavy investment of acquiring land and building houses at prevailing market rates.

The relevant literature evolves around numerous related terminology including public housing, social housing, affordable housing, low income housing, subsidised housing, community housing etc. all embodying to some extent the two foundational concepts. A more comprehensive approach for a better understanding of the mass housing concept calls for a historical developmental overview, starting as an ideology based on social reforms that unfolded into multiple scenarios with various level of success in different contexts based on a combination of influential and determining factors. An identification of these determining factors that have shaped mass housing solutions is the prerequisite step to further development's consideration.

2.2.1 Generic Challenges

Urban Population Growth and Housing Deficit. The global human population growth amounts to around 83 million annually, or 1.18 % per year. It has grown from 1 billion in 1800 to 7.3 billion as of mid-2015, with an approximately one billion added during the last twelve years only. It is expected to keep growing, with an estimated total population of 8.5 billion in 2030 and to further increase to 9.7 billion in 2050, reaching 11.2 billion by 2100 (United Nations 2015). While the resulting challenges of this fast growth are many, the urban population growth, either as a demographic growth or a result of a rural to urban migration, has triggered or remains the greatest challenge to mass housing supply. Over half (54 %) of the global population in 2014 was an urban population, up from 34 % in 1960 and is expected to keep growing (World Health Organization 2015). The growth trend is unequally distributed, as in absolute numbers over the next several decades, it is regionally concentrated in the less developed regions of the world with Africa leading followed by Asia (Fig. 2.1).

Even in these less developed countries, it is estimated that by 2017, a majority will be living in urban areas. On a historical timeline, the growth of the last 200 years is massive, affecting dramatically living standards, natural resources and the built environment. Unfortunately, only few governments in the developing regions managed to meet their rapidly growing urban population with the land, services, facilities and shelter. The results from Mexico to Jakarta, Lagos to Mumbai are cities in crisis with unregulated settlements, limited facilities, increased overcrowding and severe social ills. The rapid urban growth has created a vast shortage of urban housing worldwide. This shortage adversely affects and is more prominent in the economically weaker section, lower income group as well as the middle income group. Governments, urban planners, real estate players envisioned the development of the mass or affordable housing projects to cater to the needs of

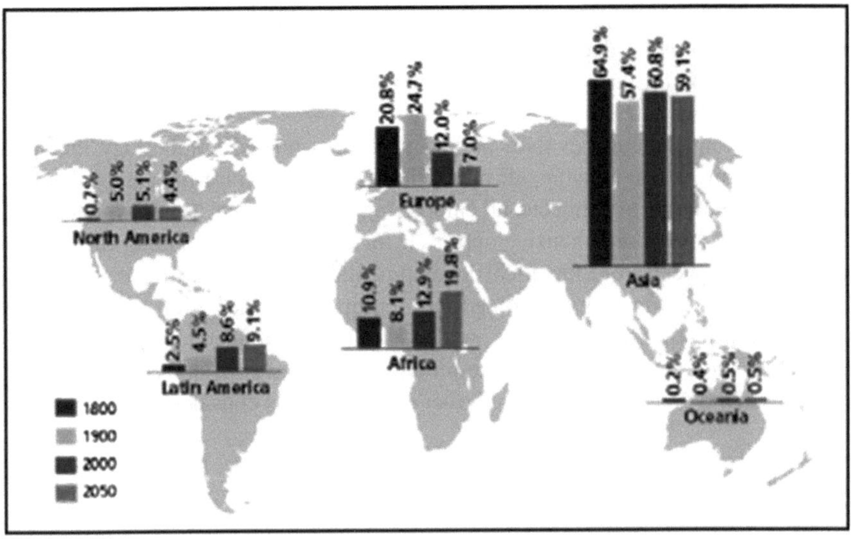

Fig. 2.1 Historical and projected population growth by continent (Sustainable Scale Project 2015)

the lower or middle income households by designing and building most often multi-storied apartment-type housing referred to as Mass Housing.

The challenges of mass housing supply in developing regions lie first in the local production capacity to meet the market demand created by an increasingly growing urban population. In developed regions, the challenges are sometimes driven by different criteria, such as an aging demographic population requiring a flexible approach to housing provision, or an existing older building stock that faces quality and performance issues (Healy 2004; Krieger and Higgins 2002; Saegert et al. 2003; Maisel et al. 2008).

Urban Land Scarcity and Resources Availability. A high population density in urban areas has triggered a massive demand for urban land, driving speculatively urban land prices and constraining affordable housing projects options. In places like Japan, Singapore and Hong Kong, a lack of developable land in urban areas makes multi-unit housing the most common mode of living for the vast majority of the residents. For these environments where high density is inevitable, there are strong supports for this type of compact city approach as a form of sustainable urban development (Wong 2010).

Institutional Framework and Major Stakeholders of Housing Delivery. Key strategic challenges, beyond population growth and land availability, lie in the legal and planning institutional framework governing the supply of housing. In most countries, a housing vision underpins the values of equality and inclusiveness, holistic and cohesive, affordable, durable and qualitative housing are stated in housing goals and materializes in housing policies and processes guiding the various actors in the housing industry. The main components enabling housing production evolve through an adequate assessment of needs, housing policy, funding

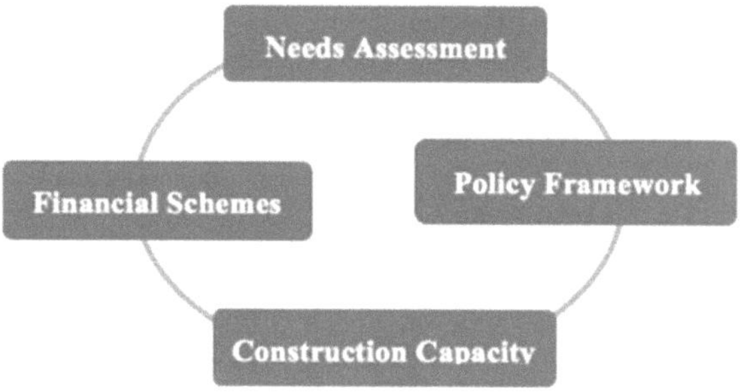

Fig. 2.2 Mass housing production general framework

arrangements and industrial and construction capacity (Fig. 2.2). Provision of housing is critically impacted by a combination of factors including political will, policy, governance, housing financing, urban land management, economy, design, technical, cultural and social factors. The high diversity and number of stakeholders involved in mass housing provision often enough result in delayed implementation (Alexander 2009). Lack of infrastructure and services are additional issues that confront citizens and policy-makers in developing countries (European Network for Housing Research 2015). Achievement of national's housing vision through key housing development strategies unfolded globally in various scenarios but remains, sometimes for different reasons, a challenge in many parts of the world.

2.2.2 Major Mass Housing Development Milestones Versus Timeline

The Industrial Revolution marked a major turning point in history, an era of per-capita economic growth that laid the path to improved standards of living while resulting in a permanent urban growth process. Social and legal reforms initiated during the Industrial Revolution, addressed the crowdedness and insalubrity that characterised the living conditions of the industrial workforce and drove the first movement of affordable housing. World Wars I and II in particular were the next major events calling for massive reconstruction and prompted extensive global mass housing production. Post-World War II era in particular witnessed a global aggressive and expanded demand for housing, a result from a combined need to rebuild of rapid population growth and urbanisation. Mass housing was the ideal choice, from the late forties to the seventies, to achieve scale and speed. The solutions materialised in several scenarios including low rise, single dwelling units but were more often in the forms of high-rise apartment blocks, located mostly peripherally from the existing city urban texture. Industrial construction as it was developed in the early twentieth century eventually became the fundamental technology for mass housing production in most countries and simultaneously rose as a stylistic principle for modern city design.

Mass Housing Timeline

1760-1840 — The Industrial Revolution

18th-19th Century

Social Reforms

- Enforce hygiene
- Support the underprivileged

- New technologies
- Improved standard of living
- Rural-urban migration

1914-1918 — World War 1

Post-War I Era

The Modernist

- Housing destruction
- Industrial construction
- Modern movement

1901-1904	La Cité Industrielle, Tony Garnier,
1920-1922	Maison Citrohan, Le Corbusier
1922-1925	Plan Voisin, Paris Le Corbusier
1926-1929	Gosstrakh & Narkomfin apart. Modular housing Moscow, **Russia**
1935	Robert Taylor Homes, Chicago, **USA**
1938	Quarry Hill social housing, demolished 1978, Leeds, **UK**

1939-1945 — World War II

Post-War II Era

Global Mass Housing

Massive Housing shortage
New technologies
Global housing solutions

- Standardization
- Prefabrication
- Customization

1941-1945	El Silencio development (Carlos Villanueva) Caracas, **Venezuela**
1946	Grindelberg houses (Bernhard Hermkes) Hamburg, **Germany**
1946	Unite d'habitation de Marseille (Le Corbusier), **France**
1947-1951	Levittown, 1st US Prototypical single detached mass housing Suburb, **USA**
1947-1952	Pedregulho and Marquês de São Vicente mass housing projects (Affonso E. Reidy), **Brazil**
1951-1956	Pruit-Igoe Homes, 2800 apartments, demolished in 1972, St Louis, **USA**
1953-Present	Mass housing initiated, Development by Housing Authority from 1973, **Hong Kong**
1959	Alton estate, one of the largest council estates in UK, Roehampton, Southwest London, **UK**
1962	Robert Taylor Homes, completely demolished in 2007, Chicago, **USA**
1960-1965	Housing Development Board, 118,000 standardised high rise units, **Singapore**
1970s	Tanji, Prefa High-rise mass housing projects, **South Korea**
1970- Present	Social housing programs, 250,000 units/year **Algeria**
1970s	Ekbatan mass housing development, Teheran, **Iran**
1975 / 1990s	Bijlmer, Built 1975/ partial demolition 1990s/ Renovation1990s, Amsterdam, **Netherlands**
2002	BedZED, Beddington Zero Energy Eco-community, Sutton, **UK**
2008-2014	Al Falah Housing Program, 5000 villas of 4-7 bedrooms, Abu Dhabi, **United Arab Emirates**
2009 – Present	Minha Casa Minha Vida Housing Program, million single-family homes and apartments, **Brazil**
2010 – Present	Energiesprong, energy efficient mass housing retrofit, **Netherlands**

Fig. 2.3 Mass housing events versus timeline

Figure 2.3 depicts a non-exhaustive timeline record list of some mass housing feature projects. The list illustrates the widespread implementation of mass housing. It spans from social reforms intents and utopian solutions, following the industrial revolution to the first experiments with modular industrialised construction and serial design during the Modernist era to finally reach a climax after the Second World War where mass housing became a global solution. While mass housing is still considered the only option in many developing countries, earlier built projects are already at a historical milestone, prompting either a total denial of the Modernists' vision resulting either in their demolition (Jencks 1977) or calling for their sustainable revitalisation (UN-Habitat 2013).

These mass housing projects took different forms with varying levels of success. It is recognised that "all over the world, local variations in mass housing relate to three factors: a diversification in the conceptual background, the internal differentiation of the respective societies, and the different architectural context in which mass housing was introduced" (Urban 2012).

The next section details the major milestones of mass housing production that have been inspired by various conceptual paradigms including the recent owner-occupier trends and the influences of user participations and comprehensive sustainability of mass housing.

2.3 Mass Housing Conceptual Paradigms

It should be firmly admitted at the beginning that the big portion of mass housing has long been chiefly associated with the state interventions in housing policies. Such adopted housing policies are usually, as Fahey and Norris (2009) claim, so frequent, varied and sometimes even intertwined. These policies often take implicit and difficult to quantify forms that reliable measurement is not available and does not provide the basis for firm conclusions about the extent of intervention at any point in time or change over time. Thus, it is difficult to draw firm separation lines between adopted housing paradigms and their policies and processes. This makes the propositioned five mass housing strategy paradigms in this section fairly indicative and are by no means expressing quite distinctive mass housing policies and processes that can be easily identifiable. Still from a strategic policy point of view, mass housing can be categorised based on the broader meaning of housing as either a service, an asset or a commodity. This would immediately produce three approaches that have been categorised here as the first three paradigms, namely, Egalitarian Policies and Social State, Welfare State and Market-Oriented. The fourth suggested paradigm reflects the impact of participatory approaches in mass housing policies and processes while the last paradigm addresses the comprehensive sustainability as an idiosyncratic approach affecting the related housing policies in general, naturally including mass housing.

This intentionally simplified conceptual framework for mass housing paradigms intended chiefly to make the adopted mass housing strategies more comprehensible for the readers from different professional backgrounds. The conceptual framework depicts each paradigm through three topics. First, the embraced policies; second, their ensuing decision-making processes and thirdly a brief review for some pertinent implementation experiences.

2.3.1 *Egalitarian Policies and Social State*

Main Policies and Processes. The best example of the egalitarian mass housing policies embraced by Social States is that developed several decades back in the former Soviet Union where state housing policy was pillared on both a socio-economical agenda that aimed at satisfying housing shortage experienced under the process of rapid industrialisation and a political agenda meant to consolidate the regime and to form what was conceived as an egalitarian communist society. One of the central ideas in the search for the right model for this 'new' society was a communal way of living and the dissolution of the traditional family model. New socialist housing ideas intended to transit the society from the family nucleus to collective living communities. In such a paradigm, mass housing was approached as a home for a wide spectrum of people, from unskilled workers to executives. By using standardised construction methods, Soviet planners were able to build mass housing in very large numbers. The initiative was to form cooperatives and build clusters for the "more equal" ones in what was assumed to be an equal society, but, according to Sutavicius (2014), things did not work actually in the way utopian communist ideas advocated. Good quality housing built before communist era was nationalised and allocated to communist party leadership. In the meantime, the vast majority of the population had to wait for long periods of time to be allocated a housing unit in an apartment block. Given their low salaries earned from state enterprises and the fact that there was no private property that could be bought, citizens actually had been left with no alternative to house themselves but to wait their turn for an attribution. Mass housing that was available for the majority, and especially the one built next to factories, was not the nicest place to live, mainly due to the size of housing units and poor building quality caused by the use of the most cost-effective and standardised buildings methods and materials (Sutavicius 2014).

Social state mass housing agenda has been defined as an unrealistic utopian notion. Sutavicius (2014) explained this situation by the combination of lack of experience, building material shortages and the theoretical problems of not knowing what exactly the new socialist city should be. These factors, Sutavicius claims, developed various ideas, which often seemed to be more utopian than realistic. With the time of relative intellectual openness in the former Soviet Union by the late 1950s, a group of Moscow University intellectuals led by Alexei Gutnov wrote 'The Ideal Communist City'. The concepts introduced in this book together with unrealistic utopian ideas presented a set of ideas, which had been reflected in the mass-housing program. Gutnov determined mass-housing design criteria encompassing, among others, the size of rooms and apartments; building typology, density; integration with green spaces; bilateral orientation, etc. In reality, Sutavicius argues, some of these design criteria had been successfully applied and some had not, such as the two-sided orientation- most likely due to either being too idealistic or due to lack of rationality.

Fig. 2.4 Narkomfin mass housing project in Moscow, 1932 (Wikimedia Commons 2016a)

In general, mass housing-related issues are similar and evident in many countries, but it is of a whole different scale in post-Soviet era countries. While, 413 million people live in Western Europe, 6 million of them live in mass housing (1.5 %) but while 293 million people live in the former Soviet Union states, 34 million of them live in mass housing (11.5 %). Thus, not only the total number of people living in mass housing in post-Soviet bloc is bigger, but also the percentage of population living in such housing is much higher. West European countries have started working on adapting mass-housing to today's demands already a few decades ago, while Eastern European countries have just started to address these same issues. Quantity of these buildings suggests seeking for a less drastic and more sustainable approach to the issue than the demolition of the existing stock (Sutavicius 2014).

Examples of Implementation. Narkomfin housing development exemplifies the mass housing pattern of the Social State (Fig. 2.4). It was designed for the People's Commissariat of Finance and was completed in 1932. The building has been in a state of dilapidation for several decades now. This example for the mass housing product depicts the low quality of the exterior and interior finishing which can be easily referred to the low cost expenditure by industries which actually did not wish to spend too much money on what was not their primary responsibility. Industry's primary goals were just getting a roof over the workers' heads so that the new plants can be constructed. Furthermore, workers were not in a position to bargain for housing quality since they had to go through long waiting lists to get an apartment (Sutavicius 2014).

Consequently, because of this natural monopoly there was no stimulus for competing for higher quality. In their attempt to reduce cost and completion time, Soviet architects adapted and developed various technologies. The earliest material used was brick. Even though production of silicate bricks enabled reaching lower price and bigger numbers but it was still too complicated, too expensive and too slow. Experts were sent to France to study the pre-cast systems and to acquire the Camus system patented by the French Engineer R. Camus in 1948. Later, new building methods using prefabricated concrete panels were developed. Their main structural typologies were:

- Precast load-bearing concrete structure assembled in situ to which solid panels of precast concrete were attached. This was probably the least successful typology, since this kind of structure was the least stable one. Even though buildings of this type were built in huge numbers, there are very few of them still standing today.
- One more type of prefabricated housing was built using large load-bearing concrete panels to create a rigid structure and joined together by welding reinforced bars and pre-installed metal plates on several points of the perimeter. Exterior panels were hollow, filled with insulating material, such as lava stone or mineral wool. Fabrication of such panels was quite complicated. As a result, it was not the most popular type of panels.
- The most popular system, which reached its peak in the 1970s, was also based on load-bearing large concrete panels that were welded together, but exterior panels as well as interior ones were not hollow. To reach bigger thermal resistance lava stone was mixed into the concrete instead of being a separate layer in the panel.

Theoretically, the expected service life span of load-bearing structures of concrete large panel houses is 125 years, and it might even reach 150 years when renovated. In addition, load-bearing structure is usually built with a reserve, meaning that additional load may be added. Many other building elements, such as roof covering, balconies, cornices, parapets, joints between panels, surface finishing and staircases, have a relatively shorter life span and should be replaced (Sutavicius 2014).

2.3.2 *Welfare State*

Main Policies and Processes. After the Second World War and driven by the massive reconstruction and welfare state development policies, housing came within the realm of public policy. This welfare focus was expressed in a wide range of ways, but it is widely believed that the compulsion towards social solidarity and equality embodied in the welfare state policies has been less realised in the housing sector than in other areas of social provision. It is claimed that in Europe and America, housing has remained the least de-commodified among social provisions. Even in countries where welfare systems are well developed, most housing services are provided by the market. It is argued that since the 1980s, housing services have tolerated the impact of cost-cutting of public welfare spending (Fahey and Norris 2009). In their analysis for the crisis of the welfare state mass housing paradigm, Swenarton et al. (2015) claim that in the 1970s the welfare state system started to unravel when three decades of economic expansion snappily came to an end with the first oil crisis and the arrival of recession. The economic crises of the 1970s revealed the limits of the welfare state and the wealthy society. These limits were not only economic but also social and political. They add that the environmental dilapidation caused by the unrestricted industrialisation and consumption raised awareness about that the western way of life was exhausting available natural resources and damaging the environment beyond regeneration. Swenarton et al. (2015) adds that a further element of the critique of the welfare state and its planning system was represented by the local actions that emerged at this time against the demolition of inner cities and historic districts as part of modernisation, slum clearance policies and functionalist planning. Jane Jacobs and 'The Death and Life of Great American Cities' (2002 edition) represents a milestone here. The costly and cumbersome bureaucracy of the welfare state, the result of a combination of Fordism and Keynesian politics designed to secure optimal redistribution of welfare for all, came under increasing attack. In Britain, the 'Winter of Discontent' of 1978–79 saw the country paralysed by a wave of public sector strikes against which the Labour government appeared powerless, precipitating the election victory in May 1979 of a new Prime Minister, Margaret Thatcher, who pledged to roll back the welfare state. The following year, Ronald Reagan was elected President of the United States, marking a definitive breakthrough of neoliberal politics in the western world, with the welfare state and its institutions now depicted as a patronizing 'nanny state' curbing individual freedom and responsibility. Swenarton et al. (2015) conclude that the welfare state was not a homogeneous phenomenon. The studies highlight the singularities of the different national contexts at different dates, allowing both a broader, and a more nuanced, portrait of the architecture of the welfare state to emerge and providing clues for further enquiry.

Fahey and Norris (2009) believe that the multiplicity of housing policy instruments that can be used across the various housing markets is one of the major complexities that hinder recognising and analysing the role of the state's welfare in housing in order to quantify their scale and impact. The housing distributive public policy adopted in welfare states was classified into either focusing on the mass housing as a 'capital' or as a 'service'. The former approach can be achieved by extending home ownership, while the later entails state intervention in the rental housing market. For example, Fahey and Norris maintain that English-speaking countries and developed south-east Asian states are commonly regarded as 'home owner societies' where public policy affords this tenure type preferential treatment. By the adoption of the mass privatisation of former state owned rented housing in the early 1990s, most of the former communist countries of central and Eastern Europe lay also within this category.

Meanwhile, northern European states generally favour state supports for rental accommodation. Nonetheless, Fahey and Norris (2009) admit that their 'capital-service' categorisation model is over-simplified. This is simply because transferals are repeatedly witnessed in some countries over time thus making them deviate from a specific type to the other.

For instance, Norway, which is commonly classified as a social democratic welfare regime, stands out for its extensive and longstanding state support for home ownership. Furthermore, in the majority of developed countries, irrespective of welfare regime, home ownership has become the dominant tenure during the twentieth century even with slower pace in northern European states due to their large social housing sectors (Fahey and Norris 2009).

Finally, Fahey and Norris (2009) point out that there is likely an upper limit to home ownership in well-functioning housing markets. This is chiefly attributed to the existence of a mobile population for whom renting is more efficient and usually also a low-income population for whom investment in home purchase is not feasible.

Examples of Implementation. An important example for the mass housing product of the welfare state adopted policies is the UK council housing (Figs. 2.5 and 2.6). A council house is a form of public or social housing built by local municipalities with the aim to supply well-built houses on secure tenancies at reasonable rents to primarily working-class people. The history of council housing development dates back to the late 19th century and reached its climax in the mid-20th century. At that time, council housing included several suburban "council estates" and numerous urban tower blocks. But since 1979, the role of council housing has been reduced by the introduction of the Right to Buy legislation, and a change of emphasis to the development of new social housing by housing

Fig. 2.5 Social housing—Perkin's Rents (Wikimedia Commons 2016b)

Fig. 2.6 Social housing in Tipper Lane (Wikimedia Commons 2016c)

associations. After all these years, a significant percentage of the population in the UK still lives in council housing. According to 2010 estimations, about 17 % of UK households still live in social housing. Additionally, 55 % of the UK social housing stock is still owned by local municipalities while 45 % is owned by housing associations (Wikipedia 2015).

Allocation of council housing has been based on the priority being given to those in greater need for housing. This yielded a long waiting list and a point allocation system according to need was utilised to define preferences for housing unit allocations where houses go to those who earn the highest number of points. Accordingly, gaining council housing property has depended on need rather than financial resources.

The council houses have two, three, four or five bedrooms, and ample back gardens intended for vegetable growing. Despite their spaciously designed living spaces if compared to the private sector housing of the same standard, council housing was tided by the rigid council rules which hindered responsiveness to the tenants' individual needs. Apartments and bungalows which were first built by local councils during the interwar years became a common pattern by the 1950s. During this decade, UK councils started building garages on new housing developments as car ownership increased. On the other hand, tower-block flats were intensively built in the 1960s. But quickly by the 1970s, the construction of these high-rise flats was interrupted mainly because of their unpopularity due to the poor insulation and structural defects. As a result, tower-block clearance schemes were becoming common by the end of the 1980s. The remaining towers have been refurbished and re-opened as "vertical warden-controlled schemes" (Wikipedia 2015).

The envisaged role that the council housing plays in helping the poor have been criticised by social policy economists mainly because it hinders labour mobility due to its system of allocating housing to those in the local area. Accordingly, working-class people are not encouraged to move across district lines as where they would be further down the waiting list for council housing in the new districts. Council housing system has been also criticised for favouring those who have already secured tenancy, even after they are no longer in need. The combination of securing tenure and subsidised rent actually does not encourage tenants to downsize from family accommodation after their children have moved out leaving behind who are in disparate need of this welfare. This rigidity in mobility and allocation are stimulated by the regulation that tenants of council housing cannot be evicted except for anti-social behaviour, serious offences committed at the premises or serious breach of the tenancy conditions, such as rent arrears. From the social point of view, council housing has been criticised for creating an isolated community where occupants usually have low aspirations and they need to have better training and more incentives to work. Finally, council estates are usually stereotyped as "problem places" where crime and welfare dependency are particularly recorded. Also, council housing residents are stereotyped as an underclass (Wikipedia 2015).

2.3.3 Market-Oriented

Main Policies and Processes. In the1990s, privatisation process in various sectors including social housing was adopted by several countries all over the world. Ample justifications have been developed by the private sector, encouraging market economy, attracting foreign capital and technologies, reducing state deficit and initiating investment programs. Mass-privatisation is the process through which the ownership of most of the state mass housing properties is transferred to the tenants. This process has been significantly implemented in post-soviet countries as their economies went through a rapid change from planned to market economy (Sutavicius 2014). This conversion from tenant occupation to owner occupation of mass social housing has been understood as a consequence of the triumph of the market and the setback of the welfare state. In the western world, the 1990s witnessed a major transformation from the tenant occupier to the owner-occupier. Nonetheless, there is still a considerable magnitude of rental in almost all western societies, but home ownership remains the general trend there.

For the meaning of the 'home ownership' Fahey and Norris (2009) argue that from the capital perspective it remains as capital i.e. a form of wealth that is accumulated and distributed through the market and is not directly controlled by the state. While welfare policies allow state's intervention in the rental market, home ownership supporting policies, by contrast, are distinguished by the market domination. Owner occupation policy enables housing as capital to be commoditised and traded in the market while still being influenced to some extent by state-imposed regulations. It is claimed that in recent years housing policies in western countries are more and more adopting an 'Anglo-Saxon' model of minimal state intervention in housing. Still, the policy instruments are so variant and strongly influenced by the sub-set of policies found in each country. Although the contraction of the social

Fig. 2.7 Bijlmer Mass Housing (Wikimedia Commons 2016d)

rented sector in many countries such as Australia, New Zealand and the UK establishes for a continuous withdrawal of state involvement in housing provision, it has usually been accompanied by intrusive policy instruments in regard to building standards, tenancy conditions and even rent controls (Fahey and Norris 2009).

Examples of Implementation. The renovation of the mass housing project of Bijlmer in Amsterdam is a good exhibit for the conversion from tenant occupation to owner occupation in mass housing as a tool for the renovation process itself (Fig. 2.7). This mass housing project, completed in 1975, was initially perceived as a town of the future as it was built in accordance with the urban principles of Le Corbusier. The project encompassed 18,000 housing units, 13,000 of which are flats in 11 storey apartment blocks. But right from the beginning, the project suffered from the lack of services and delay in infrastructure and transportation provisions. In addition to that and because of the preference for living in single family dwellings developed in the new neighbourhoods on the outskirts of small cities in the province, the residents started leaving the Bijlmer even before all the high-rise buildings had been completed. Immigrants from Surinam, the Dutch Antilles and other non-western countries quickly took over the empty homes left by the departing inhabitants (Intercultural Cities 2011).

Due to the numerous empty garages and storage spaces beneath the high-rise buildings, the Bijlmer became a notorious place for illicit activities (Fig. 2.8). Accordingly, the neighbourhood was soon labelled unsafe and dirty. To make things even worse, on 4 October 1992, a cargo plane crashed into some flats in the Bijlmer where a total number of 43 people were killed. This incident acted as the catalyst for a decision in 1992 to demolish a quarter of the flats in the Bijlmer and in the late 1990s an evaluation conducted revealed that more would have to be

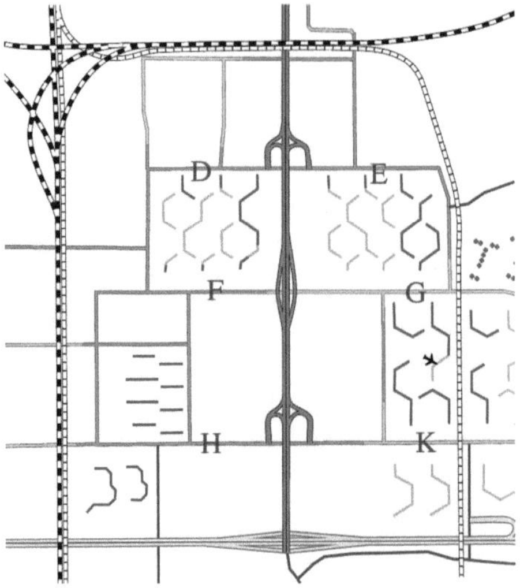

Fig. 2.8 Map of the Bijlmer Map Key: *Red* Original flats standing; *Grey* Demolished flats; *Mint* Elevated roads; *Green* Renovated roads (ground level). The plane marks the site of the Bijlmer disaster (Wikimedia Commons 2016e)

demolished. As part of the regeneration plan, new business and office premises have been built and the main shopping centre has been extended. Three major banks have built headquarters buildings on the fringes of the area, and smart new metro stations provide excellent connections to central Amsterdam. Meanwhile, the residents who wanted to remain in the area were entitled to a suitable and affordable home there. The 6500 homes marked down for demolition were replaced with 7200 new ones with about 30 % of them allocated for council housing and the majority of about 70 % allocated for the free market owner-occupied or rental sectors (Intercultural Cities 2011).

According to Bijlmer Renovation Planning Office (2008) about 1250 renovated flats have been 'repositioned' into the housing market and were eventually sold. These flats were first offered to the resident tenants, and then to other interested parties. Residents themselves initiated a project named: "Buy Your Own Bijlmer" in which 90 homes were renovated in collaboration with a group of owner-occupiers. Table 2.1 illustrates the development of the housing stock in the Bijlmer from 2002 to 2013 where it is fairly obvious that there is an increasing trend towards owner occupation. Number of social housing has declined from 19,463 to 13,701 (−22 %) while the number of owner occupier housing has increased from 2361 to 5602 (16 %) (Klundert and Bold 2014).

2.3.4 *Participatory and User-Control*

Main Policies and Processes: Mass housing programs have been criticised in the last few decades for being associated with uniformity, repetition, anonymity and uprooting. But, on the other hand, what might be called "informal" mass housing,

Table 2.1 Bijlmer housing type by owner in % of all dwellings from 2002 to 2013 divided into owner occupier, social housing and private rental housing (Klundert and Bold 2014)

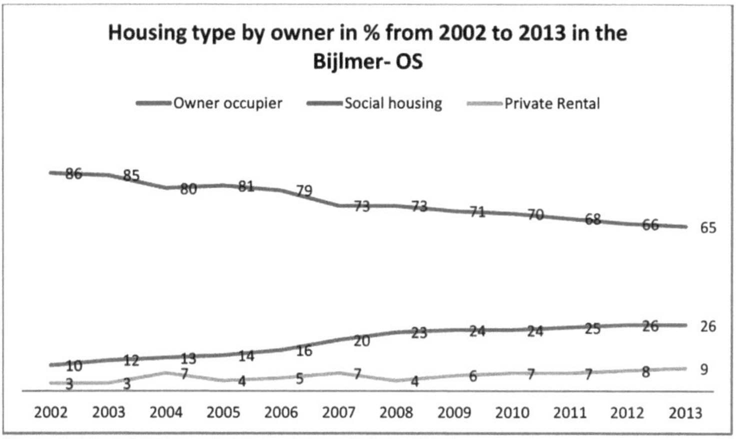

Housig type Bijlmermeer (T93/T94)	Owner occupier	Social housing	Private rental	Total
2002	2361	19,463	790	22,614
2013	5602	13,701	1937	21,240
Increase in %	16%	−22%	6%	−1374 (−7%)

or more accurately self-built housing, produces more diversity and a bigger sense of appropriation and identification of the inhabitants with their living place. Revealing the 'human' logic of 'informal' mass housing necessitates identifying the needs and the capacities of the inhabitants to intervene in the design and building processes of their future dwellings. This, for sure, would require that inhabitants be involved right from the beginning of the design process of their homes (OIKONET 2014).

Sanoff (2000) defines public participation as a direct involvement in decision-making processes whereby people share in decisions that determine the quality of their lives. The citizen participation process, Comerio (1990) maintains, is guided by two influential principles of empowerment. The first, which is mainly political, recognises the rights of all citizens to have a voice in future decisions that affect the places they inhabit, work and relax in. The second principle, which is humanistic, recognises that professionals have a responsibility to design buildings that are usable and understandable, with a sense of territoriality and an appropriate scale, and to develop neighbourhood and city plans that allow the greatest opportunity for all people to participate in the activities a city has to offer.

Participation as a process begins from the passive level, where residents do not take a real part in the decision-making processes, and ends with the most active role where they take sole control of these processes. Sherry R. Arnstein (in Julian et al. 1997) classified the levels of participation, according to the degree of power the participants have, into eight levels which she equated to eight rungs of a ladder which she called 'The ladder of participation'. The two bottom rungs (manipulation and therapy) represent non-participatory process. For example, participation on advisory boards and programs. The next three rungs (informing, consultation, and placation) are defined as precursors to participation. This includes situations in which participants either listen to, or provide advice to, traditional power brokers. The last three rungs (partnership, delegated power, and citizen control) are considered as involving a true participation level in which citizens can directly influence policy.

The current advancements in digital design methods and production technologies for mass housing brought about a renewed interest in prefabrication within construction that is based on customisation. Digital innovative technologies can help address more efficiently the current concerns with social needs and sustainability in mass housing schemes (OIKONET 2014).

Examples of Implementation. The techniques of user participation in mass housing processes may be classified into three broad categories: communication techniques, involvement in decision-making processes, and hands-on participation. First, communication techniques include disseminating information in leaflets, newsletters and exhibitions. It is also about gathering information through questionnaire surveys. Second, involvement in decision-making processes entails: holding public forums/meetings, community planning weekends and participatory workshops which aim at promoting a dialogue between the designer and user from one side, and between the different user groups from the other. Third, hands-on participation techniques that include: charrette process, open building, self-help and housing cooperatives (Towers 2000).

The charrette participatory process encompasses, in addition to a structured schedule and an open access to the process, three defined mechanisms: first, idea generation which requires knowledge transfer among all participants. Second, decision-making, which entails a dialogic discourse about the ideas presented. Third, problem-solving which provides recommendations and proposals as the outcomes of the overall process. The two main objectives of the charrette process are first, to gain the unified support of a representative cross section of citizens who are committed to implementing the resultant proposals (for mass housing project), and second, to obtain the commitment of the power structure in order to secure the required resources (Sanoff 2000).

In open building approach, a 'mass-produced' house is meant to be adaptable and easily altered in order to be 'customised' to suite various residents' needs. The main concept is that the internal living spaces of this flexible housing unit may be indeterminate so they can have many uses or might change physically with internal partitions that can be rearranged. One example of this customised mass housing is the "Living Homes" designed by Kappe, Kieran and Timberlake. Through the Living Homes project website, people can choose, within certain limitations, among several houses designs and can even customize them by adding rooms and choosing materials and finishes. Following the theory of John Habraken a flexible house may be part of a support or a permanent structure that could be a framework or

Fig. 2.9 Customisation of Adaptable Houses in the Nemausus Apartment Building, Nimes, 1987–94, Designed by Jean Nouvel (Wikipedia Commons 2016f)

infrastructure containing secondary structures, separate units of housing or in-fill built with industrial components. In a modular house design, any room can be adapted to become a bedroom, a lounge, a dining room, a study room or a playroom. Customisation entails that the house can be customised internally, externally or both (Fig. 2.9). The inhabitants could alter its façade but also could continue building, whereby the house transforms or expands (Oikodomos 2011).

Meanwhile self-help or self-build approach depends on using ordinary people, especially the unemployed, as a resource in building mass housing. In the self-build process, the professional's role is to become an enabler, rather than a dominator, who can articulate people's case and scale down the process of building design to meet the individual needs of people (Hackney 1991). In the self-build mass housing scheme in Colquhoun Street, Stirling in Scotland, the projects' land was allocated by the local authority. The beneficiaries of the projects, the unemployed people in the area, were the builders of their own houses. They were members of the Colquhoun Street Self-build Housing Co-operative. To gain the needed skills of the building profession, experts were invited to teach the needed skills such as plastering, bricklaying, electrical wiring and plumbing. In another example, Fig. 2.10 shows Michael Riggio, a mason by trade constructs a brick wall for a detached garage for his home in a Self-help Housing neighbourhood in Newville, Pennsylvania. Owners of homes in the neighbourhood helped each other with the construction. Loans for the homes in this neighbourhood were made available through the United States Department of Agriculture, Rural Development and Self-help Housing Loan Program.

Other forms of self-help mass housing are sites-and-services and core houses schemes. In sites-and-services schemes tracts of urban land are divided into serviced plots. These serviced plots could be either sold or leased to those who wish to build their own houses on them. Although authorities leave the development of the plots completely to the residents, they could develop some plot so that basic standards could be maintained (Choguill 1995). Core, unfinished or expandable houses are buildings designed by architects but whose inhabitants conclude the process, meaning they continue to build their homes. In this case the architects can

Fig. 2.10 Self-Help housing neighbourhood in Newville, Pennsylvania (Wikipedia Commons 2016g)

design cores or half-finished houses, planting the seed and leaving certain guidelines. Thus, a dwelling is not a finished product but is part of a continuous building/changing process (Oikodomos 2011).

Co-operation, with its various forms, became a key element in 'community architecture' and 'self-help' initiatives. From the 1920s, co-operative housing has prevailed in many European countries, particularly in Scandinavia and Eastern Europe. Meanwhile, in Britain it grew during the 1970s and 1980s inspired by the earliest attempts to promote rehabilitation as an alternative to large scale demolition and redevelopment. The members of co-operatives usually come together freely to work constructively with each other for their mutual benefit. A multitude of small-scale co-operatives in housing, in building and in community services were established. Most housing co-operatives are set up to provide housing for their members through self-build, part self-build, or organising and managing new construction or conversion. When construction is completed, houses are usually distributed to the members of the co-operatives. The members may own these houses and then the co-op ceases to exist (market value). Or the ownership remains with the co-op and each member becomes a tenant and in this case the co-op becomes a consumer one (par value). Or with a mixture of individual and collective ownership, the co-op becomes consumer co-op, part owning the housing (equity value) (Towers 1995).

2.3.5 *Comprehensive Sustainability*

Main Policies and Processes. Presently, it is globally accepted that the building sector, including mass housing, has the highest potential for applying sustainability measures encompassing reducing greenhouse gas emissions at a partially low cost. Despite the fact that sustainable buildings in general and sustainable housing in specific can easily achieve considerable environmental, social and economic gains, little consideration is currently given to sustainability in social mass housing projects especially in developing countries where housing provision is a major political and social priority as large segments of the population in these countries still lack access to affordable housing.

In the following section the outcomes of the UN-Habitat (2012) publication entitled Going Green: A Handbook of Sustainable Housing Practices in Developing Countries is presented with the aim to shed light over the latest debate about sustainability in mass housing especially in the developing countries where it is most needed.

From a policy point of view, there should be an adopted comprehensive approach to sustainable mass housing where it would be envisaged as a comprehensive process that considers the environmental, social, cultural, economic and institutional dimensions of sustainability. This would overcome the conventional approach of sustainability as if it means only environmental efficiency. Rather, besides considering its long-term environmental aspects, sustainable mass housing should take into account the social, cultural and economic balance of the mass housing stock and its occupants. Sustainable mass housing should be affordable and the planning and building process should be harnesses to empower communities and build residents' social and human capitals.

The assertion of the comprehensive sustainability approach simply means that a house is not going to be recognised as sustainable if it is not culturally and socially appropriate; economically affordable; located near to employment and services; and connected to a well-functioning infrastructure and service network. This integrated mass housing sustainable development should make use of indigenous methods and permits accessibility to the least empowered factions of the society as well as reducing disaster risk.

According to such a policy, the whole lifespan of a house needs to be considered at the very early stage of its design and across all the stages of its development. Thus, the whole process of building should be thought of as a sustainable long-term holistic building process. This entails: planning, construction, operation and maintenance, demolition, disposal and recycling of construction materials. On another front, it is crucial for developing countries to correlate their adopted mass housing processes with strategies of economic development, cultural preservation and social empowerment and participation of communities. Knowledge sharing, partnerships and cooperation between all actors working within the sustainable housing sector should be encouraged. Building on demonstration projects of other agencies is recommendable and scaling up sustainable housing practices of great importance.

Environmental sustainability. Environmental sustainability policies and strategies in mass housing revolve around two axes. The first, is saving energy in the whole mass housing process including manufacturing and transportation of materials with priority given to natural and recycled materials; designing and planning for achieving energy efficiency; construction. The second, is the utilisation of renewable energy and natural resources including solar energy, rain water collection and passive design techniques.

Social sustainability. Social sustainability measures that should be facilitated in mass housing strategies entail the quality of life, adequate living standards and accessibility to basic urban services and infrastructure. Social sustainability can be achieved through working for equal empowerment opportunities and participation for all gender, age and income groups of local communities in all mass housing planning, and production decision-making processes.

Economic sustainability. One of the economic sustainability dimensions in mass housing policies is income generation through the provision of opportunities for jobs and productivity. Mass housing construction itself can create a lot of jobs for locals. New employment can be created through the production of energy efficient or recycled materials and through related renewable energy and technologies. This would increase the potential for poverty reduction by building on existing skills, networks and capabilities. Another economic sustainability policy is the use of locally produced materials and local construction methods practiced by the community members as this can create local employment. Also, in order to support the long-term economic sustainability in housing tailored training workshops should be provided to local communities and professionals. This will not only increase the capacity of the residents but will develop the capacity of the housing sector in general and will secure its economic sustainability.

Cultural sustainability. Culturally sustainable mass housing is the housing that is culturally responsive and appropriate to the occupants who use it and the society in which it is located. Cultural adequacy is one of the seven criteria of 'adequate housing' as prescribed in the International Right to Adequate Housing. Cultural sustainability can be realised in mass housing projects through aspects such as the form, design, spatial layout and materials that reflect the community values, ideals and lifestyles. Culture sustainability requires that mass housing projects should focus on local community needs and allowing community-driven bottom-up approaches instead of following misallocated alignments of the top- down approaches by the governments or donors.

Example of Implementation. BedZED is a famous example of the implementation of sustainability measures in a mass housing project (Figs. 2.11 and 2.12). The development of this large-scale, mixed use sustainable community with its 100 housing units, office space, a college and community facilities was

Fig. 2.11 BedZED Eco-Village, England (by Tom Chance) (Wikipedia Commons 2016h)

Fig. 2.12 Street in BedZED (Wikipedia Commons 2016i)

completed in 2002 in Britain. The main goals of the design were to achieve big reductions in greenhouse gas emissions and water use as well as providing liveable place for people with less relying on private cars (Bioregional 2015).

The project was developed by The Peabody Trust, one of the largest and longest established providers of social housing in London, in partnership with Bioregional and designed by ZEDFactory and Arup engineers. The project has a wide range of housing types starting from one-bedroom apartments to four-bedroom houses. Half of the housing units were sold on the open market, one quarter was reserved for social (low-cost) rent by Peabody while the remaining quarter was allocated for shared ownership. BedZED is a high density development by suburban standards but still most of the houses enjoy private outdoor spaces and many have small gardens. For shared social activities, BedZED has a shared square and a large playing field. Passive solar heating was utilised for the great majority of the project's houses and flats as they were designed to have multi-storey glased sun spaces oriented to south.

While all the housing units are highly insulated they are also well ventilated through the colourful wind cowls on the roofs. This simple passive technique draws fresh outside air into the building which is pre-heated by outgoing stale air via heat exchangers. Among the passive technologies used in the project is the Thermal Mass of constructed envelop where the buildings were constructed with concrete in order to store heat in their block work, floors and ceiling slabs, helping to maintain a comfortable and even temperature night and day throughout the year (Bioregional 2015).

Sustainable systems employed in the project include a gas-fired communal boiler that supplies hot water for the entire development through an underground mini-district heating system. Each house was provided with a large hot water tank to help keep it warm in winter as well as storing hot water. Saving water was achieved through using dual flush toilets, aerated flow taps and shower heads and low water consumption washing machines. To monitor water consumption water meters were installed in locations that are easy to view in the houses. For electricity, energy efficient appliances and lighting bulbs were installed. More importantly extensive photovoltaic (PV) panels were installed and integrated on the roofs and into south facing windows. The panels managed to supply some of required electricity for the whole project. The system is designed to export surplus generated power into the local electricity grid (Bioregional 2015).

In order to reduce the environmental impacts from transport, the BedZED developers did their best to provide the construction materials from as close as possible and to make maximum use of recycled materials. Accordingly, it is claimed that over 3400 tonnes of construction materials, 15 % of the total used in BedZED, were reclaimed or recycled products. Most of the steel is reused as much of it was brought from refurbishment work at Brighton Railway Station. Reclaimed timber was used for the interior partitions and flooring. A thousand tonnes of 'sand' made from crushed glass was used under the outdoor paving slabs (Bioregional 2015). On the other hand, the bricks used on the outside walls of the project's buildings came from just 20 miles away while the timber cladding is green oak which was sourced from woodlands in neighbouring Croydon and Kent (Bioregional 2015).

Despite the high expectation about the BedZED sustainable performance as a near zero carbon development but unfortunately some of the technologies have not proved effective as anticipated. One of these techniques is the originally installed combined heat and power plant which provided carbon-free heat and electricity from local street tree thinnings never performed well so it was eventually replaced by a gas-fired boiler (Bioregional 2015).

2.4 Mass Housing Sustainability Challenges and Opportunities

For most people, owning a home is a lifelong dream, and affordability is one of the chief concerns regarding the housing industry. Not only owning a home is related to quality of life, but also gives a sense of stability, security and control over residence. Mass housing development—providing a large number of dwellings in a given area—is one of the most popular methods of building residential housing today. It involves economies of scale and, because of the high-volume of construction, it functions as an effective means of reducing housing costs when components, as well as construction processes, are standardised (Noguchi 2004). Therefore, the key factor for enabling mass housing production is standardisation. Depending on repeating the same operation, mass production is characterised by increasing productivity and outputs, in turn, contribute in lowering costs and prices without reducing quality.

Achieving sustainable development in the housing industry needs to address social, economic and environmental aspects. Housing is seen increasingly as an ongoing process rather than a final product, nor just a commodity. It is a complex process of many people, organisations and stakeholders working together or individually in order to achieve tangible goals. Demand for housing is mainly driven by demographics. However, it is also affected by other factors such as income, housing prices, cost and availability of credits, user preferences, developer preferences, price of substitutes, and price of complements. Housing, and its provisions, has long been central to our accepted understanding of quality of life. The past years have witnessed changes in many countries' housing sector mostly with governments' intention to move people to ownership through loans (Aga Khan Award for Architecture 1979). Mortgage rates contribute to housing affordability. Also, alternative-funding models can be used to assist housing initiatives driven by a partnership approach between private and public mass housing models.

Although standardised mass housing, in particular, lacks adoptability for new inhabitants versus customised one that celebrates a more responsive living space. Housing development requires sustainable solutions because building a house consumes a large amount of energy throughout the process from construction until after occupancy (Noguchi 2004). According to Hebdige (1979), standardised culture is liberating rather than limiting because the potential uses and combinations of consumer products and symbols are endless. The different roles of architects and developers make the former concerned in building for individuals rather than for people, while the latter's concern is to build for the markets rather than individuals (Groele 2013). In the past, many architects perceived the concept of mass production and prefabrication as a way to bring architecture to the masses and ordinary people and to have indisputable positive impacts (Davies 2005). In other words, the selected industrialised building systems, that determine the characteristics of prefabricated housing, can be classified into two typical types: product-oriented and process-oriented. The former type includes industrialised building systems using innovative building products (structural parts and components) to build a house at the construction site in accordance with the builder's technical specifications. The latter type includes the industrialised systems that basically use conventional materials and methods (Noguchi 2004). Therefore, the selected technique or methodology for a building system to produce housing units needs to satisfy the needs of individual consumers, as well as the society.

The role of physical design of mass housing is to better respond to social needs if integrated into everyday physical design practices. Ultimately, it offers an opportunity to engage beneficiaries and decision-makers into a dialogue to provide units that contribute to the enhancement of housing stocks. The design, construction, improvement and management of housing should be taken care of in a collaborative responsibility of local communities, not solely of the state or the market. In order to clearly perceive the intertwined responsibilities of different stakeholders, it is essential to distinguish at least three basic "levels of action" and three basic "levels of authority" in the housing sphere (Aga Khan Award for Architecture 1979). The three levels of activity can be described as the *assembly* of all the components of a dwelling environment, the supply of the *components* that are assembled, and the provision of access to the *elements* from which components and therefore assemblies are made. Consequently, environmentally improved mass housing needs to be addressed on two levels. The first is to achieve neighbourhood environmental sustainability by improving the microclimate through vegetation, urban agriculture and urban landscape elements/furniture that response to environmental needs. The second is to promote the use of local, natural and recycled building materials, passive solar design and climate-suitable design methods.

2.4.1 Potential Future Patterns of Decision-Making and Building System Processes

Mass housing approach may be seen as part of sustainability in housing development. This requires the assessment of its cycle from its planning through its occupancy. It is part of the application of industrialised homebuilding technologies

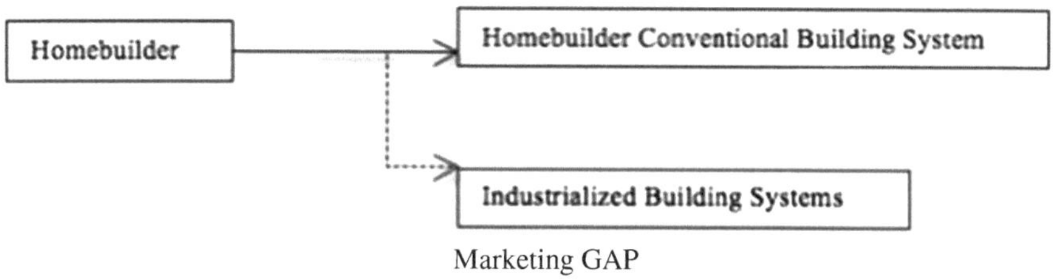

Fig. 2.13 A homebuilder's choice for a building system within the closed system mode of operation (Noguchi 2004)

to improve housing development and delivery process. It aims to provide solution to current global challenges associated with sustainable development, such as maintaining natural resources, cost and renewable energy. According to the USA and Canadian standards, the process-oriented industrialised building systems can be sub-categorised into four general types: pre-engineered, panelised, modular, and manufactured. The managerial responsibilities of leading housing development require the best use of technology to transform natural capital into human capital. In practices, few homebuilders consider such benefits as part of their managerial responsibilities within the closed system mode of operation, therefore, they are reluctant to embrace industrialised building systems and would rather use conventional operation systems (Fig. 2.13). This, according to Noguchi (2004), can be seen in modular and manufactured housing systems, whose in-factory components assembly is relatively high. Homebuilders have already partially accepted pre-engineered and panelised housing systems, which resulted in minor modifications of their production and consumption cycle, when the industrialisation technologies helped increase productivity (Fig. 2.14).

The closed system boundary

The selection of any building system needs to be made at early phases of the construction process, as it has considerable impact not only on the characteristics of the housing and its production, but also on balancing the cost-quality-time trade-offs. Such trade-offs are very much related to the environmental impact of the prefabrication cycle from construction through, or after, the occupancy. Consequently, the selected design and construction building system adopted by the decision-makers and homebuilders can very much produce sustainable mass housing approach, and housing development in general.

At the governance level, this requires a suitable balance between top-down, policy-makers, grass-root movements and local community-led interventions in providing the conditions for long-term mass housing regeneration strategies. Consistent mass housing policies are required to accommodate the participation of the community groups and beneficiaries to combine mass production for economies of scale and scope. In this regard standardisation becomes essential to mass production by enhancing possibilities and plan coordination for collective services.

As mentioned earlier, decision-making and communication in mass housing constitute a complex process that is determined by several driving forces that

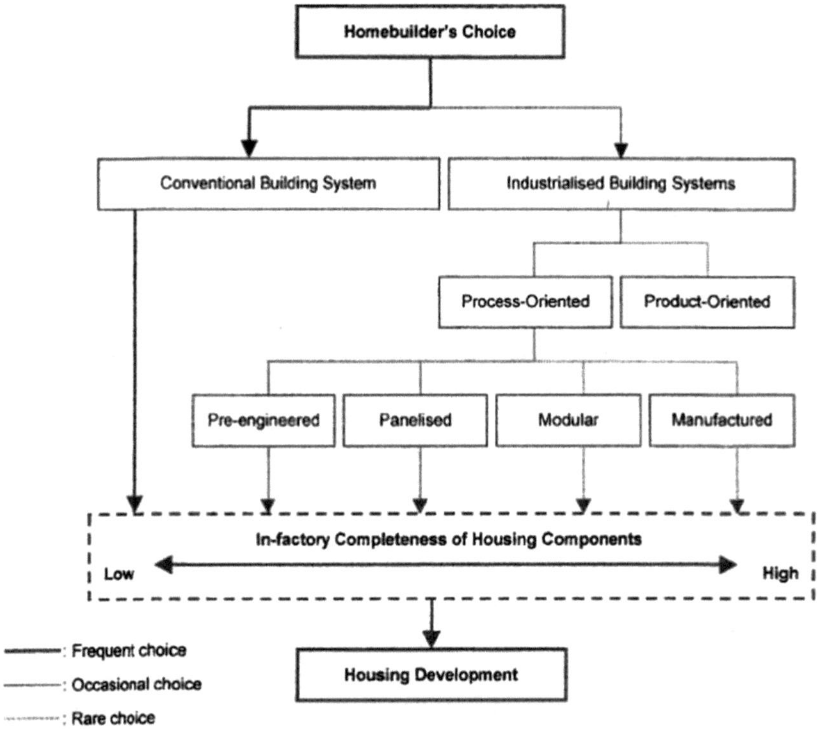

Fig. 2.14 The homebuilder choices for building systems and the level of industrialisation (Noguchi 2004)

involve obtaining information about specific products and services, evaluating alternative purchasing actions, and negotiating necessary arrangements with materials suppliers.

2.4.2 Potential Future Scenarios of Mass Housing

It is important to keep in mind that housing is more than just the physical form of the house, but is also about the embedded social trends that give more value to the form. This approach promotes the significance of this realm of architecture and helps make sensible mass housing. Mass customisation is the opposite of mass production. To mass-produce a house means to build the same model many times. However, to mass-customize a house means to manufacture many components that can be combined in various ways, where the combination of these components could be calculated and visualised using computer program and technology (Schneider and Till 2005). In this scenario, each individual may choose the house in which he/she wants to live according to his/her requirements and within a collective environment. Theories suggest that mass housing development might help drop construction cost. Also, housing manufacture helps increase product quality, as well as reduce construction time. Nowadays, consumers rely on customised end products rather than generic. The concept of quality had grown from *reliability*, popular in the 1970s, to the *individualisation* of products—initiating the present tendency for today's consumers to prefer purchasing customised products (Anderson 1997). Also, it has been argued that design and marketing are interconnected and often share the same objective: to develop the "right product, for the right market, at the right price" (Bruce and Davis-Cooper 1997). However, these ends require different skills to achieve their objectives. The tendency to move towards the individualisation of products and services may expedite current home providers to produce homes that are more customised, in response to the desires and needs of their individual clients (Schneider and Till 2005). In other words, every mass housing project focuses on the cost, quality, and time factor.

Flexible housing approach, defined by Schneider and Till (2005), is "designed for choice at the design stage, both in terms of social use and construction, or designed for change over its lifetime". They argue that flexibility is an important in housing design and should be considered if it is socially, economically and environmentally feasible. Their research suggests that the degree of flexibility is determined in two ways: the first one is the in-built opportunity for adaptability, defined as 'capable of different social uses', and the second is the opportunity for flexibility, defined as 'capable of different physical arrangements'. This principle of enabling social and physical change in housing might appear self-evidently sensible. Schneider and Till (2005) argue that the adoption of flexible housing fosters benefits on many levels. It addresses issues of finance because the flexible approach is more economic on the long term; participation as it encourages user engagement in the design process; technology in the way flexible housing involves, or is determined by, innovations in construction technology; and use in the way that flexible housing embraces to different function needs over time.

It is impossible for any administrator and professional specialist to be expert on people's needs and priorities, and impossible for them to know how that enormous multiplicity of priorities and effective demands can best be met with locally available resources, or design needs. To avoid housing burdening, people need to assert their needs and right to resources. The design, construction, improvement and management of homes and neighbourhoods are the responsibilities of local communities, not of the state or the market, and should be addressed based on communicative and collaborative approach beyond physicality and closer to sociability.

The housing industry has the same essence as all businesses—customer satisfaction (CS). Successful homebuilders focus on customer features and benefits, identifies the target market by conducting consumer research. According to adopted design approach, today's homebuilders can be categorised into three general categories: production, semi-custom and custom (Smith 1998). Production builders are structured for higher capacity construction; thus, they develop several model homes or samples, normally designed on a speculative basis, in response to the market demand. The *production* (or *speculative*) *design* approach allows homebuilders to produce a model home, or a sample, in which the buyers can study the quality and attributes of their new home in a way that paper design or blueprints alone cannot achieve—thus, helping to assure the buyers' satisfaction, or CS. The advantages of speculative design also include the reduction of the construction time and cost. The overall time to produce a standardised house is evidently shorter than that of a one-of-a-kind design, because construction personnel are familiar with the plans. Other benefits of production (or speculative) design incorporate economies of scale, selection of available design choices, practical construction methods and alteration of structural components (Smith 1998).

The builders who apply the *semi-custom design* approach are often called semi-custom builders. They usually combine features of ready-built and custom-built homes. Although they are like the production builders in terms of working with pre-existing plans or ready-design models, they are flexible when it comes to design changes, including those that need engineering and official approval. This gives more opportunities of modifications and customisation of interior and exterior finishes, structure and space of the new homes (Smith 1998). This approach has similar benefits, whenever starting with pre-existing plans, in terms of having faster construction time, less cost, and manufacturing.

Custom-builders focus on starting from a blank piece of paper to create a totally new home design according to the customer's requirements. Customers occasionally establish relationships with their architects for plan development, however, some builders sometimes act as architects or have an architect or draftspersons on staff—in such cases, these associations are called "design-build firms" (Smith 1998).

Custom design approach is seen as the optimum solution to the customisation of a new home, and it helps create one-of-a-kind homes that correspond to the consumer's individual needs. However, custom-built homes typically take the longest to complete. In addition, construction cost is higher due to having to supervise scattered site work, which is combined with the longer time that is needed to build a new custom home. Also, because of the unique design of each home, the economies of large-volume work are also lost; thus, it results in the higher prices of a new custom home (Smith 1998).

2.4.3 Mass Housing Development Towards Sustainability

The decision to adopt any development pattern of housing and community has a direct impact on the country's economic, environmental, and social future. For example, since the 1950s, the Canadian image of an urban ideal has predominantly been of a detached dwelling in a low-density community; however, today's economic, environmental and social costs challenged the sustainability of such development patterns (Quadrangle Architects Limited 2000). Accordingly, there are strong, positive correlations between the attributes of medium-density housing and current demographic social trends.

The *medium-density* housing development is defined as residential developments with a range of 12–36 dwelling units per acre (30–90 units per hectare), while low-density housing developments contain about 6 dwelling units per acre (less than 15 units per hectare) (Quadrangle Architects Limited 2000). Therefore, the townhouse is considered one of the common forms of medium-density housing development, which mostly has a simple, rectangular plan with one or two stories. The concept of a 'townhouse' has existed in the western society long ago. According to Schoenauer (1994), the origins of townhouses are found in fortified medieval cities. Townhouses are good alternatives to detached and semi-detached houses. Based on housing categories, townhouses can be indicated as row houses, where many units are attached to each other in a row. Furthermore, the models of townhouses, which are used with success in many countries, can be classified generally into four types: single-story, split-level, linked, and narrow-front townhouses.

Based on the attached units' construction, the single-story townhouse has its distinct features due to the elimination of the two facades, which has its implication on reducing the municipal service cost, shorter street exposure, and sharing walls affect internal heating and cooling cost (Schoenauer 1994).

Different from the single-story type, the split-level townhouse is not an attached version of its detached or semi-detached counterparts. It typically has an advanced plan design and arrangement of single floor façade placed on the side of a narrow pedestrian path and a two-story garden side on the private backyard.

The notion of sustainable development is based on the conversion of natural resources, and there are several design factors inherent in townhouse development that help conserve resources by reducing the amount of building materials required, and by improving the thermal efficiency of the building envelope (Friedman 2010). Thus, the medium-density housing development consisting of townhouses has the positive effect on achieving not only 'sustainability' but also 'affordability', because of its possibility of high land-use efficiency that helps reduce construction costs. More specifically, the medium-density housing, particularly single-story or narrow-front townhouses, may be appropriate to a sustainable mass housing development that produces economies of scale, when both sustainability and affordability in housing development are taken into consideration.

2.5 Conclusion

The housing sector is a sensitive issue in any given context. On one hand, when its development is sustainable, it contributes to the growth of a nation's economy, generates global competitiveness, urban success, prosperity and lies at the heart of social cohesiveness. On the other hand, housing deficiencies, of which affordable housing is a critical component, result in cities in crises, carrying economic inequality, social tension and governance challenges. Mass housing lies at the heart of this sensitive social and political balance as an attempt to meet housing needs of a global growing urban population with diverse contextual needs. This quest is further pledged with a commitment to a sustainable approach to land, materials and resources usage.

There has been a significant shift in policies and processes governing housing provision, which will be further and contextually challenged in different manners, depending on changing housing pressures, social and economic conditions including shifts in employment levels and nature of employment, demographics comprising consideration of an aging population and immigration influx. Housing stocks may have to accommodate unpredictable social changes, include flexible, adaptable and adjustable housing models as well as encourage a more effective use or reuse of the existing large stocks.

The stakeholders in the built environment in general and housing in particular, are being forced at an extensive and unprecedented pace to improve a set of conflicting objectives. A sustainable approach to meet housing needs usually calls for a collective consideration of a number of criteria. These include revisiting related laws, policies, processes and building regulations to promote its development, an improved urban land planning and utilisation, viable business models and creative financing mechanisms to motivate private involvement and public partnership.

The building industry, on the other hand, challenged by a global crisis, is currently seeking new orientation and strategies, where sustainable construction approaches and cost minimisation are a vital aspect of making mass housing projects economically viable. In this context, leverage innovative green concepts applied to construction techniques, coupled with attractive low-cost solutions and technologies including prefabrication and customisation, carry multiple benefits such as higher efficiency, lower costs, and sustainable solutions as well as meeting end users' evolving needs. These considerations can easily become the drivers for a massive change in sustainability practices and ecological products markets. The strength of mass housing future development may lay in its capacity to meet these challenges.

References

Aga Khan Award for Architecture (1979) Housing process and physical form. In: Proceedings of seminar three in the series architectural transformations in the Islamic World, Jakarta, Indonesia, 26–29 March 1979

Alexander LT (2009) Stakeholder participation in new governance: lessons from Chicago's public housing reform experiment. Geo J Poverty Law Policy XV1 1

Anderson DM (1997) Agile product development for mass customization. Irwin, Chicago

Article 25.1 (1948) United Nations Human Rights. http://www.ohchr.org/EN/Issues/Housing/Pages/InternationalStandards.aspx. Accessed 30 Aug 2015

Bijlmer Renovation Planning Office (2008) The Bijlmermeer renovation, facts and figures. Bijlmermeer Renovation Planning Office, Amsterdam

Bioregional (2015) BedZed, http://www.bioregional.com/bedzed. Accessed 14 Nov 2015

Bruce M, Davies-Cooper R (1997) Marketing and design management, 1st edn. Advanced marketing series. International Thomson Business Press, London

Choguill L (1995) The future of planned urban development in the third world: new directions. In: Aldrich B, Sndhu R (eds) Housing the urban poor, policy and practice in developing countries. Zed books, London
Comerio M (1990) Community design: idealism and enterpreneurship. In: Sanoff H (ed) Participatory design: theory and techniques. Bookmasters Inc, New Jersey
Davies C (2005) The prefabricated home. Reaktion Books, London
European housing Network for Housing Research (2015) Housing in developing countries. https://www.enhr.net/housingindeve.php. Accessed 20 Nov 2015
Fahey T, Norris M (2009) Housing and the welfare state: an overview, working paper series: Wp09/09. University College Dublin, Dublin
Friedman A (2010) Narrow houses: new directions in efficient design. Princeton Architectural Press
Groele MK (2013) Making a mark: personal customization in suburban housing. State University of New York, Buffalo
Hackney R (1991) The neighbourhood spirit. Architecture + Design v.viii (1):41–48
Healy J (2004) Housing, fuel poverty, and health: a pan-European analysis. Ashgate Pub, Aldershot, England
Hebdige D (1979) Subculture: the meaning of style. Routledge, London
Hohmann J (2013) The right to housing: laws, concepts, possibilities. Hart Publishing, Oxford
Ibem EO, Amole OO (2010) Evaluation of public housing programmes in Nigeria: a theoretical and conceptual approach. Built Environ Rev 3:88–116
Intercultural Cities (2011) The district of Amsterdam South East: intercultural profile report. http://www.coe.int/t/dg4/cultureheritage/culture/Cities/AmsterdamICProfile_en.pdf. Accessed 10 Sept 2015
Jacobs J (2002) The death and life of great American Cities. Random House Inc, New York
Jencks C (1977) The language of post-modern architecture. Rizzoli, New York
Julian D, Reischl T, Carrick R, Katrenich C (1997) Citizen participation: lessons from a local united way planning process. Am Plann Assoc J 63(3):345–355
Klundert W, Bold J (2014) Bijlmermeer regenerated: modifications of public and semi-public areas—what can we learn?. Westminster University, London
Krieger J, Higgins D (2002) Housing and health: time again for public health action. Am J Public Health 92(5):758–768
Maisel J, Smith E, and Steineld E (2008) Increasing home access: designing for visitability. AARP public policy institute http://assets.aarp.org/rgcenter/il/2008_14_access.pdf. Accessed 20 Nov 2015
Noguchi M (2004) A choice model for mass customisation of lower-cost and higher-performance housing in sustainable development. McGill University, Canada
Oikodomos (2011) Housing concepts. http://www.oikodomos.org/oikopedia. Accessed 14 Aug 2015
OIKONET (2014) Lisbon workshop: contemporary living patterns in mass housing in Europe ISCTE- University Institute of Lisbon, July 14–19th 2014. http://www.oikonet.org/public/files/Lisbon_Workskhop_A3poster.pdf. Accessed 14 Aug 2015
Quadrangle Architects limited (2000) Multiple housing for community sustainability. Ottawa: Canada Mortgage Housing Corporation, Ottawa-Carleton, & Canada. Natural Resources Canada. Mortgage and Housing. http://www.iisbe.org/abn/abn13nws.htm. Accessed 30 Nov 2015
Saegert S, Klitzman S, Freudenberg N, Cooperman-Mroczek J, Nassar S (2003) Healthy housing: a structured review of published evaluations of US interventions to improve health by modifying housing in the United States, 1990–2001. Am J Public Health 93(9):1471–1477
Sanoff H (2000) Community participation methods in design and planning. Wiley, New York
Schneider T, Till J (2005) Flexible housing: opportunities and limits. Arch Res Q 9(2):157–166

Schoenauer N (1994) Cities, suburbs, dwellings in the Postwar Era. McGill University, School of Architecture, Montreal
Smith JL (1998) Cleaning up public housing by sweeping out the poor. Habitat Int 23(1):49–62
Sustainable Scale Project (2015) Population quickfacts. http://www.sustainablescale.org/areasofconcern/population/populationandscale/quickfacts.aspx. Accessed 20 Nov 2015
Sutavicius M (2014) Mass-housing: tendencies and modernization. Unpublished research paper, TU Delft, The Netherlands. http://repository.tudelft.nl/assets/uuid:be8aab25-0ad3-4b55-b64f-92dbbc5176f2/M.Sutavicius_Research_4254791_.pdf. Accessed 23 Aug 2015
Swenarton M, Avermaete T, Heuvel D (eds) (2015) Architecture and the welfare state. Routledge, New York
Towers G (1995) Building democracy, community architecture in the inner cities. UCL Press, London
Towers G (2000) Shelter is not enough, transforming multi-story housing. The Policy Press, Bristol
UN-Habitat (2012) Going green: a handbook of sustainable housing practices. In: Developing countries, UN-Habitat, Nairobi
UN-Habitat (2013) Urban revitalization of mass housing international competition. http://www.masshousingcompetition.org. Accessed 20 Nov 2015
United Nations (2015) Department of economic and social affairs, population division. World population prospects: The 2015 revision, key findings and advance
Urban F (2012) Tower and slab: histories of global mass housing. Routledge, London
Wikipedia (2015) Council house. https://en.wikipedia.org/wiki/Council_house. Accessed 11 Feb 2015
Wikipedia Commons (2016a) File: Narkomfin Building Moscow 2007 01.jpg. https://commons.wikimedia.org/wiki/File:Narkomfin_Building_Moscow_2007_01.jpg. Accessed 4 Jan 2016
Wikipedia Commons (2016b) File: social housing-Perkin's rents-geograph.ord.uk https://commons.wikimedia.org/wiki/File:Social_Housing-Perkin%27s_Rents_geograph.org.uk_1330102.jpg. Accessed 4 Jan 2016
Wikipedia Commons (2016c) File: social housing in tipper lane—geography.org.uk—795397.jpg. https://commons.wikimedia.org/wiki/File:Social_housing_in_Tipper_Lane_geograph.org.uk_795397.jpg. Accessed 4 Jan 2016
Wikipedia Commons (2016d) File: Gooird, Bijlmer.jpg. https://commons.wikimedia.org/wiki/File:Gooiord,_Bijlmer.jpg. Accessed 4 Jan 2016
Wikipedia Commons (2016e) File: Map of Bijlmeer.svg. https://commons.wikimedia.org/wiki/File:Map_of_Bijlmermeer.svg. Accessed 4 Jan 2016
Wikipedia Commons (2016f) File: Nemausus.jpg. https://commons.wikimedia.org/wiki/File:Nemausus.jpg. Accessed 4 Jan 2016
Wikipedia Commons (2016 g) File: 09di1683-11—Flickr—USDAgov.jpg. https://commons.wikimedia.org/wiki/File:09di1683-11_Flickr_USDAgov.jpg. Accessed 4 Jan 2016
Wikipedia Commons (2016 h) File BedZED 2007.jpg. https://commons.wikimedia.org/wiki/File:BedZED_2007.jpg. Accessed 4 Jan 2016
Wikipedia Commons (2016i) File: Street in BedZED.jpg. https://commons.wikimedia.org/wiki/File:Street_in_BedZED.jpg. Accessed 4 Jan 2016
Wong JF (2010) Factors affecting open building implementation in high density mass housing design in Hong Kong. Habitat Int 34(2):174–182
World Health Organisation (2015) Urban population growth. http://www.who.int/gho/urban_health/situation_trends/urban_population_growth_text/en. Accessed 20 Nov 2015

Chapter 3
Prefabrication

Arman Hashemi, UnRyoung Kevin Kim, Pamela Bell, Dale Steinhardt, Karen Manley and Mark Southcombe

Abstract Prefabrication is a construction method where parts of a building are manufactured in a factory and transported to site for assembly. Prefabrication was an innovation in the 19th century, however it was in the 20th century that it developed into a major method of construction. This chapter covers basic and advanced knowledge on the subject. It includes background and current conditions, prefabrication fundamentals, advantages and challenges, as well as building methods, design, procurement, manufacturing and installation processes. Prefabrication is a major construction method in countries such as Sweden, the Netherlands and Japan, but is less widely adopted in countries such as Australia, the UK and US.

A. Hashemi (✉)
University of Brighton, Brighton, UK
e-mail: a.hashemi@brighton.ac.uk

U.K. Kim
Fleetwood, Crestmead, Australia
e-mail: KevinK@fleetwood.com.au

P. Bell
PrefabNZ Inc, Wellington, New Zealand
e-mail: pam@prefabnz.com

D. Steinhardt · K. Manley
Queensland University of Technology, Brisbane, Australia
e-mail: dale.steinhardt@gmail.com

K. Manley
e-mail: k.manley@qut.edu.au

M. Southcombe
Victoria University of Wellington, Wellington, New Zealand
e-mail: mark.southcombe@vuw.ac.nz

© Springer International Publishing Switzerland 2016

M. Noguchi (ed.), *ZEMCH: Toward the Delivery of Zero Energy Mass Custom Homes*, Springer Tracts in Civil Engineering,
DOI 10.1007/978-3-319-31967-4_3

3.1 Introduction

The construction process is different from most manufacturing processes. Each project and construction site is in some ways individual and unique, and requires complex production processes and involvement of many different authorities, trades and stakeholders. While a manufacturing process is most often under the control of a single management, a construction project has traditionally been the outcome of independent inputs and ideas from stakeholders including clients, designers, contractors and engineers. This situation makes the industrialisation of construction industries more complex compared to other industries (Warszawski 1999; McEvoy 1994). Historically there have been three major building methods in terms of fabrication and use of machinery and labour: traditional methods; post traditional (conventional) methods; and industrialised systems (Riley and Howard 2002) including prefabricated methods of construction.

Prefab, short for prefabrication, refers to any building part that is made (fabricated) before (pre) it is transported and assembled at a final building site. These prebuilt parts can be manufactured from any material and in a variety of sizes, from pre-cut pieces to an entire building. Prefabrication typically occurs off site in an enclosed building, a factory, or a controlled construction yard. Prefabrication may also be referred to as modular construction, offsite construction; offsite manufacturing (OSM), industrialised building, and modern methods of construction (MMC). Other common terms include pre-built, pre-configured, pre-designed, pre-planned and pre-assembled. Prefabrication refers primarily to a process, rather than a product and it is first and foremost a way, an approach, or a system of building. It does not necessarily lead to a single product outcome.

Several advantages, disadvantages and risks have been claimed for prefabricated methods of construction, and most of these have remained the same since the 20th century. The belief that prefabrication improves the quality and affordability of housing continues. For various reasons these objectives have not been achieved in many parts of the world with a few exceptions. Despite the positive outcomes associated with adoption of prefabricated construction, strong negative community perceptions of prefabricated housing have also occurred. These are driven by a historical and continuing association with temporary, emergency and low-quality housing. Across Europe, bland concrete panelised apartments are associated with periods of government-provided mass-produced housing. Australian and New Zealand residents recall demountable classrooms and mining accommodation with little fondness. Poor quality post-war housing has also tainted prefabrication's image in the UK, and in the US the prefab industry still struggles to shake its relationship with mobile trailer homes and their associated low social status (Craig et al. 2000; Daly 2009; Hashemi 2013).

While contemporary prefabricated housing industries have developed processes, production and products from their humble beginnings, they have not attracted a large mainstream market except in a few countries. Recent arguments on the necessity to increase the market share of prefabricated methods have focused on the need for improved quality, and the need to provide affordable and energy efficient sustainable homes. This chapter provides an overview, discusses prefabrication theoretical and practical issues, background and current conditions, prefabrication principles and fundamentals, advantages and disadvantages, challenges, building methods, design, manufacturing, and installation processes.

3.2 Prefabrication, A Brief History

Prefabrication was considered as an innovation during the 19th century with the first successfully prefabricated homes documented in Great Britain in 1833 (Noguchi 2012); however, it was not until the 20th century when prefabrication developed into a major method of construction. The history of prefabrication in the 20th century is linked to several issues such as the World Wars, Welfare State policies and Modern Architecture which transformed the construction industry in many respects. Massive destruction caused by World Wars caused an urgent need for reconstruction. The immense size of housing programmes were outside the capacity of traditional building methods and resources and led to the development of alternative construction methods including prefabrication. It was believed that prefabrication could change houses from an expensive to an economical and easily available product mass manufactured from major components built in the factory (Finnimore 1989; Harvey and Ashworth 1997).

Architects and building industry skilled labourers were assumed to be greatly affected by prefabrication and mass production. Skilled labourers were anxious about prefabrication eliminating the need for their skills in construction. Architects were also concerned about increased machinery and standardisation which could potentially reduce the need for design which was their main skill. While construction labourers found it difficult to adapt themselves to the new situation, architects adapted themselves more successfully (Finnimore 1989; Osbourn and King 1989).

Building industrialisation was a key tenet of Modern Movement in architecture. In 1923, Le Corbusier famously proposed that, 'we shall arrive at the House Machine, the mass-production house' (Le Corbusier 1946, p 210). He advocated the creation of a state of mind, a spirit, receptive to constructing, living and conceiving mass-production houses. This culture needed to be in place before any technological change could be accepted, understood and taken-up. This emphasis on the need for culture-change is still valid today, almost a century on. The dream of packaged kit houses purchased via mail-order catalogue or off-the-shelf has long been the holy grail of Modern architecture. After World War II, 'system building' was introduced as a theoretical idea by Modern architects raising the argument that mass production was necessary to cope with the massive housing demands (Finnimore 1989). Prefabrication of the entire building was the most idealistic outcome of these efforts (McEvoy 1994). Modernists also believed that architects should be familiar with new building sciences and technologies. The unique post-war conditions as well as massive reconstruction programmes created a great opportunity for Modernist architects to transform their prefabrication, mass construction and high-rise building theories into reality (Finnimore 1989; Burnett 1993; Harrison 2009).

The UK was one of the pioneers in developing and adopting prefabricated housing at large industrial scales. The history of prefabrication in the UK is a good example of the poor practices undertaken by many developed countries during the 20th century. The UK experienced extensive use of prefabricated methods in response to the massive housing demands caused by the World Wars and high rate of new household formation (Burnett 1993). Around one million prefabricated homes were built during this period (Ross 2002). The low quality of the design and materials, building quality and regulation led to a negative public attitude toward prefabricated methods (Harvey and Ashworth 1997; Ross 2002). Excessive use of prefabricated methods led to a widespread criticism of the industrialisation of the construction industry because quantity was valued ahead of quality. Designers were often unsuccessful considering technical issues in conjunction with social issues and aesthetics. Local authorities were also criticised for loss of tenant identity and associated social problems such as vandalism (Osbourn and King 1989). The 1960s was an era of high rise prefabricated apartments in a context of arguments against the use of prefabricated methods of construction increasing (Finnimore 1989; Harvey and Ashworth 1997). Extensive and inappropriate use of prefabricated methods, and dull and monotonous prefabricated housing estates gave rise to arguments to replace prefabricated methods with alternative methods of construction with more diversity and personalisation (Burnett 1993).

3.3 International Uptake of Prefabricated Housing

While there has been sustained interest in prefabrication as a means to improving the housing industry, and there are well-established benefits from its adoption, prefabricated construction uptake varies internationally. Table 3.1 highlights the relative uptake of prefabrication across key countries.

Most of the world's developed economies such as the United States, Australia and the United Kingdom have failed to establish and sustain significant prefabricated housing industries. The major exception is Sweden where there is a dominance of wood-based, prefabricated housing, from mostly small, weather-proof factories. 84 % of detached single dwelling houses in Sweden use prefabricated timber elements. There are clear advantages to factory-based operations which separate the construction task from the snowy, arctic climate of Sweden. While Sweden is clearly a leader in the use of prefabricated housing, attention is also drawn to other industry-leading countries such as Japan, where there is a push towards prefabricated housing with mass customisability. 15 % of new detached and semi-detached houses in Japan are prefabricated. Many of these are produced by very large prefabricated home builders that operate advanced manufacturing facilities, the largest producing more than 10 thousand houses annually. The extremely large Japanese housing industry produced more than a million new houses in 2008. These company's thrive

Table 3.1 Relative adoption levels of prefabricated housing internationally

Sweden
84 % of detached/single dwelling houses use prefabricated timber (Trä och Möbelföretagen 2015) A dominance of concrete for multi-residential housing construction
Netherlands
Estimated that 20 % of all new housing uses wood or concrete prefabrication (Clarke and Wall 2000), with a high use of prefabricated panels (Eriksson 2003)
Japan
15 % of new houses prefabricated, with dominant focus on detached/semi-detached market (Japanese Prefabricated Construction Suppliers and Manufacturers Association 2015) Prefabrication of both whole houses and components Variety of methods and materials used (Linner and Bock 2012)
Germany
9 % of new residential building permits are for prefabricated buildings, with 15 % for 1–2 dwelling buildings, and 2 % for buildings with 3 or more dwellings [G1/G2] (Federal Statistics Office Germany 2013a, b)
United States
4 % of single-family, non-mobile houses are built offsite (U.S. Census Bureau 2012b) 7 % of all dwellings are manufactured or mobile housing (U.S. Census Bureau 2012a)
UK/England
7 % of the value of the entire construction sector (including civil works) is attributable to offsite work, though with little focus on housing (Taylor 2010) 90 % of post-1990 housing is onsite masonry construction, with offsite alternatives unlikely to account for more than 5 % of all new housing (DCLG 2010)
Australia
Estimated that less than 5 % of the new housing market uses prefabrication (The C.I.E. 2013). Dominance of whole house/modular prefabrication for detached housing-majority of multi-residential projects use precast concrete (Blismas et al. 2010)
New Zealand
Prefabricated parts make up 25–28 % of all consented residential and commercial buildings (Normal and Page 2014)

on local consumers that value high quality, durability and resilience against natural disasters like earthquakes. The size of this market also facilitates ongoing development of research-intensive prefabrication companies.

Countries in mainland Europe such as Germany and the Netherlands have also developed significant manufactured housing expertise. In Germany 9 % of new residential building permits are for prefabricated buildings, while in the Netherlands 20 % of all new housing uses wood or concrete prefabrication. Other countries have a slower prefabrication uptake, with the United States, United Kingdom and Australia, having just 5 % of permanent housing with any significant prefabrication. While the US has traditionally had a substantial interest in manufactured homes, these have tended to be synonymous with low-cost trailer homes rather than mainstream housing. Unlike the Japanese market, they have been unable to harness their large population, albeit much less dense, to drive large prefabrication firms. Australia and the UK have similar low levels of annual prefabricated housing production and have industries that are dominated by traditional, craft-based, building methods. Despite significant efforts from industry groups, particularly the UK's Modern Methods of Construction movement, prefabricated housing remains under-utilised in these countries despite the evidence of the benefits.

The poor historical quality and image of prefabrication in the US (trailer parks), UK (post-war houses) and Australia (school buildings, mining camps) may be a factor in the slow prefabrication uptake in these countries, but this consideration is equally applicable to market leaders in Sweden (social housing projects) and Japan (post-war houses). Sweden has a smaller population than the UK or Australia, yet it has higher rates of housing prefabrication. This suggests that prefabricated housing activity does not necessarily need a large domestic population to sustain it. According to Steinhardt et al. (2013), these laggard nations have fallen behind because they have lacked the demand drivers of leading countries, including extreme weather, earthquakes or environmental activism.

3.4 The Fundamentals of Industrialised Building Systems

Standardisation and mass production are fundamental issues in prefabrication. It should be noted that 'Prefabrication does not necessarily imply either mass production or standardisation' (Davies 2005, p 205), although a level of standardisation is always required to reduce the costs in both traditional and prefabricated methods of construction. There were some attempts toward standardisation of building components before and after the Second World War. Theorists of the Modern Movement such as Walter Gropius and Albert Farwell Bemis carried out some work on modular coordination theory. According to Modern Movement theories, modular coordination was key to any attempts toward mass production in prefabrication (Finnimore 1989). The degree to which standardisation should be considered has historically been a major debate, particularly amongst designers. There is a competing need to avoid monotony and to make prefabricated products appealing to customers. One of the major technical challenges of prefabrication is the standardisation of individual buildings without unduly affecting aesthetics and design freedom. Long-term demand and availability of raw materials, machinery and reliable infrastructure such as transportation systems are also key issues for prefabricated construction companies.

3.4.1 Dimensional Co-ordination

Dimensional coordination is a basic requirement for standardisation and mass production. Dimensional coordination in the UK is based on a national three-dimensional grid of basic modules within which the maximum and minimum sizes of components are defined (Blanc 2014). Dimensional co-ordination creates the building's dimensional basis within which standard components can be used in a unified, interrelated way (Blanc 2014). Despite arguments around the feasibility and acceptability of dimensional co-ordination, an international basic modular dimension of 100 mm has been defined as a guide for designers and manufacturers (Blanc 2014). According to the British Standards, BS 6750:1986 and BS EN ISO 8560:1999, the basic module (100 mm), represented by the letter M, is to be used for vertical and horizontal coordination dimensions of buildings (BSI 1986; BSI 1999). British Standards recommend 3 M, 6 M, 12 M, 15 M, 30 M, and 60 M as manufacturing dimensions. Sub-modular dimensions of either 50 or 25 mm are also recommended as the first and second sub-modules, respectively. This system allows UK manufactured components to be used in conjunction with international systems and components.

It should be noted that these modular sizes are not the exact or working sizes of the components (BSI 1986). Modular components are usually required to be combined with non-modular components (Blanc 2014). Variation in component sizes is inevitable for several reasons such as manufacturing inaccuracies, high costs associated with high accuracy, locations of components on construction sites, and the physical characteristics of materials which, for instance, may cause contraction or expansion. For these reasons, building components are not designed with exact dimensions to fit into a space and there are some tolerances required in jointing areas to allow for inaccuracies (Greeno and Osbourn 2014).

3.4.2 Mass Production

Economic criteria are fundamental when establishing a prefabrication factory. In order to produce components economically, there is an optimum number of manufacturing machines used in the process and an optimum number of manufactured components produced. For standardised production systems there are still inevitably some components that are most efficiently manufactured non-standard. Non-standard components are more difficult to manufacture in systems which are more automated and therefore such components are more expensive and generally take longer to be delivered. In this respect, manufacturing methods which are capable of producing components economically and in a broader range will be more successful (Blanc 2014). The use of prefabricated building components should in theory be much cheaper due to mass production of such components however this is not always true in practice. Unlike traditional methods, prefabrication involves additional costs such as packaging, storage, transportation and assembly on site. These processes lead to the need for additional material and details for handling to protect the components against damage which may result in higher costs. For instance, precast concrete panels require extra reinforcement to protect them during transportation and assembly. They also require a rather complicated jointing system for connection with other components on site. The erection process also needs to be undertaken carefully to avoid damaging the panels, and can require special lifting and transport frames. These factors result in higher costs for onsite handling and management, so the cost of some prefabricated components increases considerably compared to similar components which are fabricated traditionally on site (Greeno and Osbourn 2014). The key advantages of industrialised methods are the quality of products and speed of construction onsite. These advantages may still result in overall cost savings despite the increased on site cost of the prefabricated components for reasons noted above.

3.4.3 Open, Closed and Flexible Systems

Industrialised building methods developed in the 20th century ended the reliance on traditional building methods. These include standardised prefabrication techniques often referred to as 'System Building'. There are two main approaches in system building (also known as industrialised building) (Riley and Howard 2002). These are Open systems, and Closed systems. Open and Closed systems have roots in construction industry inefficiency. This arises from the uniqueness of each building project and the inevitable involvement of several authorities including sponsors, designers, engineers and contractors in the construction process. The increased efficiency of all authorities involved in a project operating under a single management is highlighted by industrialised building systems. There have been two solutions to resolve this problem. The first is to merge design, production and marketing activities under the authority of one industrial system known as a Closed system. The second is to create an Open system in which standard components such as floor slabs and beams, are produced by different manufacturers. These components are applied in different building projects by any designer who is familiar with the shapes and sizes of the components (Warszawski 1999).

In an Open system building, also known as component building, different standard components, produced by different manufacturers, are used together. This allows designers to choose from a wide range of products which are installed in different types of buildings. The success of an Open system greatly depends on the degree to which products follow standardisation, dimensional co-ordination and international jointing systems. Open system building is great in economic terms however there may be some material waste. Attempts in Scandinavian countries to create Open systems based on a national standard have not been very successful elsewhere due to associated costs and aesthetic rejection (Riley and Howard 2002). Such methods of construction are much faster than traditional methods. Time saving is achieved by overlapping different tasks in the same period of time. Traditional methods are less feasible due to the linear system of design and construction. With Open systems problems can occur when two or more systems are put together. Resolving such practicality problems is the major role of designers (McEvoy 1994). Open systems also rely greatly on a range of skilled labourers that may be needed to install different applied components, whereas Closed systems usually rely on a single skill which is applicable to the entire system (Osbourn and King 1989).

In Closed system building, components are manufactured for an individual building and are not exchangeable with other buildings or systems. Closed systems are fast and efficient in assembly on site but are less adaptable to design requirements making them rather limiting when design modifications are required. Designers' choice of products is limited to what is manufactured within the system. For these reasons, Open systems are generally preferred over Closed systems (McEvoy 1994; Riley and Howard 2002). Yet, the problems of both Closed and Open systems led to the development of a new system known as a Flexible (close/open) system. Flexible systems include some fixed components which can be mixed with each other in an approved order however they offer great flexibility in layout arrangement, for example some Flexible systems are structural systems to which cladding and finishes are attached using conventional on site methods (McEvoy 1994).

3.5 Prefabricated Methods of Construction

Prefabricated buildings are wide ranging. They can be temporary or permanent, cheap or expensive, all the same or all different, small or large, with traditional or modern aesthetics, and they can be well-designed or badly designed. Prefab housing can be categorised according to the size of its parts. There are five main types of prefabrication: component (stick and sub-assembly), panel (non-volumetric), module (volumetric), hybrid (module-plus-panel) and complete buildings (box-form).[1]

Component. Component-based prefab includes stick and sub-assembly prefabrication. Stick refers to lengths of timber or steel which are pre-cut, pre-sized or pre-shaped puzzle-type pieces brought to site (Fig. 3.1). Sub-assemblies include windows and doors, fixtures and fittings, and structural members such as pre-nailed roof trusses and wall frames. The use of pre-nailed components has become an accepted part of the traditional construction process by the full range of home building companies. A common form of component-based construction is known as kitset housing.

Panel. Panelised, non-volumetric or two-dimensional prefabrication comprises manufactured panels that are transported as a flat-pack (Fig. 3.2). They can be classified as closed panels, complete with doors, windows, services, cladding or lining, or be open panels, made up of framing components. Some architects refer to closed panel systems as cartridges or cassettes.

Volumetric. Modular, sectional, volumetric or three-dimensional prefabrication refers to a three-dimensional structural unit made away from site and combined with other units or systems at site to create a whole dwelling (Fig. 3.3). The three-dimensional prefab element can be referred to as a volume, module, or section. By contrast, cores and pods refer to non-structural volumetric units often used

[1]Some parts of this chapter are adapted excerpts from 'Kiwi Prefab: prefabricated housing in New Zealand' (Bell and Southcombe 2012).

Fig. 3.1 Precut light gauge steel components assembled as house framework (Kiwi Prefab 2012)

Fig. 3.2 Timber panels assembled as house structure (Kiwi Prefab 2012)

as modules inside conventional buildings. Three-dimensional units are manufactured in controlled conditions with a high degree of services, internal finishes and fit-out installed in factory prior to transportation to site. This approach is particularly suited to highly serviced areas such as kitchens and bathrooms, which have a high added value, and cause disruption and delays on site. The modular home term came into common usage in the 1970s in the United States modular housing industry. It is a type of building that meets building codes, is factory assembled in full-dimensional units and then fixed onto a permanent foundation at site. It is a more permanent type of building than chassis-based mobile or manufactured homes which also gained popularity during that era.

Hybrid. Hybrid prefabrication is a term used for combinations of systems, such as hybrid module-plus-panel or semi-volumetric systems (Fig. 3.4). These systems use a mixture of volumetric units such as service modules or bathroom and kitchen pods for highly serviced areas and construct the remainder of the building using

Fig. 3.3 Modular volumes being set into position on site (Kiwi Prefab 2012)

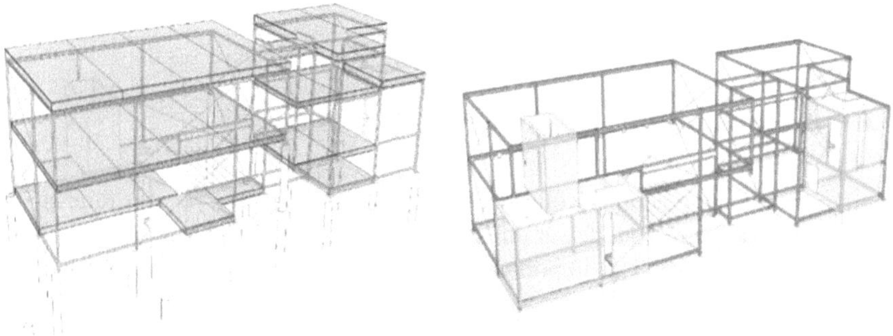

Fig. 3.4 Hybrid systems used by Kieran Timberlake in Loblolly House with panel elements highlighted above and volumetric elements highlighted below (Kiwi Prefab 2012)

Fig. 3.5 A completed building arrives at site on the back of a truck (Kiwi Prefab 2012)

panels or by another means. Hybrid prefabrication systems combine the benefits of two prefabricated construction systems, balancing construction efficiency with flexibility and consumer choice.

Complete. Box-form or complete buildings (Fig. 3.5) commonly known as portable, relocatable or transportable dwellings. These are a type of volumetric prefabrication where entire buildings are constructed in a factory or yard and then moved by a heavy haulage vehicle to site where they are attached to permanent foundations. These buildings may or may not incorporate prefabricated components, and standardised framing and sheet elements. There are subtle differences between portable, transportable, mobile and relocatable terms as used. Portable refers to a small temporary building that is light and easily moved repeatedly such as a toilet or site office; transportable refers to a larger building that is moved once from place of construction to its final site; mobile indicates a tow-able caravan-like structure on a permanent chassis which can potentially be moved repeatedly; and relocatable describes buildings that are designed to be moved several times during their lifecycle.

3.6 Prefabrication Advantages

Prefabrication can potentially offer more for less: more quality for less time at site, more known outcomes and fewer unknowns, and potentially more energy efficiency for less resource use. The importance of tangible outcomes in cost, quality and timeframe are evident through the consumer process of visiting a show-home, choosing from material samples, observing the factory manufacture, and watching the house arrive at the building site. Further prefabrication advantages include technical, social, economic and sustainability merits.

Quality: There is a common perception that prefabrication will primarily deliver a more cost-effective housing solution. However, the main advantage is increased control over manufacturing and construction conditions, creating a higher quality solution. According to research literature, improved quality and fewer defects are regarded as the principal advantages of prefabricated methods (Cook 2005; Hashemi and Hadjri 2014). This higher quality is achieved through closer coordination of labour, materials, machinery and sub-trades in controlled conditions. Quality control and resulting remedial work can be carried out before the product leaves the factory floor.

Speed of delivery: The other major advantage of using prefabrication is the increased speed of building delivery. The house can be manufactured offsite at the same time that foundations are prepared at site (Fig. 3.6). Construction tasks are simplified through the installation of prefabricated panels or modules which significantly reduces the requirement for external contractors and 'wet trades' like plastering and tiling (Poon et al. 2003). This has potential flow on effects in reducing the burden of staff management (Roy et al. 2003) and substantially increasing the overall speed of construction (Lu and Korman 2010). Prefabricated construction in a factory setting can also reduce the likelihood of construction delays due to inclement weather, one of the reasons underpinning the high adoption of prefabrication in nations such as Sweden (Bildsten 2011). This can reduce the programme time component between 30 and 60 % of a traditional construction process. The reduced onsite time can offset higher costs incurred in terms of new material development or pre-construction planning processes incorporating large prefabricated elements (Aburas 2011; Bildsten 2011). At a commercial scale, the savings are increased dramatically. However, a period of planning is needed prior to manufacture and changes cannot be made once fabrication processes commence. This presents a cultural shift from traditional linear-sequenced construction (Page 2012).

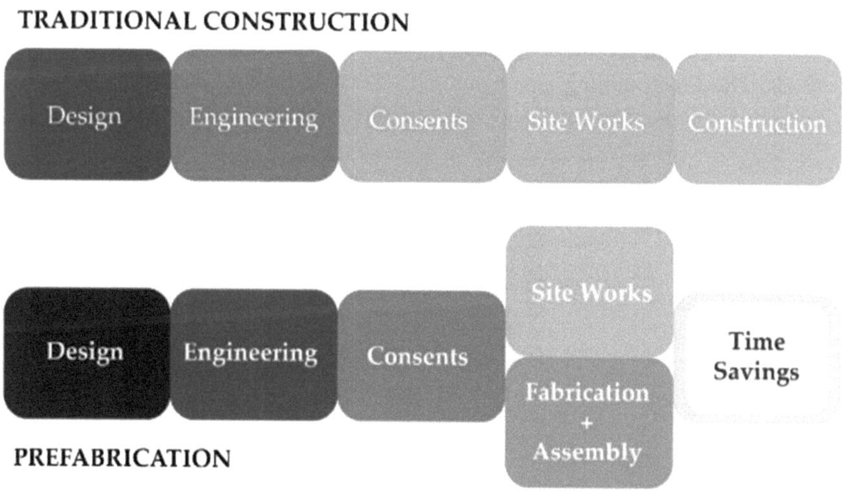

Fig. 3.6 Time savings shown in prefabrication delivery compared with traditional construction

Technical. Technical merits include tight quality controls of workmanship and materials, and the ability to test systems and prototypes within the factory. Testing, together with greater planning, accuracy and minimised tolerances, can reduce the level of mistakes and subsequent remedial work during the defects liability period post-occupancy.

Social. Social merits include being able to work under cover during inclement weather (Fig. 3.7), having tools and amenities close at hand, and improvements in health and safety. Investment in machinery and training can lead to longer-term employment stability. At site, there is likely to be less noise, dust, transportation and neighbourhood disruption than a traditional build. Prefabricated homes can be aesthetically dissimilar from each other and are often indistinguishable from conventionally constructed homes.

Economic. Economic merits include the cost savings to customers and developers from a faster delivery, reduced remedial periods and a shorter period of financial borrowing. 'Given the cost savings inherent in the construction technique, a prefabricated shell will generally be less expensive than a site-built structure of exactly the same specifications, configuration, and quality' (Buchanan 2004). Timeframes and costs can be decreased by eliminating dependence on weather for site-based construction, more efficient coordination of sub-trades in-house, reduced transportation, and price advantages from bulk ordering.

Sustainability. Sustainability merits include reduced material waste through efficient ordering, indoor protection, pre-planning and cutting. For instance, in New Zealand, construction is the 'forty percent industry', in that buildings are responsible for 40 % of energy consumption, 40 % of waste stream, 35 % of carbon dioxide emissions, and 40 % of raw material-use. The United Kingdom's Waste and Resources Action Programme (WRAP) reported that up to 90 % of waste could be reduced through the use of a variety of prefabricated manufacturing methods. United States projections are similarly optimistic, with claims that construction industry energy consumption can be reduced by 50 % by using prefabrication methods. The final building also benefits from the reduction of defects, and closer tolerances for tighter thermal and acoustic performance leading to better energy

Fig. 3.7 Off-site undercover construction (Kiwi Prefab 2012)

efficiency and lower heating bills. Prefabricated buildings can have a reduced carbon footprint through minimised transportation to site and lower energy use over its lifecycle. Deliveries to site can be reduced by 60 % for modular construction. Process benefits are from a safe, healthy and controlled environment for workers, as well as savings in labour productivity and material efficiencies. Traditionally, work at site can fall up to 50 % below potential, and an estimated 13–18 % of materials delivered are wasted from not being used properly (Gorgolewski 2005).

In its greenest form, some prefab buildings are designed for disassembly and potential future reuse of materials and components. One sustainability drawback is in the over-engineering and subsequent additional material used to brace modular units for transport, however, this over-engineering can also serve to make the house more durable and resilient once it is assembled at site. Prefabrication processes can also help tackle construction industry challenges such as a low-skilled construction workforce, increasing market demands for higher quality housing, and increasing industry regulation.

3.7 Prefabrication Challenges

Traditional housing methods still dominate the construction industry despite the clear advantages of prefabrication. There are challenges to greater prefabrication uptake. The commercial success of prefabricated housing has been inhibited for reasons ranging from prohibitive start-up costs and limited market size, to ongoing financing issues and changes in macroeconomic conditions. There are continuing challenges for designers around differing site conditions and transport box limitations. All prefabricated parts must fit on the back of a truck or in a shipping container.

Historical issues and misperceptions. Social misperceptions of prefabrication are a major challenge. These can be grouped into historical, quality, aesthetic and socio-cultural perception issues, and these are held by the public and by the construction industry. Prefabrication has been considered by many people as a cheap substitute for conventional building. It is simply a smarter way of building.

In 2001, British people still associated prefabrication with the poor materiality and bad design of temporary prefab homes and multi-storey concrete structures from the 1950s and 1960s. Despite the longevity of the post-World War II temporary prefab programme, there were negative perceptions of prefabrication from that era. ‘People have got the idea that [prefabrication] means jerry-building, tumbledown shacks, caravans, shoddy work, ribbon development, draughts and leaks and everything that’s bad in building’ (Vale 1995, p. 17). Similarly, United States citizens still judge prefabricated housing according to mobile or manufactured homes, with their connotations of being light, flimsy, temporary and cheap little boxes. Low-cost is too often confused with perceptions of low-quality. A widely-accepted myth about manufactured homes is that they are not well built, when in fact they are structurally reinforced to withstand lengthy road transportation, and typically have much higher build quality than occurs with traditional on site construction.

There is a common assumption that prefabrication results in repetitive dull outcomes. The terms prefabrication and standardisation are often confused and this causes aesthetic misperceptions. Innovative prefabricated housing has been further thwarted by the public perception that equates unconventional materials or appearance with inadequacy in performance. Most people understand a traditional housing vernacular of a pitched roof and horizontal linear cladding. By contrast, prefabricated housing that travels as a complete building on the back of a truck will be more likely to have a low or flat roof so it fits within a transport envelope.

Socio-cultural issues. A key barrier to prefabrication in, for example, the United Kingdom and New Zealand is the socio-cultural perception of the home as an economic investment and asset, rather than a consumer product. The emphasis on choice, individualisation and personal space creation has become intrinsic to home buying today. This poses a challenge because the confusion of prefabrication and standardisation dominates, and standardisation is seen as a challenge to personalisation. Some contemporary prefabricated housing businesses attempt to subvert problems caused by misperceptions around the prefabrication term, by describing it as pre-built, pre-configured or pre-planned instead of prefabricated.

Industry Acceptance. Industry conservatism is a major factor to overcome, with for example, the Australian Industry Group recently stating that there are few benefits to be gained from new construction technologies such as prefabrication (Australian Industry Group 2013). In reality there is a threat to traditional house building work patterns if automation can be applied to the production process, as this could reduce the need for unskilled or unqualified onsite labour (Dainty and Brooke 2004). Even without automation, prefabrication relies on increased standardisation to reap economies of scale. A shift to manufacturing has been noted by US-based housing manufacturers as generating a work environment that lacks employee input into customisation and individuality (Nahmens and Ikuma 2011). The relative transferability of traditional construction skills to prefabricated products or materials highlights that the technical challenges remain a lesser barrier than cultural resistance to change in the workforce (Daly 2009; Nadim and Goulding 2011). Industry acceptance can be improved by aligning with industry pressures to drive down costs, improve construction process efficiency, and improve the quality of the final house produced.

Flexibility. There is a close relationship between the standardisation of materials and processes for faster and cheaper prefabrication, and reduced design flexibility. This reduced design flexibility can cause problems for onsite builders installing prefabricated components manufactured in restricted sizes (Hofman et al. 2009). Architects who like design freedom, and consumers who have fewer choices are similarly restricted. Finding an appropriate compromise between total standardisation and total customisation is a difficult balancing act for prefabricators, but necessary for marketing success (Bertelsen 2005; Gann 1996). The end-goal of this balance is 'mass customisation'—the manufacturing of high-quality components that are adaptable to the individual choices of designers and consumers alike. Large integrated Japanese homebuilders are at the forefront of this development (Barlow and Ozaki 2005), however despite significant progress in offering both structural (floorplans) and aesthetic (fittings, appliances) customisability, the flexibility of these prefabrication companies still falls short of onsite, custom-built housing (Aitchison 2014; Patchell 2002). Customisation options in the dominant onsite UK housing market are also often limited to rearranging of internal fixtures within a predetermined, standardised structure (Roy et al. 2003). United Kingdom offsite manufacturing is also immature in terms of mass customisation (Hashemi and Hadjri 2013). Australian architects have also drawn attention to the dominant project housing market being 'reproduction designs of past styles… placed together in ways that create dull, uninspiring environments' (Johnson 2004). A major benefit of onsite built housing is that it continues to define mainstream perceptions of how housing is meant to look. Prefabrication competes against this ingrained sentiment of both consumers purchasing new homes, and the project housing builders constructing them.

Costs and quality. The most obvious issue for any construction firm is their financial viability. The promise of prefabrication is that increased industrialisation and factory-based processes can increase production efficiencies and drive down building costs. This should in turn drive down consumer costs and increase demand. The main challenges to reducing costs can be broken into the initial investment cost to establish prefabrication infrastructure (e.g. factories, supply chains), and related on-going revenue issues. The barrier of high investment capital to establish mechanised factories has been identified as a clear problem preventing smaller operators from competing in the prefabricated construction space in locations as diverse as the United Kingdom (Lovell and Smith 2010) and Hong Kong (Poon et al. 2003). This initial investment presents a significant risk because unlike traditional construction, maximum benefit is derived from prefabrication when there is continuous production and supply economies of scale (Lovell and Smith 2010). Properly evaluating the costs of prefabricated housing is complicated, as research results have pointed to both the higher upfront cost of prefabricated materials, and overall reductions once broader costs are factored in. For example, U.S. and Australian research has identified upfront material cost increases of 5–20 % for the use of Structural Insulated Panels (SIPs) (Gagnon and Adams 1999; Gurung and Mahendran 2002). Such basic analyses however overlook the potential to cut costs by up to 50 % by reducing labour and material waste, and delivering a product earlier (Miller 2010). Similarly, moving housing to a controlled manufacturing environment allows for monitoring of quality, defects, and incremental improvement of the product itself (e.g. increased insulation, improved finishes) (Gaze et al. 2007; Tam et al. 2002). Each of these improvements can also lead to cost reductions for end-users by reducing maintenance, repairs, and post-occupancy operating costs. A broader and more accurate assessment procedure would include both direct and indirect construction costs, and post-occupancy costs.

Transportation. Prefabrication introduces complications, particularly in the use of highly prefabricated, volumetric modules. Although it is often argued that prefabricated materials and transportable houses address issues of repeated travel to sites for contractors particularly on remote sites, the logistics of transporting heavy or large building modules do not compare favourably with the traditional transport of small subcomponent housing elements (Daly 2009). Prefabricated housing can therefore reduce onsite building complexity, but may shift this complexity offsite in terms of planning and transportation overheads. Transportation distances can also affect costs and sustainability of prefabricated methods in terms of the extent of CO_2 emissions.

Industry capacity. Wider systems and regulations that support prefabricated construction also need to be considered. For example, existing funding models for house building have implications for prefabrication adoption. The house construction industry has traditionally not been capital intensive, relying primarily on investment from the end user commissioning the building to finance work. The movement to prefabrication could reverse this requirement. As house-building moves to a factory-based supply model the end users can no longer be relied upon to provide the necessary operating capital (Nadim and Goulding 2011). This has particular implications for contexts where the housing market is small or geographically isolated because of the lack of a significant pool of potential investors.

The actions and policies of financial institutions, lenders and insurers are also critical. The lack of established history of quality or product lifespan often results in requirements for reassurances to lenders that prefabricated houses will last and be attractive to potential buyers (Craig et al. 2000). The lack of familiarity with prefabrication by planners, regulators, insurers and certifiers all support the current status quo, and reinforce end user uncertainty (Lovell and Smith 2010). Building codes and government contracts are also often devised without explicit consideration of offsite manufacture as a possible alternative. A specific example is Australia's 'Little Hero' apartment building in Melbourne, which relied on overseas financing from the Arab Bank rather than Australia's major banks due to a declared 'obsession with bricks and mortar' (Boyd et al. 2012). Supportive networks for innovation and new prefabrication methods assist in establishing and growing a prefabricated housing industry, however there remains need to overcome significant bureaucratic inertia to be mainstream.

3.8 Design and Procurement

Typical building procurement processes may need some adjustments in prefabricated methods of construction. There are four standard types of building procurement as follows (Garrison and Tweedie 2008):

(a) Design-Bid-Build: in this approach, an architect produces the construction drawings from which a manufacturer develops in-house fabrication drawings. This approach does not take advantage of early collaboration and may result in delays while design changes occur.

(b) Negotiate Bid: in this approach the client, architect and manufacturer team up at the beginning of the project. The manufacturer takes an advising role at early stages of design and develops manufacturing drawings at a later stage. This approach maximises collaboration by facilitating early involvement of the key stakeholders.
(c) Design-Build: in this method the manufacturer acts as the main contractor leading the design and construction processes. Architectural drawings are either outsourced or developed in-house by the manufacturer. Technical drawings for manufacture and fabrication are then completed by the manufacturer.
(d) Strategic Partnering: in this approach the manufacturer is employed by the client for an extended period of time for several projects. The design and construction documents have already been developed by the client. This method works well when a large amount of repetition is involved and therefore is not suitable for architectural projects.

Close collaboration between client, architect and manufacturer at early stages of design is crucial because manufacturer's construction systems may limit the design in certain aspects. Early collaboration to produce a complete set of manufacturing drawings helps to maximise the benefits and reduce the associated risks of prefabrication.

3.8.1 Key Check Boxes

Transportation and module sizes should be considered at very early stages of design. Module sizes are governed by the local road regulations (Table 3.2). Therefore, the transportation regulations and routes from the factory and the site should be checked prior to any decisions on module sizes. This is necessary to avoid possible issues with late design changes due to the width and height restrictions. Room sizes, building structure and transportation cost are greatly affected by the module sizes. Additional transportation increases the total costs of construction as well as increases the CO_2 emissions of the system.

Structural Integrity of each part is also a key factor in the design and construction of components, panels, modules and complete prefabricated buildings.

Table 3.2 Road regulation in Queensland Australia (QLD Gov. 2013)

No pilot or escort	Up to 3.5 m wide, 25 m long
No pilot or escort	Up to 3.5 m wide, 25 m long
1 Pilot only	3.5–4.5 m wide, 25–30 m long
1 pilot and 1 escort	4.5–5.5 m wide, 30–35 m long
2 escort + police escort	Over 5.5 m wide, over 35 m long

Height (from ground to the highest point of the module) over 4.6 m requires signs, notifications, documents etc. in/on the vehicle as per each permit document describes

Each module should be built structurally independent for transportation to site. An engineered prefabricated building therefore requires integral structural elements such as posts, beams and frames, as well as engineered joints which may limit design flexibility due to, for example, the need for easy access to the joints during installation on site. In a similar manner to traditional on site methods, issues such as fire rating, lighting, ventilation, energy efficiency, access and egress etc. need to comply with national building codes and standards; yet, detailing of prefabricated components may differ considerably from conventional methods. Such issues can have significant consequences for the design so need to be considered early during the design stage.

3.8.2 Documentation

A full construction drawing set consists of architectural, mechanical, electrical and plumbing (MEP), structural, manufacturing and installation drawing sets. A significant amount of information is constantly generated and/or requested by the stakeholders as part of all building processes. Information management is critical in prefabricated construction as the pace of prefabricated construction is considerably faster than on-site construction. Prefabricated construction processes are often carried out simultaneously in the factory line so there is no time for design changes. A well-organised design process and documentation system is essential to minimise delays. Computer Aided Design (CAD) and Building Information Modelling (BIM) facilitate offsite construction processes. BIM facilitates 'integration, documentation and exchange of information eliminating the risks of miscommunication and production of waste information'. BIM also 'reduces the risks of mistakes and late design changes by producing virtual 3D models which … help to identify any possible mistakes and clashes before the actual construction on site' (Hashemi 2014, p. 34). Application of BIM improves collaboration and documentation of information, reducing risks associated with prefabricated construction methods.

3.9 Manufacturing

The two typical types of manufacturing processes utilise Moving and Static production lines. Moving production lines utilise assembly line production where components, modules or prefabricated assemblies move through different stations while specific tasks are carried out in each station. In static production lines the components, modules or prefabricated assemblies remain in one place while staff move to the different module positions to carry out specific tasks (Fig. 3.8). Compared to static production, moving production lines typically utilise more automation, and are a more expensive setup. It is worth noting that, unlike traditional methods, modular construction is not a linear construction process and

Fig. 3.8 Supporting Pads System on the factory floor. Each module is fixed on supporting pads and stays stationary until completion (Fleetwood AU)

different construction components may be completed simultaneously and assembled together at a later stage.

A controlled factory environment reduces the possibility of defects. Defects account for up to 6 % of the total construction costs for conventional construction methods (Johnsson and Meiling 2009). Human error is the major cause of defects, although this can be reduced by training and by increasing automation which in turn increases accuracy (Burgess et al. 2013). Automation also increases the risk of unnoticed defects in prefabricated products which may be repeated several times without detection. This issue is also seen occasionally in the car manufacturing industry where production and assembly processes are highly automated.

Moving products within a factory, to temporary storage, and during transportation to site are the other major causes of defects in prefabricated methods of construction. Preventative measures such as extra bracing and temporary transit bracing during construction and transportation help to reduce the risk and extent of defects. Regular inspections and quality control procedures are also required as part of the quality control processes to identify and resolve possible defects in the factory.

3.10 Installation on Site

Local transport regulations and road conditions should be confirmed before transportation from the factory to site. Large prefabricated modules or complete buildings will likely need oversize permits from local authorities to traverse the roads to the site. Design of the logistics of the delivery, cranes, unloading and setting in place of the modules are critical in larger and more complex projects to avoid double handling, site storage, and to minimise the expense and maximise the efficiency of crane hire. Some software packages such as Catia, and NavisWorks have been developed to simulate on-site building processes. These tools can help to identify any potential conflicts during on-site construction processes (Garrison and Tweedie 2008).

On-site construction processes such as foundation work, shafts, stairwell, and ready to connect service outlets for electricity, water in/out, gas etc. are typically required prior to prefabricated element transportation and assembly on site. On site work should be minimised and prefabricated work including finishes should be maximised in the factory to ensure higher prefabricated value in relation to volume, and to maximise the benefits of the prefabricated systems. Some onsite work will also inevitably be required to finalise the project during assembly on site (Garrison and Tweedie 2008). The assembly process on-site can be divided into the following steps: (a) connecting building modules to a foundation system; (b) stitching modules together as one volume; (c) finishing external/internal mate-lines; (d) reviewing and fixing any possible defects; and (e) final inspections.

3.11 Case Study: Fleetwood Prefabricated Homes

Fleetwood is a leading prefabricated home supplier in Australia with an annual supply capacity of over 600 housing units through the West and East Coast of Australia. Fleetwood provides prefabricated homes 99 % completed from a fully controlled factory environment (Fig. 3.9). The overall house construction time is seven weeks, six weeks in the factory and one week for delivery and onsite

Fig. 3.9 A single storey 120 m^2, two bedroom house in factory, transport ready, and as delivered on site. The entire construction process from initial order to completion in the factory has been six weeks/30 working days (Fleetwood AU)

installation. Early collaboration with the client results in further optimisation of design and construction processes and reduces overall construction costs by reducing material waste and increasing labour efficiency.

The factory production line is static mimicking a typical construction site with various housing types under construction simultaneously. During construction, crew members work in groups to complete particular construction tasks at one station before moving to the next station. The construction process starts by laying down a chassis on the factory floor (Fig. 3.11). Each chassis is designed for the manufacture and transport of the homes, facilitating occasional moves from site to site. Fleetwood uses two different chassis systems: (a) a precast concrete slab chassis that minimises the threshold height and achieves in situ like construction finishes; and b) a steel chassis that combines floor structure and retractable lifters (Fig. 3.11). The number of chassis elements for a house depends on the size of the house and floor plan layout. Single dwellings are 90–120 m^2 in area and have regular square shapes. The size of each chassis is normally 4.2 m wide × 14.4 m long to minimise construction and transportation costs. All homes fully comply with Australian national construction codes and standards and pass a 6 star energy rating. The building structure is fully optimised to sustain winds up to 61 m/s. Figures 3.10, 3.11, 3.12, 3.13, 3.14, 3.15, 3.16, 3.17 and 3.18 show the design, manufacturing and installation processes for the houses.

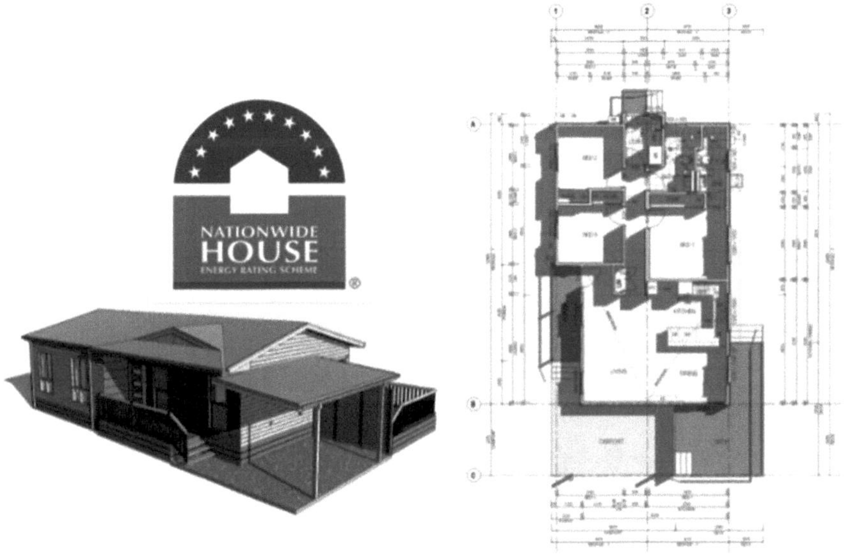

Fig. 3.10 Integrated BIM system is used to manage architectural and structural issues on one design platform

Fig. 3.11 The factory line is a static production line with each chassis blocked up from a factory floor. The *right* image shows a retractable lifting arm that accommodates bottom and top lifting mechanisms as required for the building type and site conditions

Fig. 3.12 Flooring and framing work for walls, ceiling and roof begin once the chassis is laid down. The cold formed light gauge steel frame and structure is designed to sustain cyclones up to 61 m/s

Fig. 3.13 External cladding and roofing are finished. Each building is tested by energy efficiency assessors to achieve a 6 star energy efficiency rating. The site location, building orientation, external and internal finish materials, colours, glazing, wall and roof insulation systems etc. are checked in the assessment processes

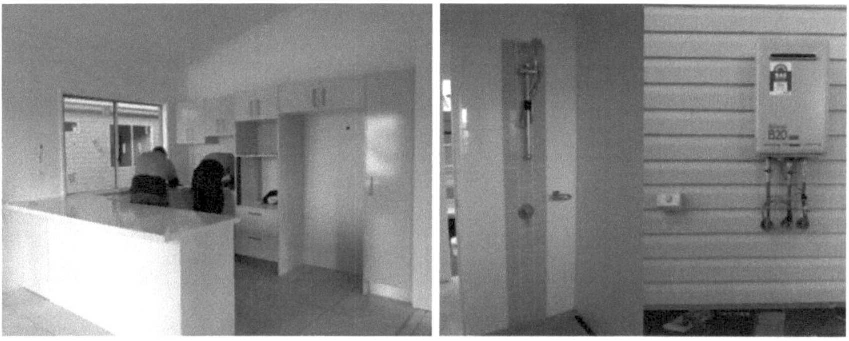

Fig. 3.14 Internal lining and finishes such as painting, tiling etc. are completed in the factory. All electrical fixtures, equipment, lightings, air conditioning system, plumbing fixtures and hot water supply unit are also installed in the factory. Module joints are the only remaining area to be completed on-site

Fig. 3.15 Each manufactured home consists of two unit modules. After completion in the factory, modules are separated and prepared for transportation to site. Ply and strap bracing are applied to the side joints to prevent damage during lifting, transportation and installation on site

Fig. 3.16 Units are loaded on a trailer. Unit sizes vary and are governed by local government road regulations. A unit size of 4.2×14.4 m is an economical dimension which can reduce material waste and transportation costs. Site location and surrounding environment, road conditions from factory to site, and local road regulations should be checked prior to design of module sizes

Fig. 3.17 Housing units installed on-site. Foundation types and connections are specified by engineers based on the soil conditions of the site installation

Fig. 3.18 Interior of a fully completed house

3.12 Conclusions

Prefabrication is not a new idea yet its innovative approach to design and construction has been slow to be adopted by the construction industry. Prefabrication needs to accommodate the needs of all key stakeholders including architects, engineers, building firms, suppliers, policy-makers and end-users together as an interactive network, not as individual actors working separately. The adoption of prefabrication necessitates restructuring of traditional construction processes. A gradual shift from traditional building processes to prefabricated methodologies has been associated with a number of advantages and disadvantages. The simplification and increased speed of onsite construction resulting from the use of prefabricated elements is well-recognised. This must be weighed against the potential for greater complexity in transport and planning. Similarly, while manufacturing offers productivity improvements over manual labour, this must be weighed against its ability to produce customisable housing that meets the market's needs. Concerns about building costs and house prices can be addressed very well by prefabrication, but more complete whole house and cost in use assessments compared with traditional building practices and housing are needed to demonstrate this objectively.

Other influences that dampen demand for prefabricated housing include a history of negative perceptions of quality and standardisation, and a regulatory environment that is not tailored to support prefabrication. Prefabrication disrupts many traditional construction norms and inevitability is compared with traditional housing produced for the ways it meets long-held expectations about construction processes, cost, aesthetics, and the individuality of housing. Key benefits of traditional construction methods include design flexibility and the ability to support small designers, developers and builders. These benefits are yet to be addressed by many prefabricated methods of construction. Mass Customisation along with recent developments in Building Information Modelling (BIM) is beginning to address some of the key issues, however the above factors will need to be collectively addressed if prefabrication is to realise its full potential more widely.

References

Aburas H (2011) Off-site construction in Saudi Arabia: the way forward. J Archit Eng 17(4):122–124

Aitchison M (2014) 20 shades of beige: lessons from Japanese prefab housing. Retrieved, from http://theconversation.com/20-shades-of-beige-lessons-from-japanese-prefab-housing-31101. Accessed 15 Apr 2015

Australian Industry Group (2013) Performance of construction index: downturn in construction industry extends to three years. http://www.businessspectator.com.au/sites/default/files/AIG%20PCI%20MAY.pdf. Accessed 18 June 2013

Barlow J, Ozaki R (2005) Building mass customised housing through innovation in the production system: lessons from Japan. Environ Plann A 37(1):9–20

Bell P, Southcombe M (2012) Kiwi Prefab: from cottage to cutting edge. Balasoglou Books, Auckland

Bertelsen S (2005) Modularisation: a third approach to making construction lean? In: Proceedings 13th international group for lean construction conference, Sydney, Australia, 19–21 Jul 2005

Bildsten L (2011) Exploring the opportunities and barriers of using prefabricated house components. In: Proceedings 19th conference of the international group of lean construction (IGLC), Lima, Peru, 13–15 Jul 2011

Blanc A (2014) Mitchell's building series: internal components. Routledge, London

Blismas N, Wakefield R, Hauser B (2010) Concrete prefabricated housing via advances in systems technologies: development of a technology roadmap. Eng. Constr Archit Manage 17(1):99–110

Boyd N, Khalfan MMA, Maqsood T (2012) Off-site construction of apartment buildings. J Archit Eng 19:51–57

BSI (1986) BS 6750: 1986: British standard specification for modular coordination in building. British Standards, London

BSI (1999) BS EN ISO 8560:1999: construction drawings—representation of modular sizes, lines and grids. British Standards, London

Buchanan M (2004) Prefab home. Gibbs Smith, Salt Lake City

Burgess JC, Buckett NR, Page IC (2013) Prefabrication impacts in the New Zealand construction industry. BRANZ study report 279. http://www.branz.co.nz/books_popup.php?id=17680

Burnett J (1993) A social history of housing 1815–1985. Routledge, London

Clarke L, Wall C (2000) Craft versus industry: the division of labour in European housing construction. Constr Manage Econ 18(6):689–698

Cook B (2005) An assessment of the potential contribution of prefabrication to improve the quality of housing. Constr Inf Q 7(2):50–55

Corbusier Le (1946) Towards a new architecture. Trans. Frederick Etchells. London Architectural Press, London

Craig A, Laing R, Edge M (2000) The social acceptability of prefabrication and standardisation in relation to new housing. In: Proceedings 16th IAPS conference: 21st century: cities. Social Life and Sustainable Development, Paris

Dainty ARJ, Brooke RJ (2004) Towards improved construction waste minimisation: a need for improved supply chain integration? Struct Surv 22:20–29

Daly G (2009) Prefabricated housing in Australia. Skill deficiencies and workplace practice. International Specialised Skills Institute Inc.

Davies C (2005) The prefabricated home. Reaktion Books, London

Department for Communities and Local Government (DCLG) (2010) English housing survey: housing stock report 2008. DCLG, London

Eriksson P (2003) Wood components in steel and concrete buildings—in-fill exterior wall panels. Nordic Timber Council, Stockholm

Federal Statistics Office Germany (2013a) Baugenehmigungen im Fertigteilbau (Neubau): Deutschland, Jahre, Gebäudeart/Bauherr. https://www-genesis.destatis.de. Accessed 22 Aug 2013

Federal Statistics Office Germany (2013b) Baugenehmigungen im Hochbau: Deutschland, Jahre, Gebäudeart/Bauherr. https://www-genesis.destatis.de. Accessed 22 Aug 2013

Finnimore B (1989) Houses from the factory: system building and the welfare state, 1942–1974. Rivers Oram Press, London

Gagnon MA, Adams RD (1999) A marketing profile of the U.S. structural insulated panel industry. For Prod J 49(7/8):31–35

Gann DM (1996) Construction as a manufacturing process? Similarities and differences between industrialized housing and car production in Japan. Constr Manage Econ 14(5):437–450

Garrison J, Tweedie A (2008) Modular architecture manual. Kullman Building Corporation

Gaze C, Ross K, Nolan E, Novakovic O, Cartwright P (2007) Modern methods of construction (MMC) in housing. BRE, Watford

Gorgolewski MT (2005) The potential for prefabrication in United Kingdom housing to improve sustainability. In: Yang J, Brandon PS, Sidwell AC. Smart and sustainable built environments. Blackwell, Oxford

Greeno R, Osbourn D (2014) Mitchell's introduction to building, 5th edn. Routledge, London

Gurung N, Mahendran M (2002) Comparative life cycle costs for new steel portal frame building systems. Build Res Inf 30(1):35–46

Harrison R (2009) Towards an archaeology of the welfare state in Britain 1945–2009. Archaeologies 5(2):238–262

Harvey CR, Ashworth A (1997) The construction industry of great Britain. Laxton's, Oxford

Hashemi A (2013) Review of the UK housing history in relation to system building. Alam Cipta 6 (1):47–58

Hashemi A (2014) Mitigating the risks of offsite manufacturing through the application of BIM. Int J 3-D Inf Model 3(4):26–35

Hashemi A, Hadjri K (2013) Code for sustainable homes: opportunities or threats for offsite manufacturing and mass-customization? In: Proceedings ZEMCH 2013 international conference, The visibility of zero-energy housing. University of Miami, Miami, USA, 30 Oct–1 Nov 2013

Hashemi A, Hadjri K (2014) Offsite construction, a potential answer to the Iranian housing shortages, In: Proceedings construction technology and management (CTM) 2014 international scientific conference, Bratislava, Slovakia, 9–11 Sept 2014

Hofman E, Voordijk H, Halman J (2009) Matching supply networks to a modular product architecture in the house-building industry. Build Res Inf 37(1):31–42

Japanese Prefabricated Construction Suppliers and Manufacturers Association (2015) Prefabricated house no genjo. http://www.purekyo.or.jp/main3-1.html. Accessed 12 Dec 2015

Johnson C (2004) Homes dot com. GAP, Sydney
Johnsson H, Meiling JH (2009) Defects in offsite construction: timber module prefabrication. Constr Manage Econ 27(7):667–681
Linner T, Bock T (2012) Evolution of large-scale industrialisation and service innovation in Japanese prefabrication industry. Constr Innov 12(2):156–178
Lovell H, Smith SJ (2010) Agencement in housing markets: the case of the UK construction industry. Geoforum 41(3):457–468
Lu N, Korman T (2010) Implementation of building information modeling (BIM) in modular construction: Benefits and challenges. Construction research congress 2010: Innovation for reshaping construction practice
McEvoy M (1994) Mitchell's external components. Longman, Harlow
Miller WF (2010) Energy and environmental benefits of structural insulated panels for residential construction in Australia. Metecno Pty Ltd, trading as Bondor, Brisbane
Nadim W, Goulding JS (2011) Offsite production: a model for building down barriers: a European construction industry perspective. Eng Constr Archit Manage 18(1):82–101
Nahmens I, Ikuma LH (2011) Effects of lean construction on sustainability of modular homebuilding. J Archit Eng 18:155–163
Noguchi M (2012) Prefabrication. In: Carswell AT (ed) The encyclopaedia of housing, 2nd edn. SAGE Publications, Thousand Oaks, pp 554–557
Normal D, Page I (2014) Prefabrication and standardisation potential in buildings. Study Report 312, BRANZ
Osbourn D, King H (1989) Components, 3rd edn. Mitchell, London
Page I (2012) Value of time savings in new housing. Study Report 250, BRANZ
Patchell J (2002) Linking production and consumption: the coevolution of interaction systems in the Japanese house industry. Ann Assoc Am Geogr 92(2):284–301
Poon CS, Ann TW, Ng LH (2003) Comparison of low-waste building technologies adopted in public and private housing projects in Hong Kong. Eng Constr Archit Manage 10(2):88–98
QLD Gov (2013) Guideline for excess dimension—vehicles carrying indivisible articles, special purpose vehicles, vehicles that require a pilot or Escort. Queensland Government, Department of Transport and Main Roads. https://www.nhvr.gov.au/files/t045-qld-guideline-for-excess-dimension.pdf. Accessed 12 Dec 2015
Riley M, Howard CA (2002) House construction Basingstoke. Palgrave Macmillan
Ross K (2002) Non-traditional housing in the UK—a brief review. Building Research Establishment, Watford
Roy R, Brown J, Gaze C (2003) Re-engineering the construction process in the speculative house-building sector. Constr Manage Econ 21(2):137–146
Steinhardt DA, Manley K, Miller W (2013) Profiling the nature and context of the Australian prefabricated housing industry, from https://www.qut.edu.au/research/research-projects/prefabricated-green-housing
Tam CM, Tam VWY, Zeng SX (2002) Environmental performance evaluation (EPE) for construction. Build Res Inf 30(5):349–361
Taylor MD (2010) A definition and valuation of the UK offsite construction sector. Constr Manage Econ 28(8):885–896
The C.I.E. (2013) Future forecasts: construction and property services skills 2016–26. Canberra: CPSISC
Trä och Möbelföretagen (2015) Husstatistik Byggfakta - Antal sökta bygglov, 2007 – 2014 (Housing Statistics from Byggfakta [Building Facts]—Number of building permit applications, 2007–2014) http://www.tmf.se/MediaBinaryLoader.axd?MediaArchive_FileID=32b3243c-4a90-41fb-975e-3283c538cad7andFileName=Byggfakta+%2B+diagram+dec+2014.pdf. Accessed 15 Apr 2015
U.S. Census Bureau (2012a) American housing survey. http://www.census.gov/housing/ahs. Accessed 5 Aug 2013

U.S. Census Bureau (2012b) Characteristics of new housing, http://www.census.gov/construction/chars/completed.html. Accessed 5 Aug 2013
Vale B (1995) Prefab: a history of the U.K. Temporary housing programme. E & FN Spon, London
Warszawski A (1999) Industrialized and automated building systems, 2nd edn. E & FN Spon, London

Chapter 4
Mass Customisation

Masa Noguchi, Carlos Formoso, Cecília G. da Rocha, María Dolores Andújar-Montoya, Victor Bunster, Rebecca Cameron and César Imai

Abstract Climate change accompanied by global warming issues necessitates that homes are more environmentally friendly in production and operation, while the drastic hike in energy costs and increasing socio-economic diversity demand housing that is more affordable and more customisable than ever before. Homes need to address social, economic and environmental sustainability issues as a whole. When the revolutionary notion of *sustainable development* was introduced the concept of *mass customisation* coincidentally emerged as a means to customise end-user products without sacrificing the production efficiency, effectiveness or low cost. This chapter will introduce the notion of mass customisation and the pragmatic application to housing design and production. This knowledge is essential to the delivery of *quality affordable homes* in which the level of performance can be determined within the economic constraints of the end-users.

M. Noguchi (✉) · V. Bunster · R. Cameron
The University of Melbourne, Melbourne, Australia
e-mail: Masa.noguchi@unimelb.edu.au

V. Bunster
e-mail: v.bunster@student.unimelb.edu.au

R. Cameron
e-mail: rejc@unimelb.edu.au

C. Formoso · C.G. da Rocha
Federal University of Rio Grande Do Sul, Porto Alegre, Brazil
e-mail: formoso@ufrgs.br

C.G. da Rocha
e-mail: cecilia.rocha@ufrgs.br

M.D. Andújar-Montoya
University of Alicante, San Vicente Del Raspeig, Spain
e-mail: lola.andujar@ua.es

C. Imai
State University of Londrina, Londrina, Brazil
e-mail: cimai@uel.br

© Springer International Publishing Switzerland 2016

M. Noguchi (ed.), *ZEMCH: Toward the Delivery of Zero Energy Mass Custom Homes*, Springer Tracts in Civil Engineering,
DOI 10.1007/978-3-319-31967-4_4

4.1 Introduction: Mass Customisation Concept

In 1987, the revolutionary notion of *sustainable development* was introduced (WCED 1987). Coincidently, the *mass customisation* concept emerged as a means to customise end-user products without sacrificing the production efficiency, effectiveness and low cost (Davis 1987). "Mass Customisation" is a complex term: for how does one combine *mass production* and *customisation*? As a technological capability, Mass Customisation was anticipated in 1970 by Alvin Toffler in his book entitled "*Future Shock*"—"Uniformity will give way to diversity" (Toffler 1970). He also asserted that maximum "individual choice is regarded as the democratic ideal" and expressed anxiety at the emergence of a more standardised mass culture and lifestyles in the future (Toffler 1970). He warned that:

> Science and technology have fostered standardization. Science and technology will advance, making the future even more standardized than the present. *Ergo*: Man will progressively lose his freedom of choice (Toffler 1970).

Toffler also directed his attention to trends in industry, indicating that companies were discovering wide variations in the wants and needs of individual customers, and they were adapting their production lines to accommodate them—"Every architect wants his own shade of green" (Toffler 1970). This implied that the rigid uniformity and long runs of identical products, which characterised traditional mass production plants, were becoming less important. According to his explanation, two economic factors encouraged this trend: first, consumers have more money to lavish on their wants; and second, as technology becomes more sophisticated the cost of introducing variations declines. However, Toffler (1970) emphasised that the latter factor is more important than the former, and he added that it is "only primitive technology that imposes standardization." Nevertheless, in this context, he seems not to be aware that standardisation for very simple items, or parts of a whole, can be accepted as offering the advantages of standardisation without limiting "creative design" (Movshin 1970). Furthermore, the rate of technological advance in the construction industry was considered to be far below that of other industries in the post-war era—or indeed even today. Toffler (1970) indicated that "construction has scarcely reached the level of mass production; it remains, in large measure, a pre-industrial craft."

In 1987, the term "Mass Customisation" was actually coined by Stanley M. Davis in his book entitled *Future Perfect* and he delineated the concept as follows:

> The world of mass customizing is a world of paradox with very practical implications. Whether we are dealing with a product, a service, a market, or an organization, each is understood to be both part (customized) and whole (mass) simultaneously….For mass customizing of products, markets, and organizations to be possible, the technology must make it economically feasible in every case (Davis 1987).

Like Toffler, Davis also stated that technology would make 'mass customisation' a practical possibility (Davis 1987). However, he knew that technology would not be sufficient by itself. Thus, he emphasised the need for the principle of mass customisation, which reflects "the paradox of the simultaneity of opposites, particularly of the whole and its parts", to be well understood so as to facilitate the widespread application of a mass customisation approach to many 'products' and 'services'.

Specifically, Davis cites the example of the Japanese prefabricated housing industry, indicating that it "has apparently learned the mass-customising message much better than their U.S. counterparts" (Davis 1987). Moreover, he briefly described Japanese manufacturers' way of mass customising their new homes as follows:

> The general sense of the public is that what you save on costs you lose in quality and in customization. If you are to buy a mass-produced house, your choices are limited to gross differences, such as two- or three-bedroom model. The Japanese, however, have applied mass customizing to housing business. Their effort began in the 1960s and has become quite sophisticated since then.
>
> The process begins by sitting down with a sales representative for a couple of hours and pretty much designing your own home on a computer screen. You choose from 20 thousand different standardized parts, like life-sized Lego blocks, and you can put those parts together pretty much as you want. Shall we add a foot to the length of living room? No problem. A small addition to add a Jacuzzi in the corner? Presto. You'd like the tea room on the other side of the house. Just press the button and the computer will make necessary adjustment in the plans and materials needed (Davis 1987).

In comparison with fully mass-customised, prefabricated housing in Japan, which usually consists of some engineered materials (e.g. a wall material, which does not burn and provides better insulation than concrete) that the manufacturers developed, Davis regarded mass-produced, prefabricated housing in the U.S. merely as "not highly esteemed" (Davis 1987). He also warned that "If U.S. home builders don't turn to mass customizing, they may end up out of business before they ever see 2001" (Davis 1987). His prediction for American homebuilders was somewhat intuitive or hyperbolic. However, the questions, which Davis posed in 1987, still remain unsolved, even today:

> Will the $6 billion U.S. housing industry adopt this [wall] material and other processes from the Japanese? Will American housing become mass customized?

It was not until 1993 that B. Joseph Pine II integrated the concept of mass customisation in his book entitled "Mass Customisation", providing a guideline for this new frontier, as well as addressing the fundamental methods of mass customisation of products and services. He regarded mass customisation as an emerging management system, which evolved from the "factory system" and the system of mass production. The factory system, common to the United States, Great Britain and the other newly industrialised nations of Europe in the nineteenth century, was a continuation of the basic idea of Craft Production—i.e. the end products themselves are not standardised. However, it allowed the production process to be standardised as the Industrial Revolution brought a general replacement of hand tools with machinery and mechanisation. According to Pine II (1993), the factory system in the United States was characterised by the following principles: interchangeable parts, specialised machinery, reliance on suppliers, focus on the process of production, the division of labour, the skills of American workers, flexibility, and continuous technological improvement. These eight characteristics differentiated the factory system of the United States from both craft production methods and the factory systems as applied in Europe (Pine II 1993).

However, in the twentieth century, these characteristics of the factory system alone become insufficient to support the growth of large enterprises that sought to meet the demands of an increasingly geographically dispersed economy in the late 1800s (Pine II 1993). At the dawn of the new century, a new system of 'Mass Production' was unveiled. Mass Production, or *Fordism* (after Henry Ford), made its first appearance in the world at Highland Park near Detroit in 1913. In order to mass-produce commodities (i.e. cars in his case), Henry Ford focused on seven principles: power, accuracy, economy, continuity, system, speed, and repetition (Nevins and Hill 1957). Ford then combined these principles into three creative components: the first was the planned, orderly, and continuous progression of the commodity through the shop; the second was the systematic delivery of the work to the mechanic, instead of bringing the mechanic to his work; and the third component was the break-down of all the operations into their constituent parts, with a suitable division of labour and material. As a consequence, the total amount of labour time spent making a single car dropped from 12 h 8 min to 2 h 35 min, when Ford's engineer introduced the assembly line in October 1913. Accordingly, sales of his car, known as the "Model T", increased tremendously (Table 4.1). This new system helped accomplish the continual lowering of its retail price and between 1908 and 1916 the price of the Model T Ford produced was drastically reduced by 57.65 %.

However, the system of Mass Production focuses essentially on producing standardised products for a *homogeneous* market. Pine II (1993) indicates that one of the primary reasons that the system of Mass Production (as well as the factory system) was developed so extensively in the United States was that the American market tended to be more homogeneous than the markets of the industrialised nations of Europe.

Table 4.1 Price and sales of Model T Fords, 1908–1916 (Pine II 1993)

Calendar year	Retail price (touring car) ($)	Change from 1908 (%)	Total 'Model T' sales	Change from 1908 (%)
1908	850	0	5986	0
1909	950	11.77	12,292	105.35
1910	780	−8.24	19,293	222.30
1911	690	−18.82	40,402	574.94
1912	600	−29.41	78,611	1213.25
1913	550	−35.29	182,809	2953.94
1914	490	−42.35	260,720	4255.50
1915	440	−48.24	355,276	5835.12
1916	360	−57.65	577,036	9539.76

Pine stated that:

> America never had the class distinctions common in Europe; therefore Americans did not have to differentiate themselves from other classes by what they purchased. Along the same lines, income distribution was also more equitable in the United States, resulting in more people clustered around similar needs and desires (Pine II 1993).

Nevertheless, there is no doubt that today's consumers have become more demanding than ever before, looking for variety and uniqueness in the products or services that they buy. They also demand high quality and low cost (Oleson 1998). Anderson (1997) suggests that "business must respond to succeed in today's turbulent climate, none is more difficult, more perilous, nor more vital than being customer-oriented." In other words, the imperative in business today is to understand and fulfil the wants and needs of each individual customer. The initiatives of most companies are usually market-focussed, rather than customer-focused, because companies also need to meet an equivalent imperative for achieving low costs—i.e. efficiency of production. Thus, Anderson (1997) concluded that "Mass Customization is the new imperative in business, one that puts the identification and fulfilment of the wants and needs of individual customers paramount within the company without sacrificing efficiency."

Pine II (1993) emphasises that, many companies no longer focus solely on producing merely standardised products or services for homogeneous mass markets. Accordingly, he suggests that a new paradigm (i.e. Mass Customisation) may help industries supply a variety of customised products or services through flexibility and quick responsiveness (Table 4.2).

In summary, the term *mass customisation* appears to be an oxymoron, but is actually a synthesis of the two long-competing systems of management—i.e. "the mass production of individually customised goods and services" (Pine II 1993). Thus, the concept of mass customisation can be explained as follows:

MASS CUSTOMISATION = ACHIEVING MASS + CUSTOMISATION

Table 4.2 Mass customisation contrasted with mass production (Pine II 1993)

	Mass production	Mass customisation
Focus	– Efficiency through stability and control	– Variety and customisation though flexibility and quick responsiveness
Goal	– Developing, producing, marketing, and delivering goods and services at prices low enough that nearly everyone can afford them	– Developing, producing, marketing, and delivering affordable goods and services with enough variety and customisation that nearly everyone finds exactly what they want
Key features	– Stable demand – Large, homogeneous market – Low-cost, consistent quality standardised goods and services – Long product	– Fragmented demand – Heterogeneous market niches – Low-cost, high-quality, customised goods and services – Short product development lifecycles

In a Mass Production system, low costs can be achieved primarily through *economies of scale*—i.e. lower unit costs of a single product or service through greater output and faster throughput of the production process. On the other hand, in a Mass Customisation system, low costs can be achieved primarily through *economies of scope*—i.e. the application of a single process to produce a greater variety of products or services more cheaply and more quickly (Pine II 1993).

4.2 Evolutions Towards Mass Customisation

As stated in the preceding section, Mass Customisation can be classified by reference to the two opposite notions of mass production and customisation. In the broadest sense, 'standardisation' is considered one of the most effective means of achieving the mass production of *products*, while 'customisation' is usually related to a process, or a *service*, that enables an end product to meet customers' individual requirements. This section focuses mainly on reviewing the two principal aspects, standardisation and customisation, in order to combine the principles of mass customisation for subsequent application to housing design and development, which are discussed later in this chapter.

4.2.1 From Standardisation to Mass Production

Standardisation is considered the key factor in establishing efficient *mass production* of products and services. However, standardisation is a discipline involving many factors (albeit not a static one since materials are continuously improved, knowledge and understanding grow and needs change over the time). According to the Oxford English Dictionary (OED), the word *standardisation* is considered "the action of standardising." However, the etymology of the word is still somewhat obscure.

Broadly speaking, a complete set of definitions for *standardisation* and *standard* covers performance, testing and certification, laboratory accreditation, and quality control. Notably, the International Organisation for Standardisation (ISO) defines the process of standardisation as:

> ... activity of establishing with regard to actual or potential problems provisions for common and repeated use, aimed at the achievement of optimum degree of order in a given context. Note: In particular, the activity consists of the processes of formulating, issuing and implementing standards (Spivak and Brenner 2001).

In this sense, Movshin (1970) emphasises that there are many sources of standards applicable to the homebuilding industry. However, there is also concern as to the effect of standards on *creativity*. Standardisation is usually accompanied by *simplification* that facilitates communication among the parties involved in a housing project and also helps to reduce the number of varieties of any given item. These factors yield economies of scale, which in turn, help lower design and production costs. According to Movshin (1970), the advantages of standardisation and simplification, from the standpoint of manufacturers, can be summarised as follows:

- Less capital needs to be tied up in raw materials, finished inventory, dies, jigs, templates, floor space, and repair parts;
- Manufacturing processes can be more economical though larger production runs—there are opportunities for introducing automated and specialised equipment, while development and experimentation costs can be reduced;
- Labour efficiency can be improved through organising the kinds of information that employees need, as well as through providing more specialised training on the products and items being produced;
- Stock depreciation and obsolescence can be reduced; and
- Improved and simplified communication can be developed throughout all phases of the production and distribution processes.

Movshin also adds some advantages from the standpoint of the user:

- As a communication device whereby information can be readily disseminated;
- As a means for ensuring that products can be readily obtained, with a known quality and at a minimum cost;
- As a way of allowing the designers to concentrate their interests and efforts on specific and significant aspects of the design; and
- As a factor in improving the maintainability and replacement characteristics of a design.

An important aspect of standardisation is the fact that it is associated with the feasibility of high-volume production—i.e. mass production.

The *Fordist* or mass production model established a system through which a unique product was created with reduced production costs and higher profits. However, since the nineteen-seventies, a change has taken place in not only the growth rates of companies and economies of scale but also in market competitiveness and changes in technology with the commencement of the information age. This resulted in an accelerated rate of change, consequent, increases in technological complexity and also changes at a sociological level (Pine II 1993). All these changes revealed more sophisticated consumers who were conscious about design and the quality and functionality of products and services (Papathanassiou 2004), meaning that standardized mass-produced products often lost their appeal (Peters and Saidin 2000). Market fragmentation has also hindered mass commercialisation, leading companies to segment their intended markets by producing and commercialising their products for segments or niches—i.e. personalized commercialization.

In this situation, organisations can reach the same number of customers as with a mass production model through mass customisation, but with a mass customisation model they also have the ability to meet the individual needs and requirements of the customers (Papathanassiou 2004; Parker 1995). However, the question that hinges on expanding standardisation in this way is: *Can standardisation permit the maximum opportunity for creativity, or customisation?*

4.2.2 From Mass Production to Mass Customisation

Proponents of Mass Customisation have discussed the meaning of *creativity* extensively. Taylor (1964) concedes the complexity of developing a single definition of creativity that suits all situations in different fields, summing up the notion as follows:

> Creativity at its highest level has probably been as important as any human quality in changing history and in reshaping the world…Man's current degree of enlightenment, particularly in certain fields, as well as his vast production of material goods, can be traced in large part to the creative performances of individuals during the course of history— to man's striving to improve his knowledge, to conquer the unknown, and to create new ideas and new, more useful things.

Taylor's description of creativity includes some interesting phrases such as "create new ideas" and "useful things". The former may be considered a generally-recognised expression of creativity, while the latter may reflect the following definition of creativity—*a novel work that is accepted as tenable or useful or satisfying by a group at some point in time* (Taylor 1964). In addition, he recognises two existing definitions of creativity as proposed by Ghiselin and Lacklen:

> Ghiselin proposes that the measure of a creation product be the extent to which it restructures our universe of understanding. Lacklen, of the Space Agency, uses the extent of the area of science that the contribution underlines: the more creative the contribution, the wider its effects (Taylor 1964).

Conclusively, Taylor (1964) admits that there is "no single definition has yet been prepared that suits all workers in the field", urging researchers "either to choose tentatively an existing definition of creativity or to develop a definition of their own that will enable them to move ahead in their work."

From the standpoint of Value Engineering, "creativity" is a behaviour that "uncovers a relationship where none previously existed; a relationship between people, objects, symbols, or any combination of these" (Dell'Isola 1997). Similarly, Parrot (2002), a Certified Value Specialist of SAVE International, indicates that the definition of "creativity" which is applicable to Value Engineering, falls into either of the following notions:

> Intellectual process resulting in the production of new and valid ideas"; or
> Innate aptitude of man in creating new combinations starting with existing elements.

The first definition of creativity above can be recognised as "to produce where nothing was before" (Oxford English Dictionary). In the homebuilding industry, a one-of-a-kind house might be of this type, in which all features of a home are customised, according to the wants and needs of individual clients, based mainly on a 'dialogue' between the user and the designer. Generically, these homes are usually called *custom homes* (Noguchi and Friedman 2002). The latter definition of creativity connotes the use of existing elements to develop new combinations that make some type of new end product. In this case, the existing elements (i.e. parts of a whole) can be standardised, while the myriad combinations of these standard parts still permit great scope for creativity. Thus, a homebuyer can directly choose standard housing components, which can be mass-produced, while the combinations of the user's 'choices' of these components make a house customised—viz. these homes can be termed *mass custom homes* (Noguchi 2001).

In contrast, homes that are customised by modifying an existing model house in response to a dialogue between the user and the designer to meet the wants and needs of individual clients, can be called *semi-custom homes* (as in this example the house is considered a whole, rather than the sum of particular parts). Homes that are totally mass-produced before they reach the market are termed *ready-built homes.*

Because of the nature of a ready-built home, the entire house itself can be standardised; thus, the level of product customisation is extremely low (Table 4.3). The characteristics of a custom home are totally opposite to those of a ready-built home. A custom home is a one-of-a-kind house, completely customised; thus, the level of standardisation in both products and processes can be considered as very low. Semi-custom homes combine the positive features of ready-built and custom homes—the model house is usually prepared on a speculative basis like a ready-built home, while, in response to the user's demands for housing, the modification of the model house allows, in part, for product customisation. On the other hand, mass custom homes can theoretically achieve a high level of standardisation for all the housing components from which homebuyers can directly select in customising their new home. The user's choice of mass-produced, standard components may actually increase the level of customisation in housing design.

Up to this point, the principles of standardisation for tangible elements of the design have been discussed. In summary, standardisation can be applied either to the whole of a complex product or to the parts. However, in terms of the scope for design customisation of a house the consequences are different—i.e. it will be a ready-built, semi-custom, custom, or mass custom home. Movshin (1970) suggests

Table 4.3 The levels of standardisation and customisation compared by housing type (Noguchi and Friedman 2002)

	Standardisation level	Customisation level
Ready-built home	High	Low
Semi-custom home	Medium	Medium
Custom home	Low	High
Mass custom home	High	High

that the "design represents an integration of *standard* and *special* components in a unique combination." On the other hand, the action of combining standard components connotes a *process*; indicating that standardisation of intangible elements (i.e. processes or services) should also be taken into account in order to maximise the advantages of standardisation for the efficient creation of an end product.

4.3 Mass Customisation Business Orientation

The idea that Mass Customisation should deliver to clients whatever they want, at anytime, anyplace, and in any form (Hart 1995) is a visionary perspective of the notion of mass customisation. This visionary perspectivism has gradually given way to a more pragmatic perspective of mass customisation. Mass customisation has been explored in several areas such as product development processes, operations management, product architecture, and supply chain management amongst others, without any real consensus on its limits and scope.

Illustrative of the absence of a consensus on the nature of mass customisation, the term has been defined as a paradigm (Huang et al. 2008; Brown and Bessant 2003; Jiao and Tseng 1999), a strategy (Piller et al. 2005; Bardakci and Whitelock 2005; Frutos and Borenstein 2004; MacCarthy et al. 2003; Salvador et al. 2002; Gilmore and Pine 2007), a theory (Bardakci and Whitelock 2005), a capability (Zipkin 2001) and an ability (Da Silveira et al. 2001). Whilst studies that define Mass Customisation as a strategy do not usually elicit the strategic planning level to which it relates, Frutos and Borenstein (2003) and Bardakci and Whitelock (2005) are two exceptions. They define Mass Customisation as a business strategy (Frutos and Borenstein 2003) and as a marketing strategy (Bardakci and Whitelock 2005). Finally, several studies (Piller et al. 2005; MacCarthy et al. 2003; Lampel and Mintzberg 1996) discuss customisation strategies, rather than a single one, emphasizing that distinct strategies are possible under mass customisation. However, Kumar recognizes there are distinct definitions of Mass Customisation and proposes two elements that are common to most of them (Kumar 2004):

- There is some degree of customisation, implying that the product delivered is similar to what the client desires; and
- There is not a proportional raise in the cost as the degree of customisation increases.

Mass customisation is understood as a concept that can be used in devising customisation strategies, which are closely related to business strategies. More specifically, mass customisation defines the competitive criteria— high product variety with reduced cost and delivery time— to be pursued at a business level. Yet, other elements such as the scope of the business, the product, the clients, and the competitors, which are not defined by Mass Customisation, need to be outlined for establishing a business strategy. In fact, it is necessary to have such elements so that the Mass Customisation concept can be tailored to and operationalized in different contexts.

This reinforces the idea that there are customisation strategies, rather than a single strategy for mass customisation. However, Mass Customisation is not conceived as a functional strategy because coordinated efforts across the different functional areas are necessary for its successful implementation. As a result, a customisation strategy needs to be supported by coordinated decisions in the areas of marketing, product design, and operations management. It is subject to external forces—competitive environment and clients—that can shape or influence such a strategy. Marketing deals with the capture and prioritising of clients' requirements, which are later translated into customisable attributes and is also concerned with the interface between clients and the organisation: namely, how these attributes are presented to clients and how clients communicate their decisions regarding the selected attributes. Product design focuses on the translation of the customisable attributes into product architecture. This requires a balance between commonality and distinctiveness across product variants, which is a key aspect to provide customisation while keeping the cost and delivery time reduced. Finally, the operations management area focuses on the production system and the supply chain and how these should be designed and managed to efficiently deliver the product variants offered.

The competitive environment can also increase or reduce the advantage yielded by a customisation strategy (Hart 1995; Pine II 1993). The timing in establishing such a strategy in relationship to rivals plays a key role: it is more successfully implemented if there are no rivals prepare to follow (Hart 1995; Da Silveira et al. 2001; Kotha 1996). The first entrant advantage in developing a customisation strategy emanates from two sources:

- Delivery of products that are able to fulfil individual needs differently from competitors; and
- Creation of a one-to-one relationship based on knowledge and trust that enables an organisation to learn more about their clients in each transaction (Hart 1995).

An organisation that initiates customisation has an advantage over its rivals, making clients them more loyal to such organisation and less susceptible to change of brands. Turbulent markets are suitable environments for customisation strategies (Pine II 1993; Hart 1995) as market instability and unpredictability increases, the potential for customised products also increases (Hart 1995).

4.3.1 Operations Management for Mass Customisation

Where efficiency in performance is required, operations *management* plays an important role in a customisation strategy. Strategic decisions within this functional area are required to provide customised products at reduced costs, with alterations to the traditional trade-off between flexibility and cost at the heart of the mass customisation approach. This change was anticipated by Toffler (1970), who stated that the pre automation or primitive technologies produce standardisation whereas advanced technologies such as computer aided design (CAD) and computer aided manufacturing (CAM) enable greater product variety to be produced. In Mass Customisation, the conflict between flexibility and cost is overcome, allowing the concept to be closely related to operations management. In some studies, Mass Customisation is even considered an exclusive attribute of operations management (although decisions in other functional areas are also require for its successful implementation). For example, Mass Customisation has been associated with the competitive manufacturing criteria of mix flexibility (Paiva et al. 2004). However, mix flexibility is not the only type of flexibility that can be adopted in pursuing a customisation strategy and customisation strategies can eventually be developed without flexible production systems, such as in the case of adaptive customisation. It is worth exploring the concept of flexibility and how it relates to customisation strategies.

Slack (1983) proposes five types of flexibility in operations management:

- New product flexibility: the capability to make something new;
- Mix flexibility: the capability to make a different mix of products;
- Quality flexibility: the capability to change the quality level;
- Volume flexibility: the capability to change the volume of the output and
- Delivery flexibility: the capability to change delivery times.

For each type, there are also three dimensions (Slack 1983):

- The range of states of the adopted system;
- The cost of moving from one state to the other; and
- The time required for moving from one state to another.

Upton (1994) defines flexibility as the ability to react or change with minor penalties of cost, effort, and performance. He proposes flexibility has a dimension of change (scope of the change required, which may be in the product, parts of it, or in a process) and a time horizon (time frame need for the change to take place). Flexibility is also defined by three elements (Upton 1994):

- Range: ability to accommodate a small or large range, measured by the number of states or the distance between extremes in a range;
- Mobility: measures the ease in moving from one state to another and can be operationalized as time or the cost of the change; and
- Uniformity: measures the sameness of the system performance when operating in different states within a range.

4.3.2 Mass Customisation and Lean Construction

From the previous section it has been deduced that in the current business environment Mass Customisation can be positioned as a strategy to ensure competitiveness in a customer-oriented market. Through Mass Customisation, value-added services are created to meet the needs and requirements of potential customers who are changing their usual way of buying and the customer becomes integrated with the process. This means than in new business models, the customer is no longer considered as a mere external entity to the processes of the organisation, but rather as an active element, determining the specific features of the desired product.

As a result, it is necessary that a new agile approach not only meets the initial needs of customers but also enables adaptation to these needs as they evolve (Barlow 1998). This approach requires the use of efficient production processes to optimise tasks and resources. The main barrier to implementation of such an approach is the complexity of management due to the diversity in the product, which may exceed the internal capacity of the organisation (Pine II 1993).

To overcome these barriers emerging techniques and tools aimed at improving the performance of production processes through the efficient management of the processes derived from mass customisation. Some of the techniques and tools with a high effect on the performance of operations are based on the philosophy of Lean Construction. In Japan, this philosophy originated in the manufacturing industry (Lean Manufacturing) (Ohno 1988), and was then adapted for the building industry as Lean Construction (Koskela 1997; Koskela and Howell 2002) to improve building execution through improvement of workflow. The main objective is to generate value for customers and to eliminate or minimise waste by reducing activities that do not add value. Likewise, this philosophy and its associated tools can reduce the cycle time, simplifying and optimizing the process by minimising the number of steps. This increases the agility and transparency of processes while focusing on the control of the whole process to reduce variability, and encouraging continuous improvement.

4.3.3 Mass Customisation in Manufacturing

The *order penetration point* (OPP) is the point where the client order first enters the supply chain, separating demand and forecast driven tasks (Yang and Burns 2003). The OPP separates the supply chain in two parts, which have distinct goals. Before the OPP, the demand is leverage and there is a limited variety of products (Naylor et al. 1999). Lean principles (e.g. elimination of waste) are applicable prior to the OPP since the demand is smooth (Mason-Jones et al. 2000). After the OPP, the demand becomes unstable and there is a proliferation of products (Naylor et al. 1999). Prior to the OPP lean principles should be applied (e.g. elimination of waste) whereas after OPP agile principles should be preferred (e.g. providing quick responses to volatile markets) (Mason-Jones et al. 2000).

The position of the OPP in the supply chain is a key decision, which requires the product type, the market, the organisation processes, and the stock characteristics to be considered (Yang and Burns 2003). Discussion on the strategic importance of the positioning of the OPP is still sparse in Mass Customisation literature

(Olhager 2003) but five classes of supply chain, based on the OPP position, have been proposed (Naylor et al. 1999):

- *Buy-to-order*: provide unique products that do not contain the same raw materials. Products are designed and produced based on demand and thus there is no risk of product obsolescence (Barlow et al. 2003), overstock, or out-of-stock. Yet, clients should be prepared to accept a long delivery time.
- *Make-to-order*: delivers different products that are made from the same raw materials. This supply chain can cope with distinct locations, volumes, and product mix. Delivery time is reduced. The supply chain is only exposed to the risk of holding raw materials and components as stock.
- *Assemble-to-order*: provides different product mix. The delivery time is reduced but there is a slight increase in the risk of overstock and out-of-stock of components.
- *Make-to-stock and ship-to-stock*: delivers standard products. Products are ordered and stocked based on forecast and ordered by clients with minimal lead-time (Barlow et al. 2003). Make-to-stock supply chains require a leverage demand of standard products that can be delivered to varied locations. Ship-to-stock supply chains provide standard products only in pre-defined locations.

Assemble-to-order supply chains allow customisation to be postponed until as late as possible (Naylor et al. 1999). In such a supply chain, product components can be modularized. In modularisation, the cost savings arise from the economies of scope and scale at the component level, since components are used across several products and produced in large volumes (Pine II 1993). Component modularisation is not possible in make-to-order and buy-to-order supply chains as they involve customisation in the product design. The similarly of the Mass Customisation approach to mix flexibility means that Mass Customisation is often related to assemble-to-order supply chains. Although the modularization is an effective strategy for customisation, it is not the only supply chain type that supports customisation. Customisation strategies can entail a myriad of types of customisation, not all of which require mix flexibility or an assemble-to-order supply chain.

Delaying the activities that differentiate a product until information about the clients' orders can be clearly identified is the logic of postponement (Yang and Burns 2003). This allows economies of scale to be explored without compromising the variety of products (Ernst and Kamrad 2000). The kind of product differentiation depends on the activity that is postponed. At least five postponement types can be highlighted (Zinn and Bowersox 1988):

- *Labelling postponement*: products are shipped in unlabelled packages to the warehouse and labels are placed according to the clients' order, allowing products to be marketed under different brands.
- *Packaging postponement*: bulk products are shipped to the warehouse and packaged based on clients' orders, allowing their storage in distinct packages sizes.

- *Assembly postponement*: a base product with a number of common parts is sold in a variety of configurations based on clients' orders.
- *Manufacturing postponement*: parts are shipped to the warehouse where manufacturing is completed according to clients' orders.
- *Time postponement*: time of shipment is postponed, which is based on actual orders, instead of forecast.

Postponing the differentiating of a product until the latest possible point in the supply chain is crucial to effective mass customising (Lee et al. 1997). This enables organisations to operate with maximum efficiency and to quickly meet clients' orders with minimum amount of stock (Lee et al. 1997). It also encourages organisations to make some strategic definitions, that is, to determine which components will be standardised and customised and which of the activities will be performed based on demand or forecast (Yang and Burns 2003). This requires a new organization and coordination among product design, production, and supply chain (Lee et al. 1997).

4.3.4 Customer Co-creation

Incorporating customer needs and preferences into the development of a product is not a new idea. In fact, many companies rely on user feedback on a regular basis as means to reduce risks before starting large-scale production or to explore new markets for their products with focus groups, surveys and prototype testing. However, the recent rise of social media and web 2.0 has enabled straightforward incorporation of the actual consumers as active agents in the development of new products. O'Hern and Rindfleisch (2008) define customer co-creation as "a collaborative new product development activity in which customers actively contribute and/or select the content of a new product offering." They argue that new product development through customer co-creation can help overcome information asymmetries that separate understanding of the users own needs, and the know-how of providers on production strategies that may be used to address those needs. Zwass (2010) further argues that co-creation challenges traditional views of customers as passive consumers of external products to understand that value is a property that can emerge through interactions between different stakeholders. Accordingly, he proposes an explanatory taxonomy of co-creation where 'sponsored' co-creation refers to strategies that enable user involvement under rules set by providers while 'autonomous' co-creation takes place independently of formal organisations. Examples of 'autonomous' co-creation include open source software communities such as Ubuntu, Gimp or Blender, whilst mass custom products that allow 'sponsored' co-creation include personalised guitar picks by companies such as ClaytonCustom© or T-shirts by companies such as Threadless©. 'Crowdsourcing' is a closely related concept in which value emerges from interactions involving large numbers of users, often organised as non-hierarchical online communities.

Examples of crowdsourcing include online platforms such as crowdSPRING© and 99Designs© for crowd-enabled logo and web design or KickStarter© and Indiegogo© from crowd funding of start-up companies.

Some precedents for advancement in customer co-creation and crowdsourcing in mass custom housing include the use of virtual catalogues for selection of prefabricated componentry such as the ones developed by companies like the Misawa Homes Group© or for active customer design of layouts through online platforms such as the one developed by Original Home Plans©. Although such platforms are not truly open to innovation, they are valuable sources of information for providers to inform new product development and improve the scope of their strategies and operations. In parallel, experimental projects such as the 'The Hemnet Home' by Tham and Videgård Arkitekter have shown that data analytics and crowdsourcing can become valuable sources of critical information for the development of customer-centric housing projects, regardless of the result being a 'one size fits all' product.

4.3.5 Pervasive Computing

With the widespread adoption of internet connectivity and the increasing computerization of current practices, the market-to-one paradigm is starting to become a reality. Widely known online companies are now able to target the specific preferences of individual customers and gain competitive advantage though personalized relationships that can enhance user experience. Parallel processing, big data, and knowledge extraction algorithms are not only changing the way in which we relate through social media, but also the ways in which we conceive objects and engage with our physical environment. Pervasive computing, also known as ubiquitous computing, refers to the vision of physical environments saturated with computing capabilities that could be gracefully integrated to everyday routines (Satyanarayanan 2001). 'Domotics' is a clear example of how such technology-enabled personalisation has started to become mainstream. Centralised cooling and space heating systems, automated illumination, home entertainment and safety systems can ease everyday life, adapt to changing environmental conditions, and even identify recurrent behavioural patterns to generate ambiences for enhancing user experience. Computing can also extend the possibilities of housing beyond personalised user experience. The increasing availability of 3D scanners, digital cameras, GPS technology, and affordable 3D printing is transforming our capacity to manage complex information as well as the inflows and outflows of knowledge among peers, including relationships between customers and providers. Technology such as the Microsoft Kinect© can collect accurate body measurements to be sent online to manufacturers to develop tailor-made jewellery and garment (e.g., Grimal and Guerlain 2014) and smartphones and digital cameras can be used to generate

3D models that can be digitally printed and delivered through platforms such as Autodesk's 123Catch© or insight3d©. In relation to housing, research initiatives such as the 'Instant House' at MIT (Sass and Botha 2006) and the 'Casa Generativa' project (García and Turkienicz 2010) (Fig. 4.1) demonstrate that

Fig. 4.1 *Casa Generativa* project, Rio Grande do Sul in 2010 (García and Turkienicz 2010)

computer-enabled manufacturing can ensure rapid development of highly customised but rather inexpensive timber frame structures, while recent innovations in digital printing led by the Chinese firm WinSun Decoration Design Engineering Co. © promise commercially feasible automated fabrication of custom housing at mass scale. Related research such as the 'Open Source Housing' (Larson et al. 2004) or the 'Wiki House' project promise to further advance housing innovation through incorporation of open source principles and technology throughout the dwelling's life cycle, including crowd sourced design, distributed manufacturing, and personalised co-creation of self-help modifications during occupancy.

4.4 Mass Custom Design System Model for Housing

One of the successful mass customisation approaches that can be applied to the homebuilding process is the modularization of housing components. The variations possible through the mass customisation process can be quantified in the light of Set Theory, a branch of mathematics developed by Georg Cantor. This allows the total number of possible ordered pairs (or combinations) of given standard housing components to be simply calculated. To bring the concept of Mass Customisation into effect, a total coordination (or systems) approach to considering products and services within the housing delivery process needs to be taken into account since to design, build and market a home requires consideration of both these aspects (Noguchi and Hernández 2005). If Mass Customisation is considered as a set of systems for designing, producing and marketing a product, both customizable products and communication services are required. If one or the other is absent, this type of Mass Customisation is not possible. To formulate the means to mass-customise homes, a system model was developed (Noguchi 2008). To discuss the potential applicability of Mass Customisation to the delivery of quality affordable homes, a generic mass custom design system model was first introduced at an internal symposium on urbanism, which was entitled *Prospectiva Urbana en Latinoamerica* held in Aguascalientes, Mexico, on 31st August 2001 (Noguchi 2001). This televised symposium was organised by the local government and over 500 audience members from the academia, construction industry and government joined the discussion.

Mass Customisation (MC) was visualized simply by making use of a conceptual analogue model as follows:

$$\mathrm{MC} = f(\mathrm{PS})$$

In this model, the *service* sub-system (S) concerns communication techniques that allow users to participate in customizing their new home, while the *product* sub-system (P) covers production techniques that aim to encourage housing suppliers to standardise housing components for mass production.

4.4.1 Service Sub-system

In mass-customizing homes, user participation is vital. Therefore, housing suppliers need to offer design support communication services to their clients. Both design-consulting staff and appropriate communication tools are required to facilitate user choice of standard components (Noguchi and Friedman 2002). These fundamental design service factors can also be integrated into a comprehensive model:

$$S = f(l, p, t)$$

In this model, the service sub-system (S) is supported by the existence of the location (l), personnel (p) and tool (t) factors.

Location factors. During the design stage, clients need to participate in customising their new home and this design consulting service normally takes place in the company's office. The company's show home may also function as an exhibition and consultation base, where experienced staff can make specific proposals concerning the external appearance and floor plans of a customised home by making use of advanced computer technology. Clients can also see and touch the samples to confirm the qualities of the company's products and they learn more about the company's suggestions regarding housing facilities. The consequence of providing such visual information, visits, and individual consultations with housing experts is that it may increase the clients' faith in the reliability of the company and its products, and thus increase the likelihood that the client will select the company when purchasing a home.

Personnel factors. In order to alleviate consumer anxiety caused by the combination of high risk and limited experience in purchasing a home, housing suppliers may strengthen their local network of housing business by locating sales staff across the country in question. These salespeople directly contact clients, in order to market, as well as design their products. In the show home, for example, the salesman may not only explain the distinguishing characteristics of the company's product, but also address the client's own requirements with rough sketches of the housing layout (Mishima 2000). Sales staff can be considered as the company's greatest assets in reaching clients and they are trained to assist clients in designing their custom homes by making use of various communication tools (e.g. catalogues and digital communication tools) that contain various standard housing components options.

Communication tool factors. Client needs are often difficult to conceptualise and articulate. However, interaction with possible prototypes can help identify these needs. Housing suppliers may be able to use various types of housing catalogues showing possible prototypes. This enhances client involvement, offering the client a great choice of housing types and components. The catalogues play a variety of roles in advertising and educating clients. In terms of client participation at the design stage, the catalogues mainly function as design tools for the manufacturers and the client. Through consultation with housing experts, clients can choose the housing type and components from the catalogues to design a custom home that meets their needs. Clearly, catalogues can be regarded as synthesised information sources that integrate and simplify extensive data through an inductive process.

The housing suppliers may start with two kinds of housing catalogues to help the client choose the housing type: a general housing catalogue and a housing style catalogue. The former contains information on all of the company's products in order to communicate general information about the company, while the latter offers more detailed information about specific types of products. Such housing style catalogues explain the characteristics of each commodity in terms of the design concept and technology.

Housing component selection catalogues correspond to the housing types, helping the client to select the various standard housing components for exterior and interior arrangements. The catalogue elaborates in detail on each component in terms of material, size, colour, texture, and functions. Such catalogues do not contain the price of each component, so that the client will choose an item according to its use—not its cost. However, a cost estimate will be offered by making parallel use of a computer with these catalogues as part of the consultation. The selection catalogues are visually integrated in order to compare the components for the client's design decision.

A physical model helps visualise the actual configurations of housing in response to stakeholders' design choices (Fig. 4.2). The construction of knowledge in the design decision making process requires the use of instruments that will enable efficient and effective communication between the user and the housing suppliers.

The use of a CAD system is important in the creation, modification, analysis, or optimisation of a design, allowing consumers to customise their choice in housing. Most housing suppliers today use a CAD system as a digital communication tool to offer flexibility in design. The benefits derived from applying a CAD system in the design stage includes the short time elapsed between the receipt of a consumer order and the delivery of the proposal, the accuracy of material and cost estimates, and the standardisation of drafting and documentation. Ease in visualising a drawing also boosts the client's comprehension of the layout of their new home. The interactive CAD system basically contributes to the creation of line drawings. However, advanced geometric modelling, such as solid modelling in three dimensions with shade and colour, helps to display more information on the graphic screen. Another advantage of a CAD system is data communication. The purpose of such data communication is to transfer data between a computer and its peripherals that can be defined as any input or output device.

4.4.2 Product Sub-system

Even though these elements are necessarily interrelated, most homebuilders and housing manufacturers have already been applying these during the design stage. An important part of Mass Customisation is that the user directly determines the

Fig. 4.2 An adjustable physical housing model for mass customisation

configuration of their home from choices given as client input during the design stage. However, this cannot be achieved without the standardisation of housing components for the volumetric, exterior and interior design arrangements. These components should be organized in a visually attractive way in a component selection catalogue that enables clients to easily choose from many options given—and the value of each component choice may also need to be explicit.

Basically, housing components can be divided into three categories: volume, exterior and interior. These can be considered the main elements of the product sub-system (P) which can be explained by the following conceptual model:

$$P = f(v, e, i, o)$$

The volume (v) components are used to configure the space of housing that determines the number and size of each room while the interior (i) and exterior (e) components serve to co-ordinate decorative and functional elements that customise a home. In addition, 'o' denotes other optional features such as air conditioning, home security system, emergency call buttons, handrails, dishwashers and other electrical appliances. Some optional features can be offered before or after occupancy with due consideration of inclusive design approaches in housing the elderly and disabled users.

When considering the concept of Mass Customisation, this product system approach can be termed *mass custom design*, which is a result of the combinations of three basic sets of design elements of housing: the volume, exterior and interior, along with any additional amenity elements (Noguchi 2001). In principle, these housing components are mass-produced, but the home itself is customised by the user's direct choices of such standard components. The exterior and interior designs include sub-categories such as the roof, walls, doors, windows, balconies, and front entrance arrangements for the exterior, as well as the kitchens, sanitary facilities, bathrooms, washrooms, toilets, storage, and finishing arrangements for the interior. The variety of sizes, materials, colours, and textures available for each component, as well as the variety of amenities offered, help expand the number of housing variations. Consequently, in order to meet individual client requirements, the manufacturers are able to provide a broad range of housing variations for their clients without producing a number of standard model homes that are usually designed on a speculative basis. The application of the mass custom design approach may have potential to reduce production costs by achieving the economies of scope, which also help customise homes.

4.5 Conclusions

Mass customisation is an oxymoron that emerged along with the revolutionary concept of *sustainable development* in 1987. From early mass production through to contemporary methods of mass custom design, mass customisation has progressively been applied as a means of holistically address social, economic and environmental sustainability. Today, mass customisation has been devised in the context of housing, where the functionality and performance of the resulting home can be pre-selected or pre-determined by user-buyers' direct choice of standardised design components proposed, which also helps define the desired and expected product quality achievable within the end-user's economic constraints—i.e. the delivery of *quality affordable homes*. This application requires the complex integration of both the product and the processes. In this chapter, a *mass custom design* system model composed of the product and service sub-systems to produce *mass custom homes* was introduced. The system illustrated includes a product sub-system encompassing production techniques that encourage housing suppliers to standardise and/or mass-produce the design components and a service sub-system where communication techniques allow the end user-buyers to participate in customising their new home. As such, mass customisation can be seen as a medium for sustainable development. In particular, it applies well to the delivery of quality affordable homes that satisfy the wants and needs of both individuals and society.

References

Anderson DM (1997) Agile product development for mass customization: how to develop and deliver products for mass customization, niche markets, JIT, build-to-order, and flexible manufacturing. IRWIN, Chicago

Bardakci A, Whitelock J (2005) A comparison of customers' readiness for mass-customisation: Turkish vs British customers. Eur Bus review 17(5):397–410

Barlow J (1998) From craft production to mass customisation? Customer focused approaches to house building. In: Proceedings of 6th annual conference of the international group for lean construction IGLC-6, Guarujá, Brazil, 13–15 Aug 1998

Barlow J, Childerhouse P, Gann D, Hong-Minh S, Naim M, Ozaki R (2003) Choice and delivery in housebuilding: lessons from Japan for UK housebuilders. Build Res Inf 31(2):134–145

Brown S, Bessant J (2003) The manufacturing strategy-capabilities links in mass customisation and agile manufacturing-an exploratory study. Int J Oper Prod Manag 23(7):707–730

Da Silveira G, Borenstein D, Fogliatto FS (2001) Mass customization: literature review and research directions. Int J Prod Econ 72(1):1–13

Davis SM (1987) Future perfect. Adison-Wesley, New York

Dell'Isola A (1997) Value engineering: practical applications for design, construction. Maintenance and operations. RSMeans, Kingston

Ernst R, Kamrad B (2000) Evaluation of supply chain structures through modularization and postponement. Eur J Oper Res 124(3):495–510

Frutos JD, Borenstein D (2003) Object-oriented model for customer-building company interaction in mass customization environment. J Constr Eng Manag 129(3):302–313

Frutos JD, Borenstein D (2004) A framework to support customer–company interaction in mass customization environments. Comput Ind 54(2):115–135

García R, Turkienicz B (2010) Generative house: exploration of digital fabrication and generative system for low-cost housing in Southern Brazil. In: Proceedings of the 14th congress of the Iberoamerican society of digital graphics (SIGRADI 2010). Bogotá, Colombia, pp 384–387, 17–19 Nov 2010

Gilmore JH, Pine BJ (2007) Authenticity: what consumers really want. Harvard Business School Press, Boston

Grimal L, Guerlain P (2014) Mass customization in apparel industry-implication of consumer as co-creator. J Econ Manag 15:105–121

Hart CW (1995) Mass customization: conceptual underpinnings, opportunities and limits. Int J Serv Ind Manag 6(2):36–45

Huang X, Kristal MM, Schroeder RG (2008) Linking learning and effective process implementation to mass customization capability. J Oper Manag 26(6):714–729

Jiao J, Tseng MM (1999) A methodology of developing product family architecture for mass customization. J Intell Manuf 10(1):3–20

Koskela L (1997) Lean production in construction. Lean construction. AA Balkema Publishers, Amsterdam

Koskela L, Howell G (2002) The theory of project management: Explanation to novel methods. In: Proceedings 10th annual conference of the international group for lean construction IGLC-10, Gramado, Brazil, 6–8 Aug 2002

Kotha S (1996) From mass production to mass customization: the case of the National Industrial Bicycle Company of Japan. Eur Manag J 14(5):442–450

Kumar A (2004) Mass customization: metrics and modularity. Int J Flex Manuf Syst 16(4):287–311
Lampel J, Mintzberg H (1996) Customizing customization. Sloan Manag Rev 38(1):21–30
Larson K, Intille S, McLeish T, Beaudin J, Williams R (2004) Open source building—reinventing places of living. BT Technol J 22(4):187–200
Lee HL, Feitzinger E, Billington C (1997) Getting ahead of your competition through design for mass customization. Target 13(2):8–17
MacCarthy B, Brabazon PG, Bramham J (2003) Fundamental modes of operation for mass customization. Int J Prod Econ 85(3):289–304
Mason-Jones R, Naylor BR, Towill D (2000) Engineering the leagile supply chain. Int J Agile Manag Syst 2(1):54–61
Mishima S (2000) Jyutaku: Nihon no Kaisya. Jitsumu kyouiku, Tokyo
Movshin J (1970) Standardization and the building industry. Ind Forum 2(1):13–18
Naylor JB, Naim MM, Berry D (1999) Leagility: integrating the lean and agile manufacturing paradigms in the total supply chain. Int J Prod Econ 62(1):107–118
Nevins A, Hill FE (1957) Ford: expansion and challenge, 1915–1933. Scribner
Noguchi M (2001) The mass custom home: vivienda personalizada masiva: optimización de la tecnología de la vivienda canadiense y japanesa en latinoamérica. Simposium internacional: prospectiva urbana en latinoamerica. Secretaría de Desarrollo Social & la Universidad Autónoma de Aguascalientes. Aguascalientes. Mexico. 31 Aug 2001
Noguchi M (2008) Structuring a choice model for mass customisation. Int J Mass Cust 2(3/4):264–281
Noguchi M, Friedman (2002) A mass custom design system model for the delivery of quality homes-learning from Japan's prefabricated housing industry. In: Proceedings of international council for research and innovation in building and construction. CIB W060–096 syllabus joint conference, measurement and management of architectural value in performance-based building. Hong Kong: CIB, pp 229–243, 6–10 May 2002
Noguchi M, Hernández C (2005) A 'mass custom design' approach to upgrading traditional housing development in Mexico. J Habitat Int 29(2):325–336
O'hern M, Rindfleisch A (2008) Customer co-creation a typology and research agenda. Wiscons innovation: thoughts on innovation from the Wisconsin School of Business University of Wisconsin, Wisconsin, USA
Ohno T (1988) Toyota production system: beyond large-scale production. Productivity Press, Cambridge
Oleson JD (1998) Pathways to agility: mass customization in action. Wiley, New York
Olhager J (2003) Strategic positioning of the order penetration point. Int J Prod Econ 85(3):319–329
Paiva EL, Carvalho L, Fensterseifer JE (2004) Estratégia de produção e de operações. Porto Alegre, Bokman
Papathanassiou EA (2004) Mass customisation: management approaches and internet opportunities in the financial sector in the UK. Int J Inf Manage 24(5):387–399
Parker MM (1995) Strategic transformation and information technology: paradigms for performing while transforming. Prentice Hall Press, Englewood Cliffs
Parrot L (2002) Value engineering workshop module 1. Unpublished course textbook. McGill University, Montreal
Peters L, Saidin H (2000) IT and the mass customization of services: the challenge of implementation. Int J Inf Manage 20(2):103–119
Piller F, Schubert P, Koch M, Möslein K (2005) Overcoming mass confusion: collaborative customer co-design in online communities. J Comput Mediat Commun 10(4)
Pine BJ II (1993) Mass customization: the new frontier in business competition. Harvard Business School Press, Boston
Salvador F, Forza C, Rungtusanatham M (2002) Modularity, product variety, production volume, and component sourcing: theorizing beyond generic prescriptions. J Oper Manag 20(5):549–575

Sass L, Botha M (2006) The instant house: a model of design production with digital fabrication. Int J Architect Comput 4(4):109–123

Satyanarayanan M (2001) Pervasive computing: vision and challenges. IEEE Pers Commun 8 (4):10–17

Slack N (1983) Flexibility as a manufacturing objective. Int J Oper Prod Manag 3(3):4–13

Spivak SM, Brenner FC (2001) Standardization essentials: principles and practice. Marcel Dekker, New York

Taylor CW (1964) Creativity: progress and potential. McGraw-Hill, New York

Toffler A (1970) Future shock. Random House, New York

Upton DM (1994) The management of manufacturing flexibility. Calif Manag Rev 36(2):72

World Commission on Environment and Development (WCED) (1987) Our common future. Oxford University Press, Oxford, p 1987

Yang B, Burns N (2003) Implications of postponement for the supply chain. Int J Prod Res 41 (9):2075–2090

Zinn W, Bowersox DJ (1988) Planning physical distribution with the principle of postponement. J Bus Logist 9(2):117–136

Zipkin P (2001) The limits of mass customization. MIT Sloan Manag Rev 81–87

Zwass V (2010) Co-creation: toward a taxonomy and an integrated research perspective. Int J Electron Commer 15(1):11–48

Chapter 5
Mass Personalisation

Victor Bunster, Masa Noguchi and Thomas Kvan

Abstract Ensuring quality in affordable housing remains a major challenge for both developed and developing countries. Residential quality can be explained as the capacity of a dwelling to meet the specific needs and preferences of its occupants, a condition that is often associated with the high costs of custom buildings. This chapter introduces the notion of *mass personalisation* and explores its potential as means to increase residential quality without breaching the limits of affordable production. Mass personalisation can be defined as a particular approach to mass customisation in which the attributes of a product or service are custom-tailored towards the implicit needs and preferences of individual users. In housing, this implies buildings capable to meet changing household requirements throughout their life cycles. This chapter reviews different techniques that may enable incorporation of such principles to housing design and development, and explores their potential with two abstract prototype implementations.

5.1 Introduction

In order to fulfill the sustainability agenda, mass custom housing should be able to meet changing occupancy patterns along the dwelling's life cycle (Noguchi and Hadjri 2010). In contrast to mass production's reliance on repetition as means to achieve cost-efficiency, mass customisation takes advantage of flexible design and manufacturing technologies to introduce diversity to industrialised development while avoinding the costs associated to craft production. However, diversity might not be enough to ensure quality in housing, as household requirements change over

V. Bunster (✉) · M. Noguchi · T. Kvan
The University of Melbourne, Melbourne, Australia
e-mail: v.bunster@student.unimelb.edu.au

M. Noguchi
e-mail: masa.noguchi@unimelb.edu.au

T. Kvan
e-mail: tkvan@unimelb.edu.au

© Springer International Publishing Switzerland 2016

M. Noguchi (ed.), *ZEMCH: Toward the Delivery of Zero Energy Mass Custom Homes*, Springer Tracts in Civil Engineering,
DOI 10.1007/978-3-319-31967-4_5

time in contrast to the rather static nature of most dwellings. This chapter presents the notion of mass personalization as a strategy to potentially enhance residential quality by addressing these dynamic conditions with the aid of information systems.

5.1.1 *Residential Quality and Satisfaction*

The properties of the residential environment are a major factor behind quality of life and general well-being (Prilleltensky 2005). In this context, UN Habitat acknowledges that housing adequacy is a complex multi-dimensional phenomenon that largely depends upon the capacity of a dwelling to meet the individual needs of its target occupants, hence calls for *user-centred* residential development:

> Adequate shelter means more than a roof over one's head. It also means adequate privacy; adequate space; physical accessibility; adequate security; security of tenure; structural stability and durability; adequate lighting, heating and ventilation; adequate basic infrastructure, such as water-supply, sanitation and waste-management facilities; suitable environmental quality and health-related factors; and adequate and accessible location with regard to work and basic facilities: all of which should be available at an affordable cost. Adequacy should be determined together with the people concerned, bearing in mind the prospect for gradual development. Adequacy often varies from country to country, since it depends on specific cultural, social, environmental and economic factors. Gender-specific and age-specific factors, such as the exposure of children and women to toxic substances, should be considered in this context (UN Habitat 1996).

In housing, the concept of personalisation is often used to refer to the different actions that a household may undertake in order to accommodate its needs and preferences to an otherwise standardised dwelling (Abu-Ghazzeh 2000; Friedman 2002), which according to authors such as Tipple (1996) can be a congruent strategy to achieve sustainable development. According to classic theories of housing adjustment, households constantly undergo a self-assessment process through which their own residential environment is judged based upon subjective norms. If a dwelling is not able to meet them, dissatisfaction motivates different adjusting behaviour including propensity to move, changes to the household structure, or modifications to the physical characteristics of a dwelling (Morris and Winter 1975; Morris et al. 1976). In this context, it can be argued that the act of modifying a dwelling might rather be the manifestation of a need or a preference that either changed over time or was not properly addressed through a personalised design process. Residential satisfaction can be formalised as the measure of the difference between the expected and perceived residential conditions of a given household (Galster and Hesser 1981; Lu 1999); hence, both occupant and design personalisation can be seen as different manifestations of a single phenomenon, which is the need for a custom fit between a household and its residential environment.

The attributes of the residential environment may be either objective or subjective and belong to the social or physical environment (Aragonés and Amérigo

1987; Amérigo and Aragonés 1997). Whilst objective attributes arguably depend upon universal needs and may therefore be standardized (e.g. ergonomics, structural soundness, thermal comfort, and ventilation) subjective attributes depend on informal perceptions that may diverge among households (e.g., aesthetic preferences, privacy, or safety). In this context, personalisation of the built environment can be seen as a strategy to achieve higher levels of residential satisfaction through its contribution to, e.g., cultural identity (Greenbaum and Greenbaum, 1981), workplace productivity (Wells 2000), and general environmental wellbeing (Heidmets 1994). However, as residential satisfaction is an uttermost subjective phenomenon that depends upon individual perceptions that may not be incorporated to design and development using objective measures, the introduction of feedback mechanisms may enable coordinating different stakeholders involved in the housing delivery process.

5.1.2 Mass Personalisation

The idea of mass personalisation is fairly recent in the specialised literature. Kumar (2008) argues that mass personalisation emerges due to technological advances that have enabled providers to effectively target the 'market of one' instead of the 'market of few'. Personalisation, he argues, is a well-known and long-dated strategy in product design and the manufacturing that has traditionally resulted in costly products. What makes personalisation unique in the mass customisation context, he argues, is its potential to achieve further levels of differentiation within costs close to the ones of mass production. Kumar (2008) also explains that although the word 'mass' has been implied along with the concept of 'personalisation' in the mass customisation literature, both concepts are not always necessarily associated and therefore calls for the reinstatement of the notion of mass personalisation as means to explicitly differentiate among approaches.

Tseng et al. (2010) similarly argue that the main difference between mass personalisation and mass customisation would be in their focuses (see Chap. 4 for a complete introduction to mass customization). Whereas mass customisation targets the needs of a group of individuals through explicit specification, mass personalisation focuses on capturing the 'latent needs' of individual users aiming to incorporate them into the value chain through engaging user experiences. Tseng et al. further observe that recent advances in e-commerce and online platforms have enabled substantial progress in the long dated objective of targeting the market of one within cost-effective frameworks due to the capacity of information systems to tailor themselves towards the implicit requirements of individual users. In this context, they argue that design for mass personalisation should take advantage of those capabilities and target the latent needs of individual customers through analysis of virtual interactions between users and products, and propose a model that targets user experience through product ecosystems (i.e., kernel), a prescribed technical framework (i.e., design context), and active user participation (i.e., engine).

Both Kumar's (2008) and Tseng et al.'s (2010) approaches to mass personalisation are influenced by recent advances in adaptive hypermedia and information systems. As explained by Sirmakessis (2006), web personalisation refers to any technique that enables tailoring web experience to the particular requirements of individual users. Brusilovsky et al. (2007) argue that these techniques emerged as a response to the increasing amounts of online information that has been made available over the past decades, the diversity of the potential users accessing that information, and the growing capabilities of the web applications. In this context, personalisation techniques enable purposeful management of the information that is presented to different users towards individualised web experience.

Although the concepts of customisation and personalisation are often used interchangeably in the field of design and manufacturing, in information systems their difference is clear. Whereas customisation intends to pursue individualized online experience through explicit user specification, personalisation is concerned with the development of systems capable of tailoring themselves towards their implicit needs and preferences. This difference has substantial implications for the way in which web services are conceptualised and for the techniques behind their implementation. While in web customisation the developer needs to transfer parts of the decision making process to the user through a number of alternatives such as, e.g., a discrete selection of 'template styles' for a graphical user interface, in personalisation approaches the system needs to be able to foresee these preferences and adapt itself in order to to match them. An example of the latter are web services such as eBay that rely extensively on recommendation systems as means to personalize the offering of products according to the user's browsing behaviour and purchase history (Ricci et al. 2011). Whereas web customisation enables differentiation through a solution space of explicit alternatives that is defined by the provider, web personalisation is concerned with the discovery of implicit preferences and needs through analysis and modelling of the user's behaviour (Brusilovsky 2001; Brusilovsky et al. 2007; Germanakos et al. 2005).

5.1.3 Towards Mass Personalised Housing

The notion of mass personalisation can potentially inform strategies for household-centred housing design and development within the limits of affordable production. As discussed above, mass personalisation can be defined as an approach to mass customisation in which a product or service is custom tailored towards the implicit needs and preferences of individual users (Kumar 2008; Tseng et al. 2010); hence, if mass customisation enables variability within costs close to the ones of mass production (Davis 1987; Pine II 1993; Tseng et al. 1996), mass personalisation should enable matching this variability with the implicit needs and preferences of

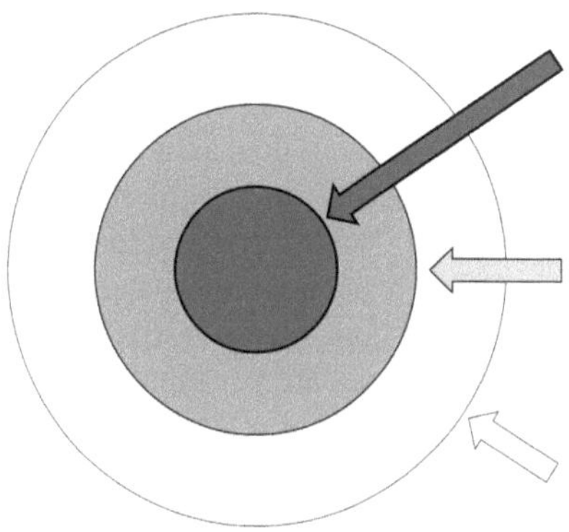

Fig. 5.1 Mass personalisation as a subset of mass customisation principles in industrialized design and manufacturing

individual users. In this conext, whereas mass custom housing relies on explicit knowledge as means to ensure coherence between households and the properties of their dwellings, in order to ensure resdential quality mass personalisation may require management of this relationship over time and at a deeper level (Fig. 5.1). Accordingly, mass personalised housing can be defined as a particular approach to mass custom housing where the implicit residential needs and preferences of the occupants are targeted along the dwelling's life cycle. After Tseng et al.'s (2010) model, it is possible to assume that this may be achieved with product ecosystems capable to inform a change in housing design and development from explicit specification towards the discovery of tacit household requirements.

As discussed by Tseng et al. (2010), products do not exist in isolation but are rather part of larger relational networks involving multiple agents, products, functionalities, technologies and services. In this context, a product ecosystem can be defined as:

> A dynamic unit that consists of all interdependent products and users, functioning together with its surrounding ambience, as well as their interactive relations and business processes (Zhou et al. 2011 p. 45).

In design and development the notion of a product ecosystem extends the way in which products are conceptualised beyond traditional unidimensional relationships between users and objects. Mobile phones, for example, no longer can be seen as a simple communication devises as their capacity to allow access to a complex network of products and services (i.e., apps, platforms, hardware, accessories, etc.) has become as important as their original functionality. In this sense, this notion acknowledges that the interactions between users and products are multi-dimensional and dynamic, suggesting that a focus on user engagement may enable more sustainable business models (Jiao et al. 2007; Zhou et al. 2011).

In the realm of housing the notion of a product ecosystem can be as powerful as in marketing research but from a different perspective. Regardless of the dynamic nature of household needs and preferences, dwellings are generally designed to be rather static end products. There are good reasons behind this assumption, as buildings are expensive and structurally complex objects that require subtantial technical expertise in order to be properly modified. Nonetheless, buildings are often modified along their life cycles in response to residential mismatches and their impacts on satisfaction. As argued by Noguchi and Hadjri (2010), mass custom housing should incorporate those changing occupancy patterns before, during and after occupancy. In mass personalised housing this might be achieved through design and development of comprehensive housing ecosystems able to support adaptive modifications in response to those implicit household requirements at different stages of the dwelling's life cycles.

5.2 Strategies for Mass Personalisation

In a recent article Salvador et al. (2009) propose that the effectiveness by which an organisation may implement mass customisation principles into its own processes can be assessed by looking at three core capabilities: (a) solution space development, (b) a robust process design, and (c) choice navigation. These three capabilities can illustrate the fundamental components behind mass personalised design and manufacturing.

The first capability, i.e., *solution space development* according to the authors, refers to the identification of attributes in a product among which the needs and preferences of users tend to diverge and therefore may offer opportunities for differentiation. If the problem of mass production is to identify the universal needs and preferences of a target group, in mass customisation, they argue, the problem would be to match the heterogeneity of their needs and preferences with the potential variability of a cost-effective solution space. In this context, mass personalisation implies a turning point for the way in which that solution space is developed and conceptualised. As argued by Tseng et al. (2010), a solution space for mass personalisation should contain alternatives capable of meeting implicit needs and preferences of individual users and, therefore, its definition requires extending the capacity of developers to foresee requirements beyond what could be achieved through direct specification. In this context, research on personalised web systems suggests that user modelling can help to capture those implicit variables through the analysis of user's behavioural patterns (Houben et al. 2009; Brusilovsky et al. 2007; Ricci et al. 2011).

The second capability, i.e., *robust product design* according to Salvador et al. (2009), refers to the problem of increasing product variability without compromising performance. Mass production enables cost-efficiency by spreading design and manufacturing costs along a large number of repeated units. In this context, any increase in variability would require major infrastructural changes and therefore act against its economic feasibility. According to Salvador et al. (2009), in mass customisation this potential increase in variability can be achieved through the incorporation, e.g., automation technologies, modular value chains or postponement in product differentiation when investment costs are justified. In this context, the capacity of a mass personalised product to meet the implicit needs and preferences of individual users may always be constrained by the potential flexibility that a value chain is able to provide whilst remaining within the limits of affordable production. In this context, the work of Tseng et al. (2010) and the experience of online platforms for the specification, production and delivery hard products suggest that mass personalisation is mostly effective when addressing the design component behind the mass customisation process. Nonetheless, emergent initiatives such as the 'open source movement' and the 'makers revolution' are extending the possibilities of personalisation strategies by narrowing the gap between industrialised manufacturing and individual users, therefore making the market of one a reality in an increasing number of products (Anderson 2012).

The third capability, i.e., *choice navigation* according to Salvador et al. (2009), refers to the problem of supporting selection of alternatives whilst avoiding information overload. This, according to the authors, might be achieved through, e.g., interactive online configurators, interfaces to facilitate understanding the impacts of interacting variables or through the incorporation of adaptive web technologies to match user preferences and the offering of alternatives. From the perspective of mass personalisation, choice navigation is fundamental for the discovery of implicit user requirements. In this context, Tseng et al. (2010) suggest that digital technologies may enable users to explore the properties of products in virtual space, while automatized assessment of such interactions may enable the discovery of tacit needs that may inform the development of personalised products. In this sense, Tseng et al. (2010) argue that personalisation extends the scope of mass customisation to incorporate user experience throughout the different stages of the product's life cycle, enabling them to participate in the definition of the actual characteristics of a product while also impacting their overall satisfaction levels.

As discussed by Kumar (2008), the capacity of mass personalisation to target the market of one is directly linked to the extent by which a product can be developed electronically. In this context, mass personalisation is a problem that goes beyond user participation. As learned from adaptive web systems, addressing the implicit needs and preferences at a massive scale requires management of variables that go beyond the cognitive capabilities of both users and providers, hence the need of predictive models capable to steer the personalisation process. The following sections present a number of techniques that may be used to inform design and development of mass personalised dwellings with the aid of information systems.

5.2.1 Co-creation and Open Innovation

Individualisation without innovation might not be enough to ensure the sustainability of mass personalised dwellings. While mass production targets universal needs that are static enough to be addressed with standardized products, the attributes targeted by mass personalisation are by definition as divergent and dynamic as the needs and preferences of individual users. This implies that design and development can be significantly demanding in terms of speed and innovation due to often shorter life-cycles (Pine and Victor 1993; Da Silveira et al. 2001; Piller 2012). In this context, recent research and development initiatives are exploring linkages between technological advancement and 'open innovation' as drivers for a 'third wave' in mass customisation (Larson et al. 2004; Piller and Tseng 2009; Rayna et al. 2014; Grimal and Guerlain 2014).

Open innovation assumes that creativity is inherently distributed and therefore external knowledge is a source of innovation that should be incorporated into the value chain of a product. In direct contrast to traditional research and development where the internal capabilities of an organisation are seen as the main factor behind the success of a product, open innovation aims to take advantage of both external and internal ideas in order to expand such capabilities and generate new opportunities and market niches (Chesbrough 2006). In this context, open innovation can help to strengthen the capacity of a product to meet customer expectations as well as providing flexible and adaptable frameworks for new product development (Piller 2012; Gandhi et al. 2013).

Active user participation in the production of products and services is a strong enabler of open innovation. Customer co-creation has been defined as "a collaborative new product development activity in which customers actively contribute and/or select the content of a new product offering" (O'Hern and Rindfleisch 2008, p. 4). As argued by O'Hern and Rindfleisch, effective design and development requires information about the user's needs but also about attributes able to meet them. In this context, customer co-creation facilitates the access to this information by transferring significant parts of the decision making process into the users, hence may help to inform design and development of mass personalised products.

5.2.2 Knowledge Discovery in Data

As explained by Han and Kamber (2006), the increasing computerization of current practices and almost ubiquitous access to the World Wide Web have resulted in an explosive growth in the available information whilst prompting an urgent demand for new techniques to make use of it. In this context, data mining—or knowledge discovery in databases—refers to the analytical process of extracting meaningful patters and relationships from large amounts of structured information. Data mining is an interdisciplinary field drawing from diverse knowledge areas such as statistical analysis, machine learning, artificial intelligence, database technology, and data visualisation, whose techniques have been implemented in fields such as fraud detection, market research, telecommunications and genetics. According to Fayyad et al. (1996) the main aim of knowledge discovery in databases is to identify potentially useful, novel, valid and understandable patterns in data, where data can be seen as a set of facts and a pattern as a high-level expression or model able to describe a meaningful sub-set of that information.

In this context, mass personalisation can be understood as a consequence of our increasing capacity to extract useful knowledge from data. Overall, data mining is a stepwise process involving data pre-processing, mining, and result validation. Fayyad et al. (1996), e.g., explain that the process starts with the (i) selection of relevant data that is (ii) pre-processed and (iii) transformed before the application of different (iv) data mining algorithms and later (v) interpretation and evaluation of the results. Similarly, Han and Kamber (2006) explain the process as involving data (i) clearing, (ii) integration, (iii) selection and (iv) transformation before the use of (v) intelligent methods to extract patterns that are later (vi) evaluated and (vii) presented to the user.

Olson and Delen (2008) explain that the Cross-Industry Standard Process of Data Mining (CRISP-DM) is one of the most widely used by the industry. The first step, i.e., *business understanding*, involves a general assessment of the current situation towards defining the goals and a particular plan for the data mining process. Then the *data understanding* step is concerned with the identification of specific data requirements that may be needed in order to pursue the data mining objectives. This step may consider data collection, preliminary assessments of the available data, and verification of the quality of that source. Later, in the *data preparation* step the gathered data needs to be cleaned and adjusted to be suitable for *modelling* step where different algorithms and visualisation methods are applied in order to identify, explore and summarize existing patterns. In the *evaluation* step the results of these models need to be evaluated in view of the initial objectives which may lead to the identification of further requirements and adaptations in the mining process. The sixth and last step consists in the *deployment* of the knowledge extracted throughout the data mining process which is done through further modelling. Similar to Han and Kamber (2006), Olson and Delen (2008) explain that this stepwise process is not rigid and may be adapted according to the specific characteristics of any given situation.

Berson et al. (1999) classify the most common data mining techniques in two major groups: classic techniques (i.e., statistics, neighbourhoods and clustering) and next generation techniques (i.e., trees, networks and rules). The group of classic techniques emerges from research on statistical analysis and, although their overlap is significant, traditionally the amount of data needed for statistical analysis is smaller and collected following a specific experimental rationale (Benjamini and Leshno 2005). Regardless of their differences in scope, (a) *statistics* is fundamental for data mining as it enables both quantification and modelling through descriptive and predictive techniques such as linear regression or correlation analysis. A second classic technique in data mining is (b) *nearest neighbour*, which is concerned with the identification of cases whose characteristics are as similar as possible in order to inform predictions of classified values using techniques such as *k*-nearest neighbour, condensed nearest neighbour, or reduced nearest neighbour algorithms among others. Similarly, (c) *cluster analysis* is concerned with the identification of groups of cases that are as similar as possible between them and as different to other groups as possible. Algorithms such as hierarchical, centroid-based, or distribution-based clustering can be used after, e.g., segmentation purposes, to identify outliers, or to automate classification procedures. Then, the group of the next generation techniques relies on methods closer to machine learning. Among them, (d) *decision trees* are models of sequential decision-making composed by a set of events (i.e., nodes) and their potential consequences (i.e., branches) that can be used after, e.g., predictive, classification and segmentation purposes. Similarly, (e) *artificial neural networks* draw inspiration from biological neuronal structures with models able to approximate functions that depend on a large number of input factors by processing information through a number of layers of interconnected nodes (i.e., neurones), each containing an activation function that can alter the weights of these interconnections using a learning rule. Then, (f) *association rules* work by assessing the prevalence of 'if-then' relationships in a given dataset in order to identify recurrent patterns. A common use of this technique is in assessing consumer behaviour; e.g., if purchase records show that a large number of individuals buy milk when buying cereal, then association rule mining can provide the probability of this purchase or identify hidden purchase patterns.

Knowledge discovery in data or data mining provides the means to explore and quantify recurrent patterns in relational datasets. These techniques can be used to inform mass personalisation approaches to housing by assessing implicit needs and preferences before and during occupancy. However, in order to relate these patterns with individual users they need to be formally incorporated into the data workflow. This can be done based upon techniques such as user modelling and profiling.

5.2.3 User Modelling and Profiling

In adaptive web systems and human-computer interaction, *user modelling* is a process concerned with the development of formal representations of individual users towards which the information flow can be custom tailored. As explained by Gauch et al. (2007), most web personalisation approaches strongly rely on data instances of user models as means to specify individualised delivery of web content. These models are known as user profiles, and may contain any type of information that is collected from the users either explicitly (i.e., through direct input by the user) or implicitly (i.e., through automatized assessment of the user's online activities). Gauch et al. further explain that user profiling often consists of three main phases, i.e., data collection, profile construction, and technology or application, which can be approached using different techniques according to the type of information, structure and objectives of the personalised output. In the realm of mass personalised housing, these models should focus on the changing needs and the preferences of the target occupants (i.e., a household) over the dwelling's life cycle rather than on short term interactions between a user and a platform.

User models can range from simple static profiles based on socio-demographic characteristics to be explicitly specified up to complex dynamic representations of their cognitive capabilities to be inferred from implicit interactions and browsing behaviour (van Rijn et al. 2011). Attempts to categorize such different approaches include those focused on the way in which the users interact with a system in order to achieve a certain objective versus those focused on modelling cognitive traits, and models based on stereotypes that need to be built a priori versus those focused on the actual features that may be found through user interaction (Rich 1979; Biswas and Robinson 2010), or those focused on either the knowledge or the interests of the users (Brusilovsky and Millán 2007).

Rich (1983) argues that the scope of user models can be characterised using a three dimensional space where models can (a) represent a single canonical (i.e., stereotypical) user or different individuals, can (b) be explicitly defined by the developer of a system or inferred from the behaviour of the users, and can (c) address long term characteristics such as demographic factors or short term ones such as their current objectives. Rich (1983) further explains that the move from canonical to individualised models enables higher levels of personalisation but also implies greater technical challenges as in many cases explicit customisation driven by the users is not adequate or inconvenient. Accordingly, she suggests that such systems able to build models through inferring the characteristics of the users that can be the most appropriate. Rich argues that in order to provide effective responses, information systems often need understanding of the users at a level beyond their current objectives or preferences. In this context, although some of these long term user patterns might be inferred based upon factors such as demographic characteristics, their history of interactions can help to infer properties beyond immediate preferences and, therefore, help to inform the personalisation process.

User modelling can guide design and development of mass personalised housing. Together with customer co-creation and data mining techniques, such models may help to infer the long term needs and preferences of individual households and thus guide custom tailoring of product ecosystems. Nonetheless, user modelling might not be enough to ensure a custom fit between a household and the characteristics of a dwelling if users are not aware of the long-term concequences of their decisions on the quality of the residential environment. In this context, recommender systems may enable the introduction of feedback mechanisms to steer the exploration of the solution space towards adequate alternatives.

5.2.4 Recommender Systems

Recommender systems can help to avoid information overflow and establish effective feedback mechanisms for design and development of mass personalised housing. In general terms these systems rely on patterns of user behaviour as means to identify items that although have not yet been experienced, are likely fit their needs and preferences. Accordingly, in mass personalised housing recommender systems may enable identification of alternatives able to meet implicit household requirements. Attempts to categorize such systems often distinguish between (a) collaborative filtering, (b) content-based filtering, (c) knowledge-based filtering, and (d) hybrid recommendation approaches (Jannach et al. 2011). These categories are described in the following sections.

(a) *Collaborative filtering* stands on the assumption that similar individuals tend to share similar preferences and, therefore, if the history of two users tends to overlap their future actions are most likely to be similar. Consequently, if these patterns can be found in a large number of similar users, then the possibilities of accurate recommendations tend to increase significantly. Collaborative filtering is widely used in mainstream retail platforms such as the ones of Amazon or eBay, where it enables automatized offering of items according to browsing behavior. One of the key advantages of collaborative filtering is its simplicity, as it does not need formal representations of the attributes of an item in order to provide accurate recommendations. As explained by Jannach et al. (2011), recommender systems based on pure collaborative filtering are content-agnostic as they use a rating matrix of users and items as single input to (i) predict the probabilities by which a given user may like or dislike an item and (ii) generate a ranked list of recommended items (Table 5.1). Over this matrix different data mining techniques can be used to identify patterns of user preferences including Nearest Neighbour analysis, Pearson correlation analysis, and Association Rule mining among others.

(b) *Content-based filtering* uses representations of the item's attributes in order to assess the available information and generate recommendations. In contrast to collaborative filtering approaches, content-based filtering requires formal representations of qualitative variables for each item and a user profile that contains information regarding the preferences of individual users. Table 5.2, e.g., shows some of the contents that may be included in an online food delivery dataset where the attributes of different restaurants are specified in

Table 5.1 Example user-rating ranked matrix

	User 1	User 2	User 3	User 4
Item 1	4	5	4	3
Item 2	3	4	1	1
Item 3	1	?	1	4
Item 4	4	1	2	?

Table 5.2 Example user-rating ranked matrix

ID	Name	Cuisine	Location	Minimum order
1	Barbalu's	Italian	Hampton	$18
2	Thaicoon	Thai	Brighton	$30
3	Roti Boti	Indian	Hampton	$45
4	Win Ho	Chinese	Sandringham	$25

rows according to different categories (i.e., name, cuisine, location, and minimum order) assigned to an identifier (i.e., ID). Based upon this type of information, content-based filtering can use different data mining techniques to identify recurring patterns among users and these attributes in order to generate recommendations. An example of content-based filtering is the 'Music Genome Project', which collates music information according to more than 400 categories (e.g., music genre, gender of vocalist, year, etc.) to be used by the Pandora online music platform to generate personalised recommendations according to the interests of individual users. According to Jannach et al. (2011), a key advantage of content-based filtering is its capacity to (i) provide accurate recommendations without the need of massive user communities like in collaborative approaches and to (ii) automatically recommend new items after categorized. Nonetheless, these advantages come with significant challenges as formalising and categorizing qualitative attributes is often not and straightforward process.

(c) *Knowledge-based filtering* tackles problems whose characteristics make collaborative or content-based filtering inadequate, and therefore benefit from the incorporation of users' reasoning as part of the recommendation process through some sort of explicit interaction. Examples of such problems are the recommendation of products that might be acquired only once or few times in a life time (i.e., durable goods such as a house or a car) and therefore might not enable access to a purchase history. Jannach et al. (2011) explain that two main approaches to knowledge-based filtering are case-based (e.g., Felfernig and Burke 2008) and constraint-based (e.g., Burke 2000) recommendation systems. While both approaches require explicit input by the users in order to identify alternatives that may potentially satisfy their own requirements, case-based systems are concerned with identifying items based upon similarity criteria while constraint-based systems rely on explicit recommendation rules. Burke (2000) explains that the main advantage of knowledge-based filtering lies in its capacity to assist users through the process of exploring a solution space while helping them to identify alternatives that may fit their individual preferences without necessarily understanding their complex attributes. Furthermore, although complementary to other types of system, knowledge-based filtering does not require large datasets or formalisation of subjective attributes in order to provide accurate recommendations.

(d) *Hybrid recommendation systems* combine more than one of the former approaches aiming to overcome the limitations or improving the performance of other filtering strategies. Netflix, e.g., uses attributes of both films and user profiles as well as community ratings as criteria to inform its recommendations. Burke (2007) explains that hybridization may overcome limitations such as the need of ratings at the beginning of the process in collaborative filtering (i.e., the 'ramp-up problem'). Accordingly, he categorizes different hybridization strategies in seven groups as: (i) *weighted*, when the score of an item is computed on the basis of all the results of different recommendation systems, (ii) *switching*, when the recommendation strategy can be switched according to a certain criterion, (iii) *mixed*, when a number of recommendations can be produced and combined to identify alternatives, (iv) *feature combination*, when the features on one technique are incorporated into another, (v) *feature augmentation*, when a new feature is generated through incorporation of techniques from other recommendation technique, (vi) *cascade*, when the recommendations are structured in a hierarchy where scores set at a high level cannot be overrun by the ones of lower levels, and (vii) *meta-level* hybrids when a given recommendation technique is used by a different one. Hybridization has been shown to increase the performance of recommendation engines when the information available does not fit other paradigms (Burke 2000; Jannach et al. 2011).

5.3 Prototype Implementations

The mass personalisation principles presented in this chapter were tested in two parallel experiments that used the problem of defining low-cost energy-efficient dwellings under the Chilean social housing program as case study context. While the first experiment focused on the definition of a context-general information workflow to assess the feasibility of using the proposed data mining techniques to develop personalised incremental housing configurations, the second focused on communication between different stakeholders through a prototype online configurator. The prototype implementations are presented in the following sections aiming to encourage further explorations on the development of information systems to support design of mass personalised dwellings.

5.3.1 Incremental Configurations

The first experimental implementation of the principles and techniques presented above focused on the definition of an abstract solution set to represent the development of incremental dwellings in response to the implicit needs and preferences of 180 households. Accordingly, the CASEN 4R, CASEN 2013 and MovieLens100k datasets were used as proxies to specify the demographic conditions and subjective

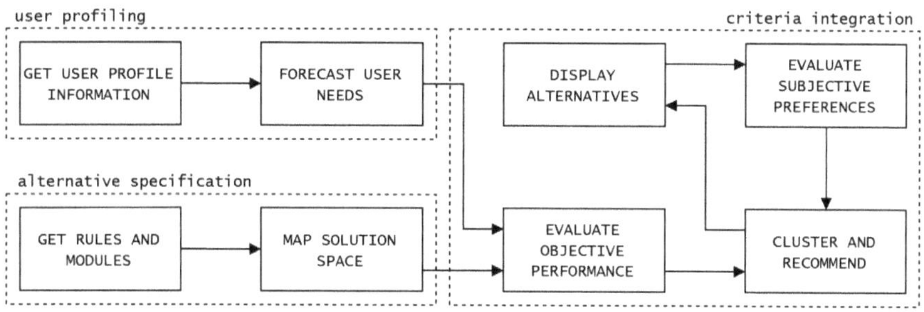

Fig. 5.2 Information workflow behind the first prototype implementation

biases of the target population (i.e., household needs and preferences), while the thermal performance and building costs of the prototypes was used as main development criteria (Fig. 5.2).

Stage 1: Mapping the Solution Space

The first stage of this prototype was concerned with the co-creation of modular product ecosystems in response to the implicit needs and preferences of the target users within the potential customisation capabilities of componentry providers. Accordingly, the first step of the personalisation process consisted in the definition of a solution space based on procedural recombination of modular building components, which were represented by five identical 27 m^3 cube-shaped modules for building 135 m^3 incremental assemblages (i.e., 5 modules) over a period of 20 years (i.e., 4 × 5 year time steps) on a 144 m^2 base surface. The aggregations were obtained through 5 rules: (a) only one module could be added on each iteration, (b) new modules had to be attached to one of the faces of the module added in the previous iteration, (c) growth was only allowed over the x and y vectors, and (d) within the limits of the base surface with (e) no overlaps between modules (Fig. 5.3).

The solution space generated through this growth algorithm was mapped using brute force methods. Although the base surface allowed 16 potential starting points, only one 'seed' and its subsequent 'branches' were included aiming to further constraint the output (Fig. 5.4). This initial mapping allowed a first layer of differentiation among the resulting prototypes as, regardless of having the same total surface area, differences in number of corners and vertical elements can be assumed to have a direct impact on building materials and therefore on construction costs (Table 5.3). From this starting point, the analyses focused on the identification of a discrete number of incremental growth alternatives able to respond to the needs and preferences of different households within the limits of affordable production.

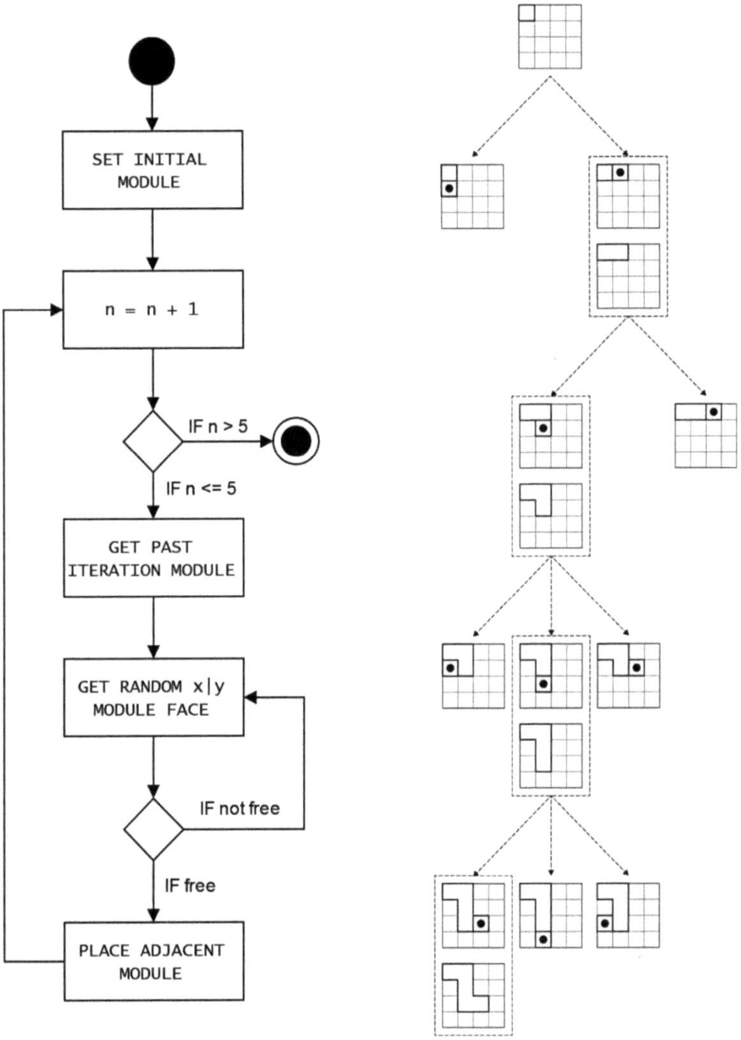

Fig. 5.3 Incremental growth algorithm (*left*) and example output (*right*)

Stage 2: Profiling User Needs

The second stage of this prototype implementation was concerned with gathering and processing critical household information to enable identification of implicit requirements and inform the definition of a discrete solution set able to maximise residential quality. This was done based upon the proxy demographic characteristics and browsing behaviour of the target population. Accordingly, 180 households were randomly selected from the CASEN 2013 dataset, a socioeconomic characterization survey conducted periodically by the Chilean government

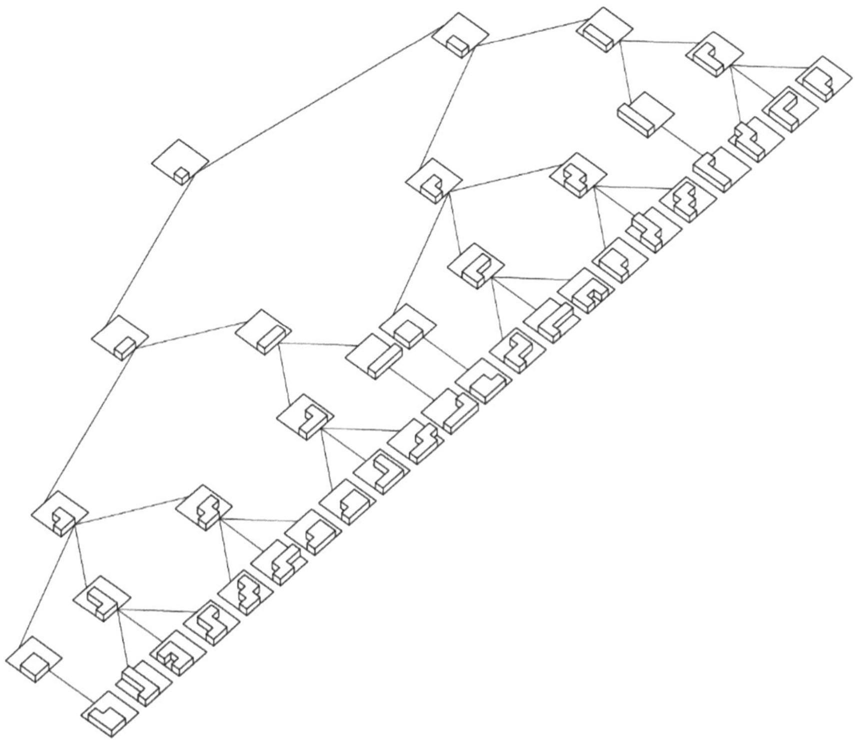

Fig. 5.4 Combinatorial tree displaying 22 incremental growth alternatives from a single generative 'seed'

(mean household size 4.73, with a minimum of 1 and maximum 10 users), and their changes over time were forecasted using a linear regression model extracted from the longitudinal CASEN 4R dataset using data mining methods (Fig. 5.5). The mean change in household size over this time period was −0.488 users with a maximum increase of +3 and decrease of −4 for a reduction of the mean household size from 4.73 to 4.24 over the complete time series (i.e., 20 years).

This information was then used to assess the thermal performance of the prototype configurations developed in step 1 as shown in Fig. 5.6. The simulations were conducted with EnergyPlus using default construction materials and schedules, no artificial HVAC systems or artificial internal gains and the climatic file of Concepción, southern Chile. While household size and simulated mean air temperature were almost perfectly correlated ($r = 0.995$, $n = 202$, $p < 0.005$) and one-way ANOVA confirmed statistically significant differences between household size groups, the influence of household size on temperature decreased over time ($F(9210) = 2.682\text{E}{+}33$, $p < 0.005$ for step 1, while $F(9210) = 2013.819$, $p < 0.005$ for step 5). The mean annual air temperature at step 1 was 20.826 °C with 17.702 °C

Table 5.3 Cost proxy based on total number of vertical elements and corners of each alternative

Alternative number	Vertical elements	Corner elements	Proxy cost
1	10	6	16
2	12	6	18
3	12	8	20
4	12	8	20
5	12	10	22
6	12	8	20
7	10	6	16
8	10	6	16
9	12	6	18
10	12	8	20
11	12	6	18
12	10	6	16
13	12	8	20
14	12	6	18
15	12	8	20
16	10	6	16
17	12	8	20
18	12	10	22
19	12	6	18
20	12	8	20
21	12	6	18
22	10	6	16

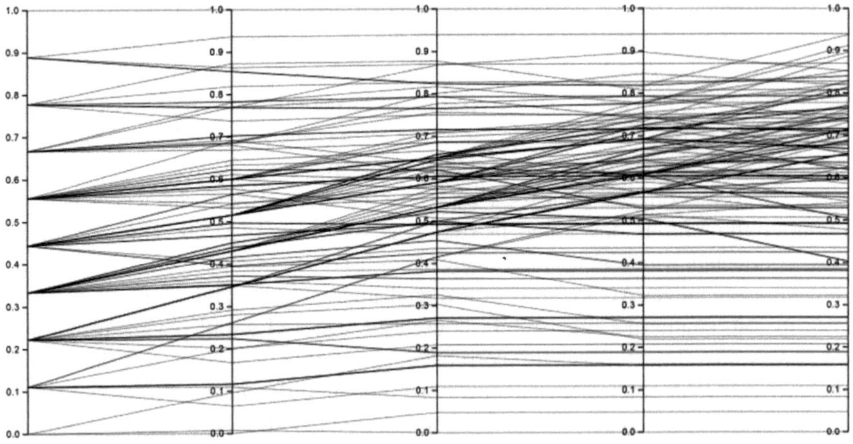

Fig. 5.5 Forecasted household size for the 180 cases over a period of 20 years (*left to right*)

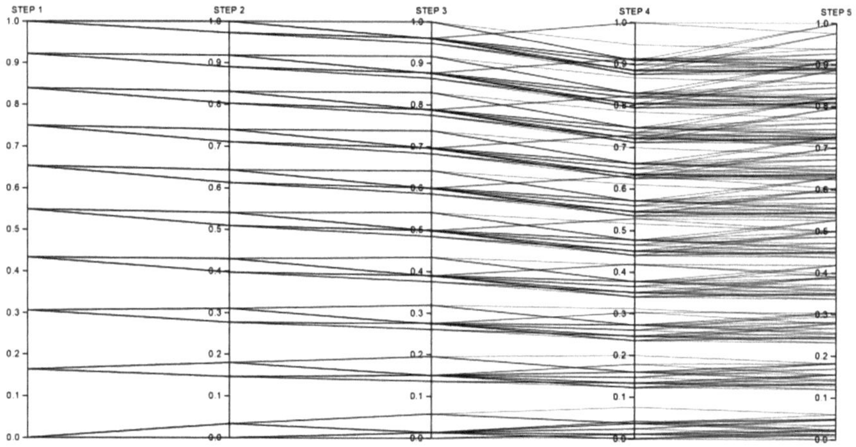

Fig. 5.6 Simulated thermal performance of the alternatives available in the solution space over a 20 year time period (*left to right*) according to case study household sizes (*y* axis)

for one user and 23.454 °C for ten, while at step 5 it was 3.612 °C with 3.347 °C for one user and 3.847 °C for ten.

Then, the MovieLens100k dataset was used to generate proxy ratings for each one of the 22 growth paths available in this solution set. MovieLens is a virtual community focused on movie recommendations developed and maintained by GroupLens, University of Minnesota. MovieLens100k is an open dataset that contains 100,000 ratings of 1700 users for 1000 movies. The first 22 movies of the dataset were used to represent each of these growth paths and the ratings given by the first 180 users were assigned to each one of the target households. This resulted in a dataset containing 690 five-star ratings (where 1 is the lowest and 5 the highest) with a mean rating of 3.9, and a minimum of 3 and maximum of 81 rating instances per item. Significantly, the resulting dataset was incomplete (i.e., not all alternatives are rated by all users) as one of the main problems that recommendation systems aim to address is identification of resemblance patterns to propose alternatives that have not yet been rated (i.e., movies in the MovieLens case).

The outcome of stage 2 of this prototype test is a relational dataset containing structured information regarding the implicit needs (i.e., household size) and preferences (i.e., ratings) of the target households as well as the performance of the alternatives available in the proposed solution space (i.e., construction costs and thermal performance). In this context, personalisation implies that each household should be provided with the alternative that best suits its specific needs and preferences; however, this level of individualisation was assumed to be unfeasible for acting against scale economies. Stage 3 was concerned with the exploration of resemblance patterns among the target households towards the definition of a discrete and cost-effective mass personalised output.

Stage 3: Defining a Discrete Solution Set

The third and final stage of this prototype was concerned with the identification of a discrete solution set through matching the properties of available alternatives with the implicit needs and preferences of target users. Towards this objective, two complementary strategies were used: (i) exploration of optimum clustering options and (ii) automatized feedback for stakeholders with a recommendation system. Whilst clustering was used as black-box method for the identification of solution sets able to respond to heterogeneous user, provider and developer-specific criteria, feeding this information back to the stakeholders enabled management of satisfaction through participation in the decision making process within a transparent and cost-effective framework. Although this approach enables dynamic identification of resemble patterns, this prototype was limited to a single iteration. Firstly, *k*-means clustering was used to classify the 180 target households according to forecasted changes in their size (i.e., needs) over the complete time series. This resulted in 5 statistically significant groups that were used to compute the thermal performance of the 22 growth alternatives (i.e., 3960 five dimensional variables as shown in Fig. 5.7), which later enabled identification of the best growth paths over the complete time series (Table 5.4; Fig. 5.8).

At this stage of the mass personalisation process the developers may explore different clustering alternatives and assess their impact on the performance, feasibility and costs of potential solutions sets. Nonetheless, personalisation requires incorporation of subjective attributes that may not be accessed through simulation; hence, in the prototype these potential solutions were fed back to the target occupants. Accordingly, user preferences were incorporated using ratings obtained through the MovieLens100k dataset and manipulated through a recommender system. The possibilities behind this strategy can be illustrated by analysing the potential impacts of feedback information on a single household.

After mapping the solution space, gathering subjective preferences and storing this information in a dataset through explicit five-star ratings, the next step consited in the identification and offering of clustered options that should fulfil diverse performance criteria with a recommendation system. In this context, household 145, e.g., had 3 members, the age of the head of household was 32 years old, and their autonomous income was CL$100,000 at the moment of the survey. According to the data mining models developed in stage 1 there was 45.5 % of chances for this household to grow in one member over the following 5 years and a second one during the next 15. This implied that this household belonged to cluster 3 and the best incremental growth alternative to satisfy its thermal performance needs was number 18. However, according to the proxy rankings dataset household 145 gave a moderate 3 out of 5 rating to this particular option (Table 5.5).

Moreover, addition of normalised thermal performance, preferences and costs scores showed that alternative 18 was among the worst ranked for household 145 (0.111 where the mean was 0.373), while alternative 7 was the best available option

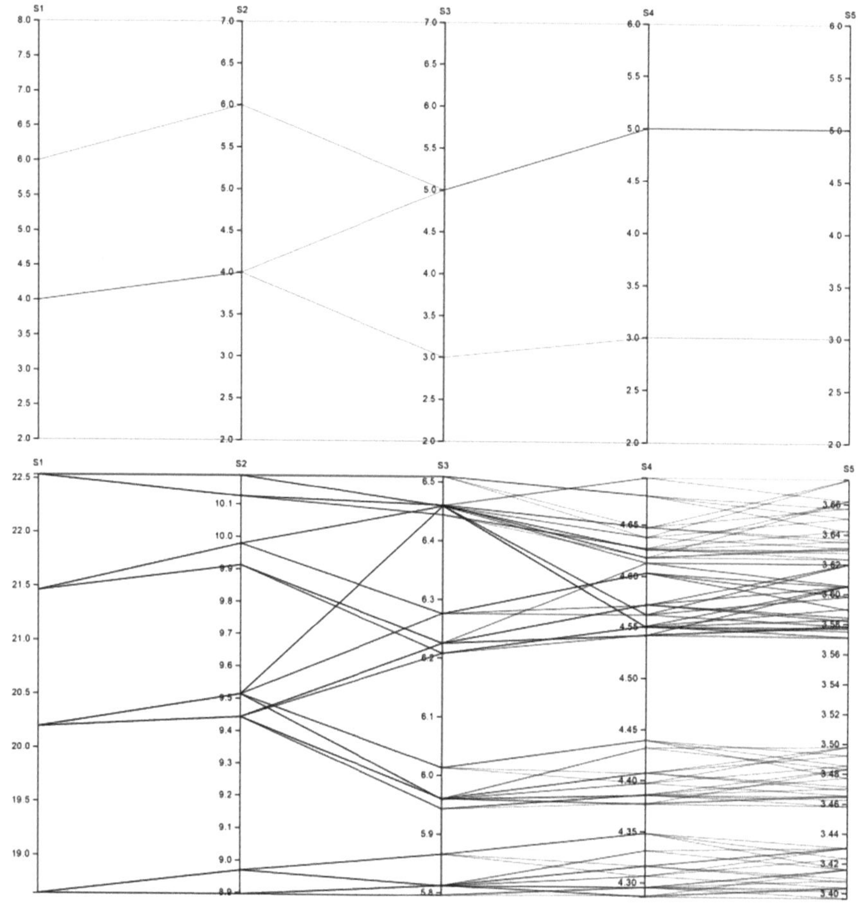

Fig. 5.7 *K*-means clustering results for the 180 households (*upper series*) and the thermal performance of the 22 available growth alternatives over the studied period (*lower series*)

(Fig. 5.9). Similar situations emerged when other households were analysed individually as there is no direct correlation between thermal performance and subjective preferences. In this context, feeding this information back to the users should enable finding alternatives able to maximise objective and subjective attributes.

Accordingly, a *k*-nearest neighbour collaborative filtering recommendation system was used to identify alternatives that should suit the subjective preferences of household 145 based on the MovieLens100 k dataset (accuracy of R^2 0.959). These alternatives were later sorted according to their performance and cost attributes to be fed back to the target households and inform further recommendations (Table 5.6). In parallel, this prototype enabled exploring different clustering

Table 5.4 Clustered alternatives ranked according to their thermal performance over the time series (upper to lower rows)

	Cluster 1	Cluster 2	Cluster 3	Cluster 4	Cluster 5
01	19	12	12	19	19
02	20	16	16	20	20
03	22	17	17	21	21
04	21	13	13	22	22
05	12	14	14	18	18
06	18	15	15	12	12
07	16	19	19	16	16
08	01	20	20	17	17
09	17	21	21	13	01
10	13	22	22	14	13
11	14	18	18	15	14
12	15	01	01	01	15
13	02	02	02	02	02
14	04	04	04	04	04
15	03	03	03	03	03
16	07	07	07	07	07
17	08	08	08	08	08
18	09	09	09	09	09
19	05	05	05	05	05
20	06	06	06	06	06
21	10	10	10	10	10
22	11	11	11	11	11

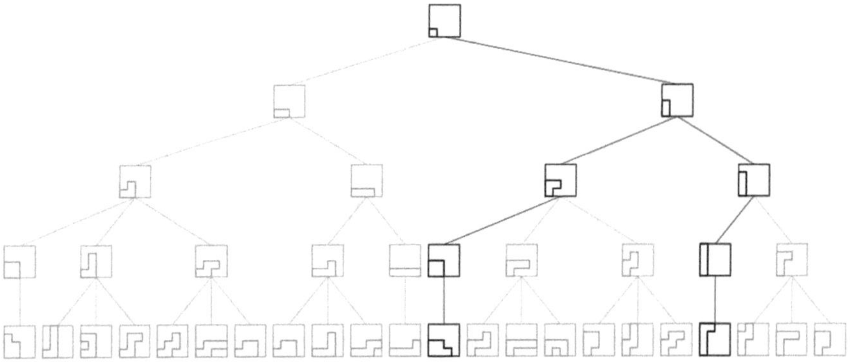

Fig. 5.8 Best growth alternatives for the five household clusters, where alternative 12 (*left*) ranked the best for clusters 2 and 3 while alternative 19 (*right*) was the best for clusters 1, 4 and 5

Table 5.5 Subjective preferences for household 145 according to proxy ratings dataset

	01	02	03	04	05	06	07	08	09	10	11	12	13	14	15	17	18	19	20	21	21
145	3	–	3	–	3	–	5	–	2	–	5	5	5	–	2	3	3	–	–	–	5

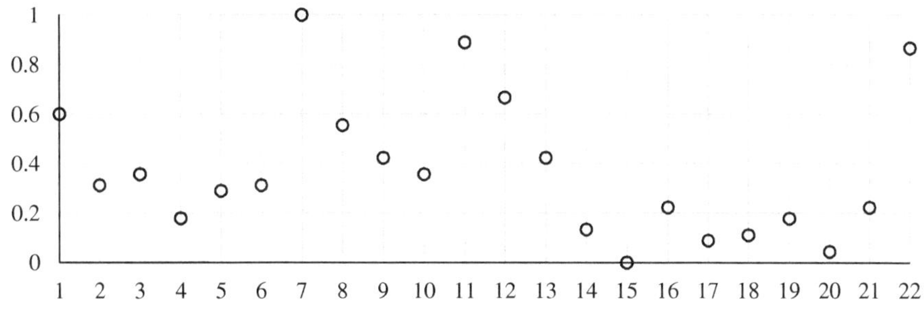

Fig. 5.9 Normalised scores for household 145 through addition of thermal performance, preferences and construction costs for each of the 22 alternatives

Table 5.6 Output of the proposed recommendation system showing best alternatives according to forecasted preferences (a), objective attributes (b), and final display order (c) (left to right)

(a) Preferences	(b) Performance − cost	(c) Final display order
22	0.75	08
13	0.04	22
08	1.00	10
04	0.39	02
14	0.32	06
02	0.61	21
21	0.46	16
10	0.68	04
20	0.18	19
19	0.39	14
18	0.04	20
16	0.46	13
06	0.61	18

alternatives until finding solution sets able to fulfill the objectives of different stakeholders involved in this proxy housing delivery process.

Although this prototype was highly abstract and limited to a single round of recommendations, 22 alternatives and one household, it illustrates how a mass personalisation process may enable identification of solution sets able to combine objective and subjective attributes of multiple househonds and alternatives within a cost-effective framework.

5.3.2 Knowledge-based Configurator

Online configurations are among the most widely used strategies to incorporate the needs and preferences of individual users into the value chain of mass custom products (Felfernig et al. 2014; Blecker et al. 2004). Beyond enabling straightforward customer co-creation, these systems provide the means to generate, gather, and process information that may help in the establishment of long term relationships between customers and suppliers through virtual platforms and product ecosystems. Accordingly, the second prototype focused on the development of a web configurator aimed to assist the exploration of a solution space towards the definition of a discrete set of alternatives in response to both individual household preferences and performance criteria.

One of the main objectives of product ecosystems is to engage user experience beyond one-dimensional interactions between users and products. In this context, this second prototype mass personalisation test was intended to explore a double incentive scheme for different stakeholders to be encouraged to participate in the definition of modular housing assemblages. On the provider side, this knowledge-based configurator is intended to enable housing developers access to information regarding the needs and preferences of the target users that may help them to orient their operations, while on the user side, this same configurator should enable households to personalise their dwellings while been guided towards solutions that were expected to exceed expectations beyond what might be achieved through explicit specification. Specifically, the search was focused on housing configurations able to meet spatial, energy, and cost-efficiency criteria within predefined building codes and regulation (Fig. 5.10).

Accordingly, the constraints defined by Chilean social housing regulation were used to define a discrete set of 18 modular building components, whose total combinatorial possibilities were explored using brute force methods to then generate envelopes based on the minimum bounding box that fitted the resulting assemblies (Fig. 5.11). Regardless of these limitations, the resulting housing prototypes provided a large-enough solution space to enable exploration and discovery in a proxy design and development situation. The performance of each configuration was measured according to their total hours of exposure to solar radiation and total thermal discomfort hours in both summer and winter seasons including only household size in the simulations (EnergyPlus). In parallel, a cost factor was assigned according to the total volume of the envelopes. The overall performance of each solution was calculated according to its proximity to the Pareto front using multivariable optimisation search methods.

An online configurator was developed aiming to assist the exploration of the resulting search space and identification of prototypes that should fit both personal preferences and performance criteria (Fig. 5.12). This configurator required explicit input of demographic factors as means to forecast energy consumption patterns of individual households, and provided a user interface to explore and rate different

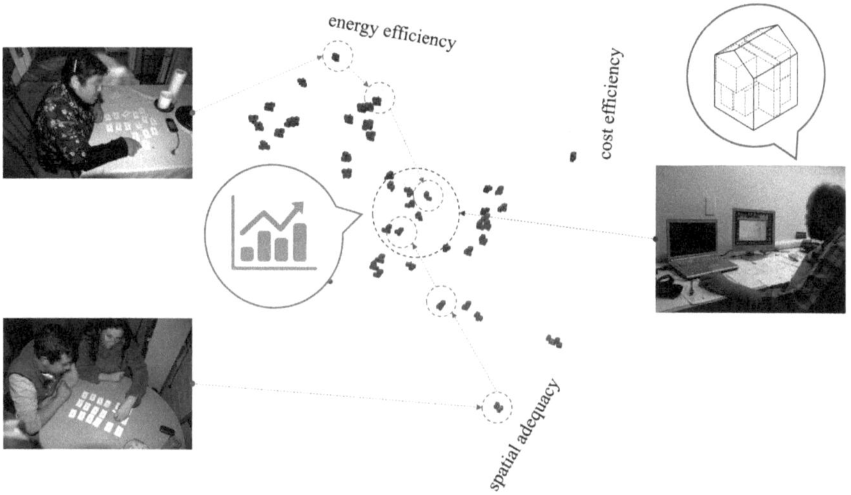

Fig. 5.10 Exploration of housing configurations using minimum surface areas defined by Chilean social housing regulation and the preferences of different stakeholders as constraints, plus energy, cost and spatial adequacy as performance criteria

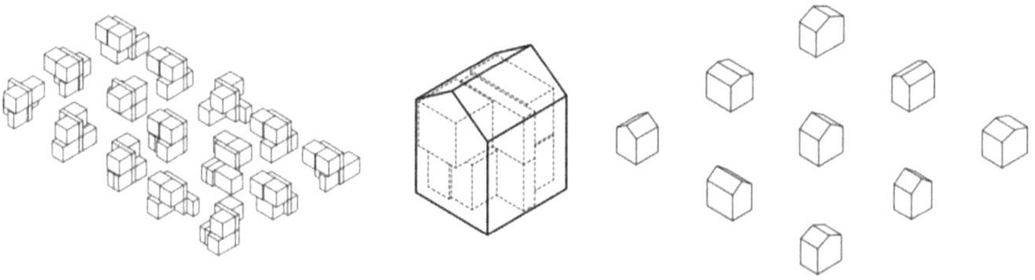

Fig. 5.11 Example two storey housing prototypes obtained through combinatorial addition of modular building componentry based on Chilean regulatory standards

alternatives through visualisation of their thermal performance (i.e., objective attributes) and geometry (proxy for aesthetic properties, i.e., subjective attributes) using a five-star rating system. In the background of this configurator, a database stored ratings and used collaborative filtering to recommend housing alternatives according to resemblance attributes linking different households.

Although this working prototype enabled only limited interactions between different stakeholders involved in the housing delivery process, it illustrates how mass personalisation can inform housing design and development based upon computational techniques such as knowledge-based configurators. Further work should focus on automated forecasting of changing household needs and preferences, real time performance analysis and visualisation, as well as on the incorporation of further modular housing componentry.

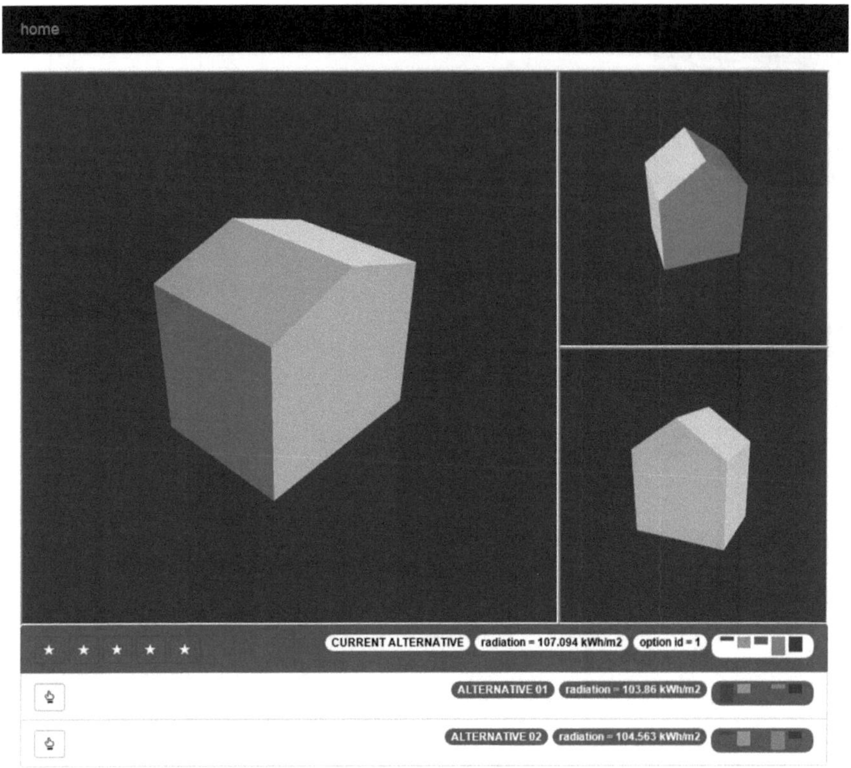

Fig. 5.12 Screen caption of the prototype web configurator showing the current alternative (*left*), recommendations (*right*), an interactive five-star rating system (*lower left*) and summarized performance information (*bottom right*)

5.4 Conclusions

This chapter introduces the notion of mass personalisation and discussess its potential to inform design and development for residential quality. It is argued that mass personalisation can be understood as a particular approach to mass customisation in which products or services are custom tailored towards the implicit needs and preferences of the target users. In housing, this may be pursued through the development of product ecosystems capable to meet the changing residential requirements of individual households over their life cycles. Mass personalisation emerges in the context of an increasing computerization of products and services that has enabled the development of techniques to target the latent user requirements through information modelling and data mining. Accordingly, this chapter presents a number of methods that can be used to inform design and developmet of mass personalised dwellings and explores their potential with two prototype implementations. The learning outcomes of this chapter are (a) a general introduction to the notion of mass personalisation and (b) its potential for design and development in housing, as well as (c) a review of some of the techniques that may help to incorporate such principles in the design of future dwellings.

References

Abu-Ghazzeh TM (2000) Environmental messages in multiple-family housing: territory and personalization. Landscape Res 25(1):97–115

Amérigo M, Aragonés JI (1997) A theoretical and methodological approach to the study of residential satisfaction. J Environ Psychol 17:47–57

Anderson C (2012) Makers. The New Industrial Revolution Crown Publishing Group, New York

Aragonés JI, Amérigo M (1987) Satisfacción residencial: Un concepto de calidad de vida. Documentación Social:133–154

Benjamini Y, Leshno M (2005) Statistical methods for data mining. In: Maimon O, Rokach L (eds) Data mining and knowledge discovery handbook. Springer, London, pp 565–587

Berson A, Smith S, Thearling K (1999) Building data mining applications for CRM. McGraw-Hill Professional, London

Biswas P, Robinson P (2010) A brief survey on user modelling in HCI. Paper presented at the proceedings of the international conference on intelligent human computer interaction (IHCI 2010), Allahabad, India, 16–18 Jan 2010

Blecker T, Abdelkafi N, Kreuter G, Friedrich G (2004) Product configuration systems: state-of-the-art, conceptualization and extensions. Paper presented at the Proceedings of the Eighth Maghrebian conference on software engineering and artificial intelligence (MCSEAI 2004), Sousse, Tunisia, 9–12 May 2004

Brusilovsky P (2001) Adaptive hypermedia. User Model User-Adap Inter 11:87–110

Brusilovsky P, Millán E (2007) User models for adaptive hypermedia and adaptive educational systems. In: Brusilovsky P, Kobsa A, Nejdl W (eds) The adaptive web: methods and strategies of web personalization. Springer, Berlin, pp 3–53

Brusilovsky P, Kobsa A, Nejdl W (eds) (2007) The adaptive web: methods and strategies of web personalization. Lecture notes in computer science, vol 4321. Springer, Hannover

Burke R (2000) Knowledge-based recommender systems. Encyclopedia of library and information sciences, vol 69. CRC Press, Boca Raton

Burke P (2007) User models for adaptive hypermedia and adaptive educational systems. In: Brusilovsky P, Kobsa A, Nejdl W (eds) The adaptive web: methods and strategies of web personalization. Springer, Berlin, pp 377–408

Chesbrough H (2006) Open innovation: a new paradigm for understanding industrial innovation. In: Chesbrough H, Vanhaverbeke W, West J (eds) Open innovation: researching a new paradigm. Oxford University Press, New York

Da Silveira G, Borensteinb D, Fogliatto FS (2001) Mass customization: literature review and research directions. Int J Prod Econ 72(1):1–13

Davis SM (1987) Future perfect. Addison-Wesley, Reading

Fayyad U, Piatetsky-Shapiro G, Smyth P (1996) From data mining to knowledge discovery in databases. AI Mag 17(3):37–54

Felfernig A, Burke R (2008) Constraint-based recommender systems: technologies and research issues. Paper presented at the proceedings of the 10th international conference on electronic commerce (ICEC '08), Innsbruck, Austria, 18–22 Aug 2008

Felfernig A, Hotz L, Bagley C, Tiihonen J (2014) Knowledge-based configuration: from research to business cases, 1st edn. Morgan Kaufmann, Waltham

Friedman A (2002) The Adaptable house: designing homes for change. McGraw Hill Professional, New York

Galster GC, Hesser GW (1981) Residential satisfaction: compositional and contextual correlates. Environ Behav 13(6):735–758
Gandhi A, Mangar C, Roberts R (2013) How technology can drive the next wave in mass customization. McKinsey Bus Technol 32:2–9
Gauch S, Speretta M, Chandramouli A, Micarell A (2007) User profiles for personalized information access. In: Brusilovsky P, Kobsa A, Nejdl W (eds) The adaptive web: methods and strategies of web personalization., Lecture notes in computer scienceSpringer, Berlin, pp 54–89
Germanakos P, Mourlas C, Isaia C, Samaras G (2005) An optimized review of adaptive hypermedia and web personalization: sharing the same objective. Paper presented at the proceedings of the 1st international workshop on web personalization, recommender systems and intelligent user interfaces (WPRSIUI 2005) of the 2nd international conference on E-business and telecommunications networks (ICETE2005), Reading, England, UK, 3–7 Oct 2005
Greenbaum PE, Greenbaum S (1981) Territorial personalization: group identity and social interaction in a Slavic-American neighborhood. Environ Behav 13(5):574–589
Grimal L, Guerlain P (2014) Mass customization in apparel industry: implication of consumer as co-creator. J Econ Manag 15:105–121
Han J, Kamber M (2006) Data mining: concepts and techniques. Morgan Kaufmann, New York
Heidmets M (1994) The phenomenon of personalization of the environment: a theoretical. J Russ East Eur Psychol 32(3):41–85
Houben GJ, McCalla G, Pianesi F, Zancanaro M (eds) (2009) User modeling, adaptation, and personalization. Lecture notes in computer science. Springer, Heidelberg
Jannach D, Zanker M, Felfernig A, Friedrich G (2011) Recommender systems: an introduction. Cambridge University Press, New York
Jiao RJ, Xu Q, Du J, Zhang Y, Helander M, Khalid HM, Helo P, Ni C (2007) Analytical affective design with ambient intelligence for mass customization and personalization. Int J Flex Manuf Syst 19(4):570–595
Kumar A (2008) From mass customization to mass personalization: a strategic transformation. Int J Flex Manuf Syst 19(4):533–547
Larson K, Intille S, McLeish TJ, Beaudin J, Williams RE (2004) Open source building: reinventing places of living. BT Technol J 22(4):187–2000
Lu M (1999) Determinants of residential satisfaction: ordered logit vs. regression models. Growth Change 30:264–287
Morris EW, Winter M (1975) A theory of family housing adjustment. J Marriage Family 37(1): 79–88
Morris EW, Crull SR, Winter M (1976) Housing norms, housing satisfaction and the propensity to move. J Marriage Family 38(2):309–320
Noguchi M, Hadjri K (2010) Mass custom design for sustainable housing development. In: Piller FT, Tseng MM (eds) Handbook of research in mass customization and personalization, vol 2., Applications and conceptsWorld Scientific Publishing, London, pp 892–910
O'Hern MS, Rindfleisch A (2008) Customer co-creation a typology and research agenda. Wiscons innovation: thoughts on innovation from the Wisconsin School of Business. University of Wisconsin, Wisconsin, USA
Olson DL, Delen D (2008) Data mining process. In: Olson DL, Delen D (eds) Advanced data mining techniques. Springer, New York
Omar EOh, Endut E, Saruwono M (2012) Personalisation of the home. Proc Soc Behav Sci 49:328–340
Piller FT (2012) Mass customization: a strategy for customer-centric enterprises. In: Lyons A (ed) Customer-driven supply chains: from glass pipelines to open innovation networks. Springer, London
Piller FT, Tseng MM (eds) (2009) Handbook of research in mass customization and personalization, vol 2. World Scientific, Toh Tuck Link
Pine BJ II (1993) Mass customization: the new frontier in business competition. Harvard Business School Press, Boston
Pine J, Victor B (1993) Making mass customization work. Harvard Bus Rev 71(5):108–117

Prilleltensky I (2005) Promoting well-being: time for a paradigm shift in health and human services. Scand J Publ Health 33(66):53–60
Rayna T, Striukova L, Darlington J (2014) Open innovation, co-creation and mass customisation: what role for 3D printing platforms? Paper presented at the proceedings of the 7th world conference on mass customization, personalization, and co-creation (MCPC 2014), Aalborg, Denmark, 4th–7th Feb 2014
Ricci F, Rokach L, Shapira B, Kantor PB (eds) (2011) Recommender systems handbook. Springer, London
Rich E (1979) User modeling via stereotypes. Cogn Sci 3(4):329–354
Rich E (1983) Users are individuals: individualizing user models. Int J Man Mach Stud 18(3): 199–214
Salvador F, Martin-de-Holan P, Piller F (2009) Cracking the code of mass customization. MIT Sloan Manag Rev 50(3):51–78
Sirmakessis S (2006) Adaptive and personalized semantic web, vol 14., Studies in computational intelligenceSpringer, Netherlands
Tipple G (1996) Housing extensions as sustainable development. Habitat Int 20(3):367–376
Tseng MM, Jiao J, Merchant ME (1996) Design for mass customization. CIRP Ann Manuf Technol 45(1):153–156
Tseng MM, Jiao RJ, Wanga C (2010) Design for mass personalization. CIRP Ann Manuf Technol 59:175–178
UN-Habitat (1996) The habitat agenda: goals and principles, commitments and the global plan of action. UN-HABITAT, Istanbul, Turkey
van Rijn H, Johnson A, Taatgen N (2011) Cognitive user modelling. In: Vu KPL, Proctor RW (eds) Handbook of human factors in web design, 2nd edn. CRC Press, New Yersey, pp 527–542
Wells M (2000) Office clutter or meaningful personal displays: the role of office personalization in employee and organizational well-being. J Environ Psychol 20(3):239–255
Zhou F, Xu Q, Jiao RJ (2011) Fundamentals of product ecosystem design for user experience. Res Eng Des 22(1):43–61

Chapter 6
Inclusive Design

Karim Hadjri, Yasemin Afacan and Tulika Gadakari

Abstract This chapter will explain and discuss the principles, role and importance of Inclusive Design particularly in the context of an ageing society. It will review the changing and complex user needs and requirements through case studies and current work of leading organizations. Current standards used in the UK and elsewhere will be reviewed to establish whether they need to take into account sensory and cognitive impairments into consideration. So far, these have not been fully accepted by industry and practice and more needs to be done by policy makers. Findings of recent research on users' needs and requirements will be reviewed to highlight the needs for more inclusivity in the design of the built environment. Additionally, barrier-free design and Inclusive Design will be further examined to assess the use of technology in embedding accessibility during the design stage. This chapter will allow students, lecturers and designers to understand the value and purpose of Inclusive Design and its potential to provide an accessible and age-friendly built environment.

K. Hadjri (✉) · T. Gadakari
University of Central Lancashire, Preston, UK
e-mail: khadjri@uclan.ac.uk

T. Gadakari
e-mail: tgadakari@uclan.ac.uk

Y. Afacan
Bilkent University, Bilkent, Ankara, Turkey
e-mail: yasemine@bilkent.edu.tr

© Springer International Publishing Switzerland 2016
M. Noguchi (ed.), *ZEMCH: Toward the Delivery of Zero Energy Mass Custom Homes*, Springer Tracts in Civil Engineering,
DOI 10.1007/978-3-319-31967-4_6

6.1 Defining Inclusive Design

6.1.1 What Is Inclusive Design?

Inclusive Design applied to environment, services and products would allow people of all ages and abilities to use them. Inclusive Design is also referred to as 'design for all', universal design or accessible design (Burton and Mitchell 2006).

Inclusive Design is a development of Universal Design that prioritizes the needs of building users in the design process. It aims to remove barriers from the built environment in order to provide most users with easy mobility. Its effective implementation can enhance the design of open spaces and buildings and can have a positive effect on users' wellbeing.

Historically within the built environment disciplines Inclusive Design was concerned with disability and mobility. However, recent demographic trends, and more specifically an ageing population, have led to the emergence of new health issues namely visual and cognitive impairments such as dementia. Hence policy makers, designers and architects need to consider this shift in user needs and requirements.

6.1.2 Principles of Inclusive Design

Design guidelines have been developed in many countries to improve inclusivity and equality in design such as Lifetime Homes Standards in the UK. Laws and acts on inclusivity and equality in the UK have been proposed since 1984.

Some of the principles of an inclusive environment are:

- Easily used by a majority of people.
- Able to offer freedom of choice in relation to access and use.
- Legible, enjoyable and of high quality.
- Flexible, safe and convenient (Langton-Lockton 2004).

The key principles of Inclusive Design were highlighted by UK Commission for Architecture and the Built Environment (CABE) in 2005:

- It places people at the centre of the design process.
- It is equitable, enjoyable and convenient to use by all.
- It is simple in its design, legible and easy to understand by the user.

Further design principles are concerned with perceived accessibility which are more challenging such as those caused by sensory and cognitive impairment which affects wayfinding abilities and decision making (Fig. 6.1). Design guidelines related to this condition are explained at the end of this Chapter.

Fig. 6.1 Tactile floor for visually impaired people

6.2 Importance of User Needs, Requirements and Expectations for Inclusive Design

The process of selecting the right set of user requirements becomes significant in terms of Inclusive Design. Since all requirements cannot be equally satisfied, designers need to involve users to support their decisions, meet their needs, prioritise them, create designs and evaluate the final alternatives with them. User involvement methods and prioritization techniques are discussed in this section to highlight the importance of presenting the diverse user needs. The primary aim of assessing the relative importance to each requirement is to derive the overall requirement priorities for all solution alternatives in order to determine the ideal solution that satisfies the best priority values. The benefits and challenges of both user involvement and requirement prioritisation techniques are significant in terms of finding an appropriate technique that fits well into the Inclusive Design problem-solving process.

AGE Platform Europe proposes a series of recommendations to make 'age-inclusive places' a reality. It can be noted that the recommendations have been interpreted based on user needs, requirements and expectations:

- Participation
- Motivation
- Inclusive Design
- Organisation
- Desirable and shareable urban places
- Health and well-being in our publicly used buildings
- Sustainable neighbourhoods for all ages
- Wellness in Sustainable Housing
- Seamless public transport
- Responsive and integrated personal transport (Bond et al. 2010)

6.2.1 What Is a User Requirement?

For designers, it is impossible to design built environments and/or interior spaces without having negotiations or discussions with customers and/or users on their needs, requirements and expectations. A user requirement is an architectural feature/attribute of a built environment/product/service that should have or how it should perform from the user perspective (Afacan 2008). User requirements should be concerned with the architectural standards, building codes and regulations, and municipality and governmental policies. Although requirement elicitation process seems straightforward, it is complicated by the following three endemic syndromes (Leffinwell and Widrig 2000):

1. The "Yes, But" syndrome;
2. The undiscovered ruins syndrome; and
3. The user and the developer syndrome.

The 'Yes, But Syndrome' is a part of the human nature and is related to user inability to express real requirements concerning the usability, accessibility and affordability. The user attitudes guide designers to conduct earlier evaluation and construct a rapid prototype. The 'Undiscovered Ruins Syndrome' is related to the undiscovered requirements. Users do not always know what they want (Clarkson et al. 2011). They think they know what they need until designers provide them with what they said they required. The third syndrome arises from the communication gap between the user and designer (Afacan and Erbug 2009; Afacan and Demirkan 2010).

In this respect, the process of selecting the right set of user requirements becomes significant in terms of Inclusive Design. A good user requirement specification should be solution independent, specific, objective, and quantified where possible, measurable and testable, traceable, accurate in their representation, complete and well structured (Clarkson et al. 2011). Thus, it requires the simultaneous assessment of multiple Inclusive Design principles. This multi-parameter decision making during Inclusive Design process complicates the conceptual design phase, where designers have to deal with most of the conflicting design decisions simultaneously. Since all requirements cannot be equally satisfied, designers need to involve users to support their decisions, create designs and evaluate the final solution alternatives with them.

6.2.2 *User-Centred Design and User Involvement Methods*

User-centred design (UCD) is a discipline for collecting and analysing user requirements (Jordan 1998). The aim of UCD is that the built environment should suit user, rather than making the user suit the environment. Although many studies have been conducted to develop methods and techniques for UCD and user involvement process, the representation of user needs during the design process requires taking a broader user perspective as defined in the Inclusive Design process. Thus, Inclusive Design is consistent with current practices in UCD with focus on user need specifications to design considering all groups (Smith-Jackson et al. 2003). Smith-Jackson et al. (2003) stated that "when integrated with accessibility, Inclusive Design practice should focus on the contexts of use, operational limitations and capabilities, and preferences of persons with disabilities". Paciello (2005) stated that UCD supports Inclusive Design principles and suggests the use of usability testing methods to broaden the scope of user inclusiveness to engage users with disabilities.

> User participation in the design process is considered to be a good mechanism for increasing levels of satisfaction with a complicated building and is a particular concern when people with varying mobility, visual, cognitive and hearing ability may use a building (Luck et al. 2001).

Next section introduces a literature review on major studies on user involvement grouped under the three methods as suggested by Damodaran (1996). First method is the informative, in which users are serving as information sources answering specific questions that can arise during the design process (Ryd 2004; Olsson 2004). Second method is the consultative method. It deals with user opinions and suggestions on a predefined design (Lahti et al. 2004). Third method is the participative involvement. Participation in user involvement can be defined as a general concept covering different forms of decision making by a number of involved groups (Sanoff 2000).

Informative Form of Involvement. Barki and Hartick (1994) defined informative form as deciphering the relationship between the activities performed by users and developed design systems. Responses from users can be considered as shared sense of purpose (Sanoff 2000; Weiss et al. 2004). Identifying the problems and analysing the requirements of the users can be achieved by feedback at every day usage. Considering user experiences as the central part of research, Taylor (1999) laid emphasis on information and communication from the perspective of users. Sanoff (2000) stated that "Participants need to share information and identify additional information required". Information obtained from users can be considered as clear statements of specification written by the users (Imrie 1999; Luck et al. 2001; Ryd 2004). According to Olsson (2004) users are active agents without the power that can be involved in the design process as information sources.

Typical methods for informative form of involvement include architect-user conversation, observation and interviews (Kujala 2003; Olsson 2004). Developing a program or providing a brief according to the user requirements related to functions, connections, area needs and technical systems can be also considered as the informative form of user involvement (Ryd 2004). Ryd (2004) emphasized the growing awareness toward involving user needs throughout the entire cycle from concept to implementation and stated that "this holistic view means that the clients must, at an early stage, create a platform for a clear understanding of user needs and ensure that final product meets these wishes".

Consultative Form of Involvement. In the consultative form of involvement, coordination, cooperation and collaboration are the key issues while sharing user opinions on design problems (Lahti et al. 2004). During the coordination process, users are carrying individual tasks whereas in the cooperation users actively communicate and work together with designers (Hennessy and Murphy 1999). Consultation can be interpreted as a multidisciplinary teamwork between users and designers. Schaik (1999) emphasized the significance of the user consultation and stated that simply asking users is not enough in consideration of design process. Since users are not designers, they could not give suitable responses and identify their desired functionality and system requirements. Therefore, in the consultative form of involvement users are defined as active empowered partners (Olsson 2004).

Various methods can be employed in user consultation. Older et al. (1997) emphasized the task allocation methods. The user centred design is also emphasized in relation to the consultative form of involvement (Gulliksen et al. 2003). Task analysis, prototyping and usability evaluations can be used in consultative form of involvement (Kujala 2003).

Participative Form of Involvement. There are numerous benefits to participative form of user involvement in terms of successful outcomes in developing information systems as well as architectural designs. User participation is defined as 'a behavioural construct' performed by users and their associated activities during the design process (Lin and Shao 2000). According to Olssen (2004), users are considered as passive subjects to be observed during the participative form of involvement. User participation is studied broadly by Wulz (1986) and Sanoff (2000) and the meaning is extended covering not only user groups but also communities. Reich et al. (1996) mentioned participatory design as an antithesis for the traditional design where "participation is often side-stepped by reducing the user to a databank". Evaluating user needs throughout the entire life cycle requires user participation. Bühler (1996) described user involvement with respect to user experiences based on the tests on prototypes with users.

Usability testing, prototyping, workshops and heuristic evaluation involve participation activities (Beyer and Holtzblatt 1999). There are also qualitative and quantitative methods for investigating designer and user relationships. In this sense, unstructured interviews, sketching, videotaping, brainstorming and scenario building are the techniques that can be used in user participation (Sanoff 2000; Demirbilek and Demirkan 2004).

6.2.3 The Process of Selecting the Right Set of Requirements

Having obtained user needs through user involvement methods, the designer must determine the relative importance of one requirement to another (Afacan and Demirkan 2010). The choice of candidate requirements for implementation and determination of their strength and importance degrees are primary determinants of user satisfaction (Karlsson and Ryan 1996). Thus, the Inclusive Design requirements need to be prioritized to resolve the conflicts between the parameters and to support the inevitable trade-off in decision making. Despite the extensive literature on requirements prioritization in software development, requirements engineering and product design fields (Lin et al. 2008), there is a limited amount of research on the systematic specification and prioritization of Inclusive Design requirements in architectural design context (Afacan 2008). Similar to requirements in engineering studies, architectural design process also needs analysis and prioritization of the requirements. An Inclusive Design requirement priority is needed, not just to ignore the least important requirements but also to guide designers in coping with conflicts and trade-offs between multi-attribute requirements simultaneously (Wiegers 1999).

Because of the challenging and complex nature of prioritizing process, there exist a number of different techniques for requirements prioritization in the literature that can be analysed under two categories with respect to their usage of ordinal scale or ratio scale (Karlsson et al. 2007). The first category includes prioritizing techniques that result in priorities on an ordinal scale and provide a ranked order among requirements, e.g. the Numerical assignment, the Planning Game (PG), the Quality Function Deployment (QFD), the Bubblesort, and the Binary Search Tree (Beck 1999). The techniques in the second category provide the results on a ratio scale and provide information on how much more important one requirement is compared to another. Examples of this category are the Analytical Hierarchy Process (AHP), the Wiegers' method and the 100$ test (Leffingwell and Widrig 2000; Saaty 1980; Wiegers 1999). Karlsson and Ryan (1996) stated that the techniques based on a ratio scale are more accurate and informative than the ones based on an ordinal scale. Each of these three techniques provides an ordered priority list of requirements as an outcome.

The suitability of a prioritization technique to an application that can help in coping with the challenges of prioritization during conceptual design phase is essential in terms of analysing the trade-offs between requirements and assigning a local priority to each requirement with respect to others and in setting global priorities. According to Karlsson and Ryan (1996), an efficient and accurate prioritization technique should give a designer the following advantages:

- a clear means for selecting the right set of requirements for implementation;
- support to resolve the conflicts between requirements and
- support to evaluate the alternative design solutions.

In this respect, in addition to the ranks of the requirements, the decision maker also needs to know the relative distance between the ordered requirements to achieve an effective trade-off.

6.3 Inclusive Urban Environments

Over the last few decades the design of the urban environment has not had the same interest as building design. But this is changing. Recent work is pointing towards the importance of the outdoor setting and its role in providing inclusive and enabling environments.

Burton and Mitchell (2006) found that there are many benefits to using the outdoor environment for older people; these are:

- Freedom of autonomy.
- Dignity and sense of worth.
- Fresh air and exercise (physical health).
- Psychological wellbeing and enjoyment (mental health).
- Social interaction.

They also highlighted familiarity, legibility, distinctiveness, comfort, and safety as essential characteristics of outdoor environments (Fig. 6.2).

Physical Aspects. The outdoor environments present significant challenges to older people and people with disabilities. It hinders their mobility and enforces exclusion. However there have been some significant developments in relation to wheelchair user.

The physical aspects alone are not the answer to the increasingly complex problem of inclusivity and accessibility in outdoor environments. More research is needed on how to remain independent and age-in-place for as long as possible.

Sensory and Cognitive Aspects. The importance of sensory and cognitive aspects of disability is gaining momentum amongst researchers, practitioners and policy makers. This is due to the impact of an ageing population. The work of Burton and Mitchell (2006) highlighted the need to design outdoor environments that are dementia-friendly. People with dementia are particularly vulnerable and have been marginalized due to age and cognitive impairment (Lubinski 1991). A significant development in the UK law is the inclusion of sensory and mental conditions together with physical as part of the UK Disability Discrimination Act (DDA) 2005. As a result architects, urban design and planners have the obligation

Fig. 6.2 Tactile pavement and ramps for easy access and mobility

to design built environments using Inclusive Design principles so that people of all ages and regardless of physical, mental and sensory disabilities, can access these without impediment.

Social Aspects. As the awareness of disability grows more, anti-discrimination legislation has been produced such as the UK Disability Discrimination Act (2005), and the US Americans with Disabilities Act 1990. Nowadays the social model of disability is acknowledged, where the built environment is seen as limiting accessibility and capability of many user groups (Burton and Mitchell 2006). The social model supports the design of products and environments that meet the needs of all users, and to minimise disability (Imrie 2001).

The outdoor environment also presents problems to older and disabled people in particular those who generally perceive it as unsafe and unsecure compared to the home environment (Imrie and Kumar 1998). Social inclusion and social cohesion are also key components that ensure sustainable communities and neighbourhoods are achieved.

Sustainability. Sustainability policies across the world have been targeting sustainable development through efficient land use planning, housing policies, and achieving sustainable communities. The UK government for instance has encouraged brownfield redevelopment, high density housing, and compact city policies (Burton 2002; Urban Task Force 1999). Sustainable communities within sustainable settlements aim to have high quality design as a key component. High quality design in this context refers to place identity and successful outdoor areas.

A socially sustainable environment is desirable because it leads to good quality of life, is safe and easy to use, is accessible and pleasant, and is well serviced (Burton 2003).

6.4 Building Standards for Inclusive Design

6.4.1 Accessibility Standards (UK Part M)

The UK Part M building standard was devised in the 1980s as a reaction to inadequate self-regulation within the construction industry (Imrie 2003). The Disability Discrimination Act 1995 influenced this standard by making wheelchair access compulsory for all new dwellings (Madigan and Milner 1999). Part M aims to assist occupants with ambulant impairments and to make visitability easier. Visitablity features are concerned with making the home more accessible particularly for people who want to visit the occupants but have mobility impairments.

Part M includes measures such as level or gently sloping approaches to properties from car parking spaces, level entrances to properties, adequate heights for switches and sockets, and a WC at entrance level. A common criticism of Part M is that the building regulations were too interpretive. The requirements of Part M are not comprehensively targeting minimum standards, and primarily aimed towards visitability features only (Imrie 2006).

6.4.2 US ADA Standards for Accessible Design

The US ADA Standards for Accessible Design 2010 "2010 Standards" are enforceable accessibility standards. In 2012, compliance with these Standards was required for some new construction and alterations and for accessibility and barrier removal. ADA standards are concerned with accessibility and mobility features such as accessible sleeping rooms, accessible routes throughout, and turning spaces and kitchen, and work spaces that comply with its standards. Additionally, all doors and doorways in residential units should provide user passage, and platform lifts to connect different levels within the building. Accessible parking is also required, mail boxes to be within easy reach, play areas and swimming pools when available should provide mobility features. Guidance for grab bars for WCs, bathroom and kitchen design, communication systems, accessible routes, and fire alarm system are also provided (US Department of Justice 2010)

6.4.3 UK Lifetime Homes Standards

The UK Lifetime Homes Standards (LTHS) were developed by the Helen Hamlyn foundation in the 1980s. These are a set of 16 criteria used to design homes that are inclusive of disabled people and older people (Department for Communities and Local Government 2008). LTHS criteria are designed to enable people to 'age in place' and reduce the need to move to more adequate housing. LTHS were developed in response to the need for more Inclusive Design for housing environments. LTHS homes have the potential to allow residents to be independent for longer (Hanson 2001). They are however focused on physical impairments and further research should be conducted to include sensory and mental requirements (Rooney 2014).

LTHS include the following criteria:

- Access (car parking widths; approach to dwelling from parking; approach to all entrances; entrances; communal stairs and lifts).
- Inside the home (internal doorways and hallways; circulation space; entrance level living space; potential for entrance level bed-space; entrance level WC and shower drainage; WC and bathroom walls; Stairs and potential through-floor lift in dwellings; Potential for fitting of hoists and bedroom/bathroom relationship; bathrooms).
- Fixtures and fittings (glazing and window handle heights; location of service controls) (Fig. 6.3).

6.4.4 Other International Standards

Most countries have now introduced standards to align with inclusion requirements and disability acts. There are currently a number of standards for inclusive design from all around the world, including the developing countries. Listed below are a select few best-practice building standards.

The Americas. Accessible Design for the Built Environment, Canada: It is a national technical standard developed by the Canadian Standards Association and contains requirements for making buildings and other facilities accessible to persons with a range of physical, sensory, and cognitive disabilities (CSA 2004). National Building Code of Canada, Canada: Developed by the National Research Council Canada, the Building Code details the minimum provisions acceptable to maintain the safety of buildings, with specific regard to public health, fire protection, accessibility and structural sufficiency (NRC 2010).

Recomendaciones de Accesibilidad (Accessibility Recommendations), Mexico: Developed by the Representative Office for the Promotion and Social Integration of Persons with Disabilities, Presidency of the Republic in 2001. They cover recommendations on various areas such as urban environment and open spaces,

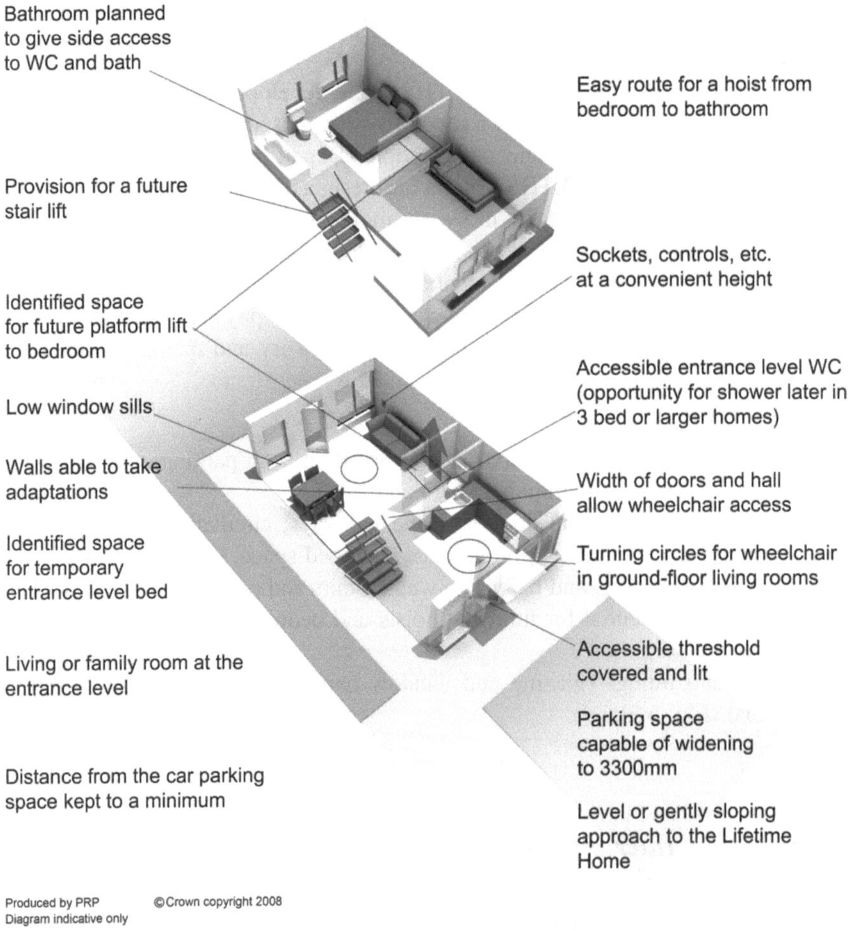

Fig. 6.3 Lifetime Homes Standards. *Source* Habinteg Housing Association (http://www.lifetimehomes.org.uk/)

architectural environment and covered spaces and signage (Presidencia de la República 2001).

Europe. Most of the European countries have their separate National Building Standards to design for accessibility for the disabled and older people such as Building for Accessibility: Austrian Standards, Austria; *Guía técnica de accesibilidad en la edificación*, Spain; and Building Regulations: Technical Guidance Document M—Access for People with Disabilities, Ireland (CHRC 2007). 'The Older Persons' Housing Design: A European Good Practice Guide' has been comprehensively developed as part of the EU-part funded project—Welfare Housing Policies for Senior Citizens (Wel_hops) which involves partners from Italy, Sweden, Spain, Hungary and the UK (WEL_hops 2007; Housing LIN 2012).

Australia and New Zealand. Australian Standard—Design for access and mobility, Australia: Has been approved by the Council of Standards Australia in 2001 and provides building designers and users with the minimum design requirements, to enable access for people with disabilities (ANUHD 2015). New Zealand Standard—Design for Access and Mobility, Buildings and Associated Facilities, New Zealand: This Standard sets out requirements for the design of buildings, facilities within buildings, driveways, car parks, passages and any associated landscaping and access ways for use by people with disabilities (Standards New Zealand 2000). Lifemark Design Standards and 5-Star Rating System, New Zealand: Lifemark Design Standards are best practice for building an adaptable, accessible and safe home while the Lifemark 5-Star rating system provides information about adaptability, safety and ease of use to achieve a 3-, 4- or 5-star rating (Lifemark 2014).

Asia. Code on Accessibility in the Built Environment, Singapore: The Code, since its inception in 1990, has been an important driver behind improvements to the accessibility in Singapore. It places great emphasis on universal design concepts to benefit a wide spectrum of people, including older persons, persons with limited mobility, and parents with infants (BCA 2013). Malaysian Code of Practice on the Accessibility and Mobility of Persons with Disabilities, Malaysia: Developed by the Government of Malaysia in 1991 (CHRC 2007; Hussein and Yaacob 2012). Guidelines and Space Standards for Barrier Free Built Environment for Disabled and Elderly Persons, India: These access standard codes have been developed to ensure that the physically disabled and older persons will have equal access to everyday life in the city (CPWD 1998). Accessibility for the Disabled—A Design Manual for a Barrier Free Environment, Lebanon: Developed by the Lebanese Ministry of Social Affairs; National Committee for the Disabled and United Nations Economic and Social Commission for Western Asia (ESCWA) (UN 2004).

Africa. The South African National Standard—Facilities for persons with disabilities in buildings and the built environment, South Africa: This standard was approved by the Council of the South African Bureau of Standards (SABS 2011).

6.5 Other Considerations for Inclusive Design

6.5.1 Designing for an Ageing Population

Half of the global population now lives in cities (UNPF 2007). By 2030, 60 % of the global population will be living in urban environments. Hence the challenge will be building sustainable age-friendly cities (UNDESA 2006). Population ageing and urbanisation are the result of human development, and are major global challenges for the next few decades.

The increasing number of older adults in the UK has triggered research and technological development programmes that aim to benefit the ageing population (Medical Research Council 2013). Over the last decade, governmental and non-governmental initiatives have been implemented, including research into healthy ageing and how to improve the quality of life of older people at home. Some findings highlight that there is growing evidence of the links between the built environment, health and wellbeing (Croucher et al. 2007).

In the long term, the ageing population will be required to live in homes that are age-friendly or barrier-free and environmentally comfortable. If homes are to be designed with adaptability in mind, then universal and Inclusive Design approaches should be examined and tested. These methods ensure a product is inclusive and usable with the majority of the population, and therefore potentially suitable. Additionally, with prospective economic development in mind, demographic change most notably implies the maintenance of innovation capacity and productivity within an ageing society.

There is evidence to suggest that the built environment creates significant challenges to older people and those with disabilities. Recently, there has been significant developments to create a more enabling built environment through the age-friendly cities concept and initiatives. Active ageing requires an environment that is supportive offering opportunities for participation, and promotes health and wellbeing. An age-friendly city is capable of offering these opportunities while also being inclusive and accessible to older people.

In order to compensate for physical and social changes associated with ageing, older people will increasingly require supportive and enabling living environments for both indoor and outdoor spaces. "Making cities more age-friendly is a necessary and logical response to promote the wellbeing and contributions of older urban residents and keep cities thriving" (WHO 2007).

6.5.2 Multi-generational Housing

An ageing population and economics of intergenerational transfers have led to the appearance of new living arrangements (Palloni 2002). Living arrangements for older people are driven by the need for family, community or institutional support, and health and life satisfaction (Heiss et al. 2003; Börsch-Supan 1989). This has required the adaptation of housing environments to enable ageing-in-place and multi-generational living.

For example, "multigenerational houses are a key part of Germany's ageing population plan" where multigenerational houses were established since 2003. Intergenerational living appears to be the next step following multigenerational socialising, which would bring nurseries and nursing homes together. Additionally, the student-style housing typology for older pensioners have become very popular, by also achieving a good balance between young families and older people (Oltermann 2014).

This new phenomenon will require an innovative approach to Inclusive Design which combines the needs and requirements of a broad range of users.

6.5.3 Designing for Dementia: Wayfinding Guidelines

Wayfinding (how people orient themselves in the physical environment) outdoors for people with dementia has been investigated and has led to design guidelines for dementia-friendly neighbourhoods in the UK (Mitchell and Burton 2010). People with dementia need to remain active through daily walks and activities. Access to the outdoor environment when safe should be encouraged and maintained. This is important to keep their minds and bodies active and alert. Recent research on the design of dementia environments suggests that people with dementia generally cannot recognise or misinterpret modern building designs such as use of sliding or revolving doors. This also extends to modern street furniture (Burton and Mitchell 2006).

A decline in wayfinding ability, can have a negative psychological impact in those with cognitive impairment by causing confusion and agitation or even aggression. Wayfinding abilities are affected by reduced cognitive function, and this is why the design of the physical environment can impact on dementia residents by affecting their spatial orientation (Hadjri et al. 2012).

The Dementia Services Development Centre's (DSDC) Design for Dementia Audit Tool checklist has a section on 'General Design Principles' which provides a valuable basis for use as a briefing tool for designers of new facilities, and constitutes the essential criteria identified in the audit tool (Cunningham et al. 2008). Essential criteria are those which are evidence based in research. The audit tool uses the concept of 'Inclusive Design' and follows NHS Scotland design guidance and assessment tools used for designing nursing and care homes for dementia patients. The aim of this tool is "to ensure that the built environment does not present insurmountable barriers to those who use it" (Dementia Services Development Centre 2007).

In terms of architectural design, building layout, hierarchy of spaces and their organisation should provide free movement and access throughout (compact layouts, no dead-ends or kinks in corridors). Outdoor areas should form an integral part of the overall building design and should be accessible all year. Wayfinding success can be enhanced by the use of landmarks and signage (Faith 2014).

6.5.4 Designing for Visual Impairment

Research highlights that indoor and outdoor space is an important consideration when designing homes for visually impaired people, and that given people's change in needs and health, flexible use of space is critical. Emerging requirements for

Inclusive Design comprise: adequate lighting levels and quality; aspect (views); glare and controls; colour; tactile domain; safety; layout and wayfinding (Rooney 2014).

As a result, there should be new recommendations to improve guidelines, policy and building standards including Lifetime Homes that are more inclusive of the needs of visual impairment (Rooney 2014) (Figs. 6.1 and 6.2).

6.5.5 The Role of Smart Homes

Smart homes can be defined as housing with the integration of automated building controls and assistive technologies such that all essential equipment in the home is connected via a network that can receive and transmit information to provide a better quality of life. More specifically smart homes for older people will simplify, monitor and proactively assist in the completion of everyday tasks thus allowing them to stay in their own house for as long as possible whatever their health condition (apart from critical) by focusing on safety, security, care and comfort (Berlo 2002; Helal et. al. 2004; Bassi et al. 2011; Lushai and Cox 2012; Thomas et al. 2014).

Smart care spaces rely on sensors for collection of data from the environment. Sensors can collect physical, chemical and biological data based on measures including temperature, movement, light, pressure, electric fields, glucose in blood, blood gases, volatile organic compounds in body fluids such as breath, urine or faeces, and bio-markers in blood for a disease (Lushai and Cox 2012; Thomas et. al. 2013). In a Smart home interoperability is key, so that greater efficiency can be achieved by sharing resources across multiple building systems to improve ease of use, better human factor capability, greater choice, international economies of scale, and hence lower unit costs (Edge et. al. 2000; Bassi et al. 2011). Home Sensor Networks comprise of sensors placed anywhere in the home using wireless technology, including on moving objects, and also to be worn by the patient as they move around the home (Lushai and Cox 2012; Thomas et al. 2014).

Smart homes may also be connected to Telehealth systems which act as a virtual link between clinicians and patients. The system includes a central communication platform which incorporates a user interface with large screen, large buttons and voice control—to communicate with a clinician or an automated service to ask the patient questions about symptoms; and vital signs monitoring devices with sensors to monitor blood pressure, glucose, weight, pulse rate and blood oxygen levels. Telehealth systems are typically designed to manage long-term conditions such as COPD, heart failure and diabetes from the comfort of the home (Empirica and WRC 2010; Lewin et al. 2010; Thomas et al. 2013; Berlo 2002; Helal et al. 2004; Empirica and WRC 2010; Lewin et al. 2010; Bassi et al. 2011; Ageing Well Network 2012; Lê et al. 2012; Thomas et al 2013, 2014).

Some of the benefits of 'Smart Homes' highlighted by the literature includes providing an environment that is monitored, safe and secure; assistance with

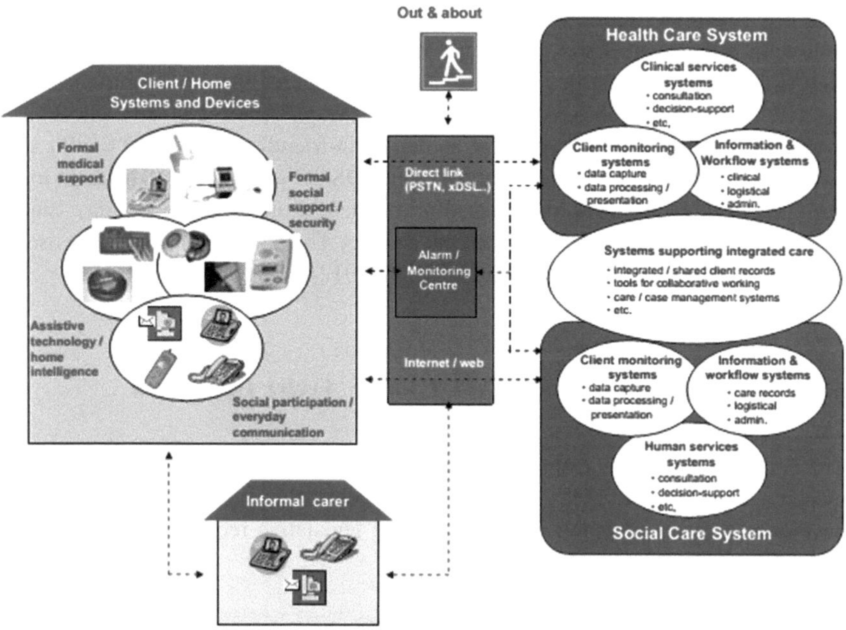

Fig. 6.4 Various types of technologies available in a smart home. *Source* Empirica and WRC (2010)

day-to-day tasks through automation; saving time and cost through fewer visits to a hospital outpatient department; allowing users to take greater control of their medical condition and improve their fitness; facilitating in the rehabilitation of individuals; enabling and empowering the user by retaining some independence; improving level of confidence and feeling like less of a burden on carers; enabling greater social contact with friends and family; providing opportunities to pursue hobbies and interests in a social network setting; and establishing links with the local community (Edge et al. 2000, Helal et al. 2004; Lewin et al. 2010; Ageing Well Network 2012; Lê et al. 2012; Thomas et al. 2013, 2014) (Fig. 6.4).

6.6 Emerging Practice in Inclusive Design

6.6.1 Dementia Services Development Centre (DSDC) Design Audit Tool

The Design for Dementia Audit Tool developed by the DSDC at the University of Stirling (Cunningham et al. 2008) is based on the work of Marshall (2001). It is a briefing guide for designers of new facilities and is used to ensure that the built environment is inclusive and "does not present insurmountable barriers to those who use it" (DSDC 2007). It is designed to aid organizations develop an action plan following assessment. Essential criteria must be met fully in order for a facility to gain Gold Standard accreditation by DSDC. DSDC guidelines are acknowledged by the UK National Health Service. The DSDC Design Audit Tool is a checklist to evaluate and certify an environment as dementia-friendly. An official certificate is awarded by an independent assessor following DSDC's guidelines. The audit tool is designed to include newly built and refurbishment projects as well as day centres, dementia wards, care homes and medical centres. The audit tool can also be used to improve the design of existing areas (DSDC 2012).

6.6.2 People Project: A CAD Tool to Assist Designers

The PeopleProject is the result of a PhD study completed by Nicholas Humes at the Queen's University Belfast in 2012. The project followed a systematic software development process which responded to the legislative requirements of Lifetime

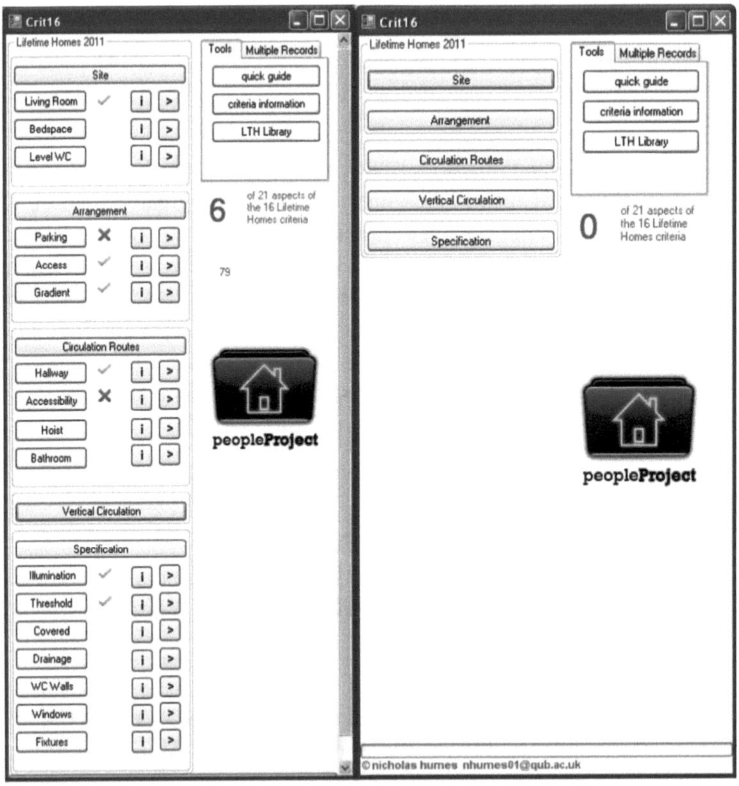

Fig. 6.5 Views of the main PeopleProject page (Humes 2012)

Homes Standards. This research developed a software that is robust by following best practice techniques. The code allows for future development. The software is simple to use, this is because of the graphical layout, the use of only four necessary windows or the low level of technical skill and input required from the architect. The PeopleProject framework has the capability to be developed further, and is adaptable, extensible and scalable framework to which different principles could be applied (Fig. 6.5).

6.7 Conclusions

This chapter reviewed Inclusive Design by first defining its meaning, need, principles, approaches and applications. It then discussed the importance of user needs, requirements and expectations in this area of design. It also highlighted user involvement methods and prioritisation techniques, and the importance of presenting diverse user needs. The primary aim of assessing the relative importance to each requirement is to derive the overall requirement priorities for all alternatives in order to determine the ideal solution that satisfies the best priority values. The benefits and challenges of both user involvement and requirement prioritisation techniques are significant in terms of finding an appropriate technique that fits well into the Inclusive Design problem-solving process. Although each of these UCD methods and prioritisation techniques help designers in the systematic formulation of necessary calculations of Inclusive Design needs, they do not contribute to the time-consumption and extensive work of comparisons. Since the requirements in Inclusive Design process are complex, volatile, vast and multi-faceted, a manageable prioritization process, which can handle increasing number of requirements, should be considered of high importance (Ozkaya and Akin 2006). Any Inclusive Design decision should include the careful consideration of the prioritised set of Inclusive Design requirements so that the assigned priorities can also be checked for consistency and certainty.

Inclusive urban environments are also key elements of an all-inclusive built environment particularly in the context of an ageing population, where social, physical, sensory and cognitive difficulties are all combining to create very challenging design problems to achieve an inclusive environment.

International Building Standards were also reviewed briefly to highlight the importance of these tools in supporting the implementation of Inclusive Design principles. Additionally, a number of considerations have been put forward to raise awareness about the challenges facing designers, policy makers, and all stakeholders involved with the design and production of inclusive environments.

Better solutions could be developed if real users and business needs consider stronger evidences of user involvement techniques. Responding to the diversity of ageing requirements, engaging with disabled people and achieving their physical, social and cultural needs have become a prominent part of Inclusive Design.

References

Afacan Y (2008) A computer assisted universal design (CAUD) plug-in tool for architectural design process. Dissertation, Bilkent University, Ankara, Turkey
Afacan Y, Demirkan H (2010) A priority-based approach for satisfying the diverse users' needs, capabilities and expectations: a universal kitchen design case. J Eng Des 21(2–3):315–343
Afacan Y, Erbug C (2009) Application of heuristic evaluation method by universal design experts. Appl Ergon 40(4):731–744
ANUHD (2015) Australian network for universal housing design. www.anuhd.org. Accessed 8 Dec 2015
AWN (2012) The new agenda on ageing: to make ireland the best country to grow old. Ageing Well Network, Dublin
Barki H, Hartick J (1994) Measuring user participation, user involvement, and user attitude. MIS Q 18:59–62
Bassi R, Quillin K, Warner P, Maher D, Snook K, Belmonte L (2011) A guide for assisted living: towards lifehome 21, 1st edition, 3D Reid. BRERIBA, Marylebone
BCA (2013) Code on accessibility in the built environment. Building and Construction Authority, Singapore
Beck K (1999) Extreme programming explained. Addison-Wesley, Reading MA
Berlo AV (2002) Smart home technology: have older people paved the way? GeronTech J 2 (1):77–87
Beyer H, Holtzblatt K (1999) Contextual design. Interactions 6:32–42
Bond R, Georgantzi N, Kucharczyk M, Mollenkopf H, Rayner P (2010) Towards smart, sustainable and inclusive places for all ages. Age Platform Europe, Brussels
Börsch-Supan A (1989) Household dissolution and the choice of alternative living arrangements among elderly Americans. The economics of ageing: 119–150
Bühler C (1996) Approach to the analysis of user requirements in assistive technology. Int J Ind Ergon 17:187–192
Burton E (2002) Measuring urban compactness in UK towns and cities. Environ Plann B 29 (2):219–250
Burton E (2003) Housing for an urban renaissance: implications for social equity. Hous Stud 18 (4):537–562
Burton E, Mitchell L (2006) Inclusive urban design streets for life. Architectural Press, Oxford
CABE (2005) Better neighbourhoods: making higher densities work. Commission for Architecture and the Built Environment (CABE), London
CHRC (2007) International best practices in universal design: a global review. Canadian Human Rights Commission, Ottawa
Clarkson PJ, Coleman R, Hosking I, Waller SD (2011) Inclusive design toolkit, 2nd edn. www.inclusivedesigntoolkit.com. Accessed 8 Dec 2015
CPWD (1998) Guidelines and space standards for barrier free built environment for disabled and elderly persons. Central Public Works Department- Ministry of Urban Affairs and Employment, New Delhi
Croucher K, Myers L, Jones R, Ellaway A, Beck S (2007) Health and the physical characteristics of urban neighbourhoods: a critical literature review. Glasgow Centre for Population Health, Glasgow
CSA (2004) B651-04 accessible design for the built environment. Canadian Standards Association, Mississauga
Cunningham C, Marshall M, McManus M (2008) Design for dementia: audit tool. University of Stirling, Stirling
Damodaran L (1996) User involvement in the systems design process: a practical guide for users. Behav Inf Technol 6:363–377
Demirbilek N, Demirkan H (2004) Universal product design involving elderly users: a participatory design model. Appl Ergon 35:361–370

Department for Communities and Local Government, Department of Health and Department for Work and Pensions (2008) Lifetime homes, lifetime neighbourhoods: a national strategy for housing in an ageing society. Communities and Local Government Publications, Yorkshire
DSDC (2007) Dementia design checklist, version 1 in collaboration with health facilities Scotland. University of Stirling, Dementia Services Development Centre, Stirling
DSDC (2012) Audit and accreditation: the design audit Tool. http://dementia.stir.ac.uk/design/audit-and-accreditation. Accessed 8 Dec 2015
Edge M, Taylor B, Dewsbury G, Groves MA (2000) The potential for 'Smart Home' systems in meeting the care needs of older persons and people with disabilities. Senior's Hous Update 10 (1):6–7
Empirica and WRC (2010) ICT and ageing: European study on users, markets and technologies—final report. European Commission, Brussels
Faith V (2014) Designing for Dementia: an assessment of the physical environment on wayfinding success for residents in long term care settings. Dissertation, Queen's University Belfast, Belfast
Gulliksen J, Göransson B, Boivie I, Blomkvist S, Persson J, Cajander A (2003) Key principles for user centered design. Behav Inf Technol 22:397–409
Hadjri K, Faith V, McManus M (2012) Designing dementia nursing and residential care homes. J Integr Care 20(5):322–340
Hanson J (2001) From sheltered housing to lifetime homes: an inclusive approach to housing. In: Winters S (ed) Lifetime housing in Europe. Katholieke Unversiteit Leuven, Leuven, pp 35–57
Heiss F, Hurd M, Börsch-Supan A (2003) Healthy, wealthy, and knowing where to live: predicted trajectories of health, wealth, and living arrangements among the oldest old. NBER working paper no. 9897
Helal A, Lee C, Mann WC (2004) Assistive environments for individuals with special needs. In: Cook D, Das S (eds) Smart environments: technology, protocols and applications. Wiley, Hoboken, pp 361–383
Hennessy S, Murphy P (1999) The potential for collaborative problem solving in design and technology. Int J Technol Des Educ 9:1–36
Housing LIN (2012) Older persons housing design—a European good practice guide. http://www.housinglin.org.uk/Topics/browse/Design_building/AccessibleDesign/CaseStudies/?&msg=0&parent=8577&child=1666. Accessed 9 Oct 2015
Humes N (2012) A CAD based tool to assist design decision making and the assessment of inclusive design. Dissertation, Queen's University Belfast, Belfast
Hussein H, Yaacob NM (2012) Malaysian perspective on the development of accessible design. Asian J Environ-Behav Stud 3(8):101–115
Imrie R (1999) The role of access groups in facilitating accessible environments for disabled people. Disabil Soc 14:463–482
Imrie R (2001) Inclusive design: designing and developing accessible environments. E & FN Spon, London
Imrie R (2003) The impact of Part M on the design of new housing. Access Des 96:6–8
Imrie R (2006) Independent lives and the relevance of Lifetime homes. Disabil Soc 4:359–374
Imrie R, Kumar M (1998) Focusing on disability and access in the built environment. Disabil Soc 13(3):357–374
Jordan PW (1998) An introduction to usability. Taylor & Francis Ltd., London
Karlsson J, Ryan K (1996) Supporting the selection of software requirements. In: Proceedings of the 8th international workshop on software specification and design (IWSSD' 96), pp 146–149
Karlsson L, Thelin T, Regnell B, Berander P, Wohlin C (2007) Pair-wise comparisons versus planning game partitioning: experiments on requirements prioritization techniques. Empirical Softw Eng 12(1):3–33
Kujala S (2003) User involvement: a review of the benefits and challenges. Behav Inf Technol 22:1–16
Lahti H, Hakkarainen PS, Hakkarainen K (2004) Collaboration patterns in supported collaborative designing. Des Stud 25:351–371

Langton-Lockton S (2004) What is inclusive design? [Great Britain]. Access Des 101:9–11
Lê Q, Nguyen HB, Barnett T (2012) Smart homes for older people: positive aging in a digital world. Future Internet 4(2):607–617
Leffinwell D, Widrig D (2000) Managing software requirements-a unified approach. Addison-Wesley, Boston
Lewin D, Adshead S, Glennon B, Williamson B, Moore T, Damodaran L, Hansell P (2010) Assisted living technologies for older and disabled people in 2030. Final report to Ofcom. Plum Consulting, Sagentia, Loughborough University, Aegis, London
Lifemark (2014) Lifemark standards and 5 star rating system. http://www.lifemark.co.nz/about-lifemark/lifemark-standards.aspx. Accessed 8 Dec 2015
Lin TW, Shao BMB (2000) The relationship between user participation and system success: a simultaneous contingency approach. Inf Manag 37:283–295
Lin M, Wang C, Chen M, Chang A (2008) Using AHP and TOPSIS approaches in customer-driven product design process. Comput Ind 59(1):17–31
Lubinski R (1991) Learned helplessness: application to communication of the elderly. In: Lubinski R (ed) Dementia and communication. Singular Publishing Group, San Diego, pp 142–151
Luck R, Haenlein H, Bright K (2001) Project briefing for accessible design. Des Stud 22:297–315
Lushai G, Cox T (2012) Assisted living technology: a market and technology review. Report for project funded by European regional development fund. Life Sciences-Healthcare and the Institute of Bio-Sensing Technology for the Microelectronics and Biomedical iNets, Bristol
Madigan R, Milner J (1999) Access for all: housing design and the disability discrimination act 1995. Crit Soc Policy 19(3):396–409
Marshall M (2001) Environment: how it helps to see dementia as a disability. Care Homes Dement: J Dement Care 6:15–17
Medical Research Council (2013) A strategy for collaborative ageing research in the UK. Lifelong Health and Wellbeing (LLHW). https://www.mrc.ac.uk/documents/pdf/llhw-a-strategy-for-collaborative-ageing-research-in-the-uk/. Accessed 8 Dec 2015
Mitchell L, Burton E (2010) Designing dementia-friendly neighbourhoods: helping people with dementia to get out and about. J Integr Care 17(6):11–18
NRC (2010) National Building Code of Canada 2010. http://www.nrc-cnrc.gc.ca/eng/publications/codes_centre/2010_national_building_code.html. Accessed 8 Dec 2015
Older MT, Waterson PE, Clegg CW (1997) A critical assessment of task allocation methods and their applicability. Ergonomics 40:151–171
Olsson E (2004) What active users and designers contribute in the design process? Interact Comput 16:377–401
Oltermann P (2014) Germany's 'multigeneration houses' could solve two problems for Britain. Guardian newspaper. http://www.theguardian.com/world/2014/may/02/germany-multigeneration-house-solve-problems-britain. Accessed 8 Dec 2015
Ozkaya I, Akin O (2006) Requirement-driven design: assistance for information traceability in design computing. Des Stud 27(3):381–398
Paciello M (2005) Enhancing accessibility through usability inspections and usability testing. In: Proceedings of center on disabilities technology and persons with disabilities conference 2005. California State University, California
Palloni A (2002) Living arrangements of older persons. Population Bull Unit Nations Spec Issue 42(43):54–110
Presidencia de la República (2001) Recomendaciones de Accesibilidad. Oficina de Representación para la Promoción e Integración Social para Personas con Discapacidad, Mexico
Reich Y, Konda SL, Monarch IA, Subrahmanian SNL (1996) Varieties and issues of participation and design. Des Stud 17:165–180
Rooney C (2014) Blind spot: an investigation into lifetime home standards and visual impairment. Dissertation, Queen's University Belfast, Belfast
Ryd N (2004) The design brief as carrier of client information during the construction process. Des Stud 25:231–249

Saaty TL (1980) Multicriteria decision making: the analytic hierarchy process. RWS Publications, Pittsburgh
SABS (2011) South African national standard—The application of the national building regulations part S: facilities for persons with disabilities. SABS Standard Division, Pretoria
Sanoff H (2000) Community participation methods in design and planning. Wiley, New York
Schaik PV (1999) Involving users in the specification of functionality using scenarios and model-based evaluation. Behav Inf Technol 18:455–466
Smith-Jackson TL, Nussbaum MA, Mooney MA (2003) Accessible cell phone design: development and application of a needs analysis framework. Disabil Rehabil 25:549–560
Standards New Zealand (2000) New Zealand standard—design for Access and mobility, buildings and associated facilities. Standards New Zealand, Wellington
Taylor G (1999) Empowerment, identity and participatory research: using social action research to challenge isolation for deaf and hard of hearing people from minority ethnic communities. Disabil Soc 14:369–384
Thomas AM, Moore P, Shah H, Evans C, Sharma M, Xhafa F, Mount S, Pham HV, Wilcox AJ, Patel A, Chapman C, Chima P (2013) Smart care spaces: needs for intelligent at-home care. Int J Space-Based Situated Comput 3(1):35–44
Thomas AM, Moore P, Evans C, Shah H, Sharma M, Mount S, Xhafa F, Pham HV, Barolli L, Patel A, Wilcox AJ, Chapman C, Chima P (2014) Smart care spaces: pervasive sensing technologies for at-home care. Int J Ad Hoc Ubiquitous Comput 16(4):268–282
UN (2004) Accessibility for the disabled: a design manual for a barrier free environment. Manual by the urban management Department of the Lebanese Company for the development and reconstruction of Beirut Central District (SOLIDERE) in collaboration with the United Nations Economic and Social Commission for Western Asia (ESCWA), the Ministry of Social Affairs and the National Committee for the Disabled. http://www.un.org/esa/socdev/enable/designm/. Accessed 8 Dec 2015
UNDESA (2006) World urbanization prospects: the 2005 revision. United Nations Department of economic and social affairs, population division, New York. www.un.org/esa/population/publications/WUP2005/2005wup.htm. Accessed 8 Dec 2015
UNPF (2007) Urbanization: a majority in cities. United Nations Population Fund, New York. www.unfpa.org/pds/urbanization.htm. Accessed 8 Dec 2015
Urban Task Force (1999) Towards an urban renaissance, final report of the urban task force chaired by Lord Rogers. Urban Task Force, London
US Department of Justice (2010) 2010 ADA standards for accessible design http://www.ada.gov/2010ADAstandards_index.htm. Accessed 8 Dec 2015
Weiss JC, Williams K, Heerwagen J (2004) How to design for humans. Architecture 93:39–41
WEL_hops (2007) Older persons housing design: a european good practice guide. Project part funded under the European Union INTERREG IIIC programme. Welfare Housing Policies for Senior Citizens (WEL_hops)
WHO (2007) Global age-friendly cities: a guide. World Health Organization
Wiegers K (1999) Software requirements. Microsoft Press, Redmond
Wulz F (1986) The concept of participation. Des Stud 7:153–162

Chapter 7
Energy Use in Housing

Hasim Altan, Liangxiu Han, Karl Wagner, Jose Luis Gomez Ortega and Akash Deep

Abstract This chapter mainly focuses on energy use in housing and covers areas related to energy use breakdown, demand side management, fuel poverty, occupant behaviour and smart metering. It also touches on importance of reducing energy use and its associated carbon emissions by discussing best practice benchmarks. Some of the learning outcomes anticipated as follow: (1) Overview of the energy use in housing including its breakdown and usage trends. (2) Understanding the relationships between the demand side management and its associated carbon emissions as well as the supply side perspective. (3) Understanding the correlation between energy use and fuel poverty both locally and globally. The idea is to provide an overview of, and a basic ability to understand the energy use in housing in order to help design and retrofit housing projects toward reduction of energy demand and/or energy end-use efficiency. Energy benchmarks and best practice guidelines have been discussed as well as fuel poverty and interventions for

H. Altan (✉)
Department of Architectural Engineering, University of Sharjah (UOS), Sharjah, UAE
e-mail: haltan@sharjah.ac.ae; hasimaltan@gmail.com

H. Altan
Welsh School of Architecture (WSA), Cardiff University, Cardiff, UK

L. Han · J.L.G. Ortega
Manchester Metropolitan University, Manchester, UK
e-mail: l.han@mmu.ac.uk

J.L.G. Ortega
e-mail: jose.l.gomez@mmu.ac.uk

K. Wagner
Universiti Kuala Lumpur, Kuala Lumpur, Malaysia
e-mail: profwagnerkarl1@gmail.com

A. Deep
GRIHA Council, The Energy and Resources Institute (TERI), New Delhi, India
e-mail: akashdeep@grihaindia.org

© Springer International Publishing Switzerland 2016

M. Noguchi (ed.), *ZEMCH: Toward the Delivery of Zero Energy Mass Custom Homes*, Springer Tracts in Civil Engineering,
DOI 10.1007/978-3-319-31967-4_7

adopting energy conscious behaviour. In addition, integration of renewables and smart technologies will be touched on to support demand side management including operation and maintenance.

7.1 Energy Demand in Housing

This section will detail on the energy used by occupants in various activities. These activities vary from lighting, space conditioning equipment usage, etc. The section compares the different demand loads on basis of usage and geographical locations in housing only. In addition, the section will illustrate strategies on the various measures, which can be adopted to manage the demand and optimize the consumption. Energy is one of the major commodity used worldwide, with the growing worldwide crises, the only way to manage any commodity is to have set benchmarks. The section presents an energy-measuring unit, which can provide worldwide set benchmarks in comparing same typology of building in respective climate zone (Fig. 7.1).

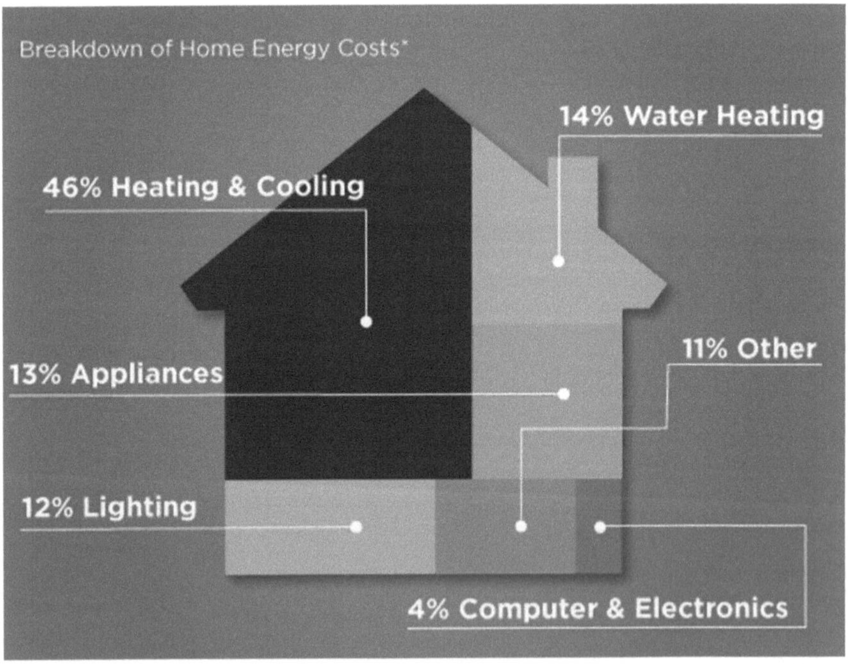

Fig. 7.1 Breakdown of home energy costs (GME 2015)

7.2 Energy Demand Breakdown

Lighting

For doing any kind of work we need specific light levels to perform that activity without stressing our eyes. The term associated with proper lighting is called visual comfort. The unit for measuring light intensity is Lux (lx). In a building, the lighting source can be of two types:

Natural (Daylighting)

In a residential building integration of daylighting plays key role in energy optimization. Every room in the house is generally planned with daylight integration. Daylight integration can reduce major lighting load in parking and common areas in apartment buildings (Fig. 7.2).

Artificial Lighting

Artificial lighting in a house is used in outdoors and indoors. Different types of outdoor lighting as per usage are—street lighting, bollard lighting, landscaping lighting, facade lighting and architectural lighting. Different types of indoor lighting as per usage are—regular lighting, task lighting and mood lighting. There are many types of fixtures available in the market. These fixtures vary with the wattage (W) requirement and the lumen (lm) output provided with respective fixture. Depending on the usage and activities to be performed in specific room, the occupants can select types of fixtures. Also with growing human visual comfort requirement nowadays, there is mood lighting available with different manufacturers. These lighting system change colours of the room with lighting fixtures to set different environment as per personalization (Fig. 7.3).

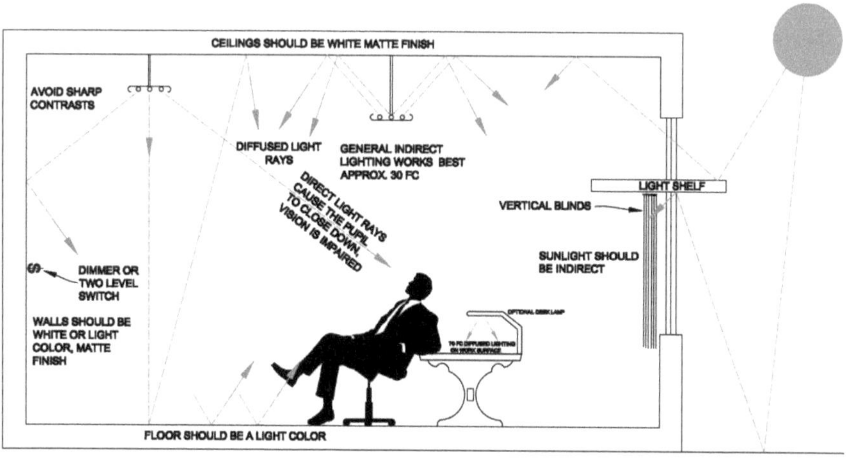

Fig. 7.2 Lighting in building spaces (TERI 2011)

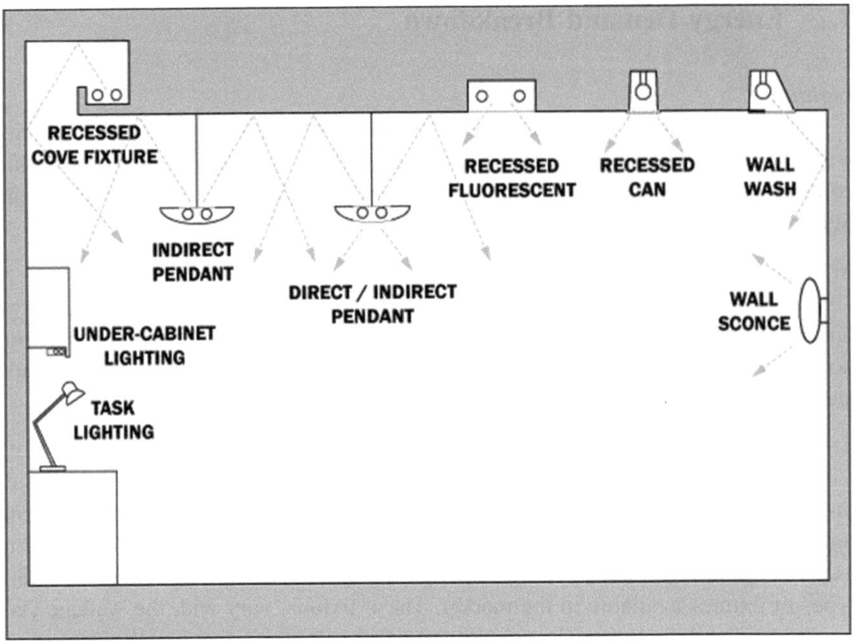

Fig. 7.3 Artificial lighting in building spaces (Autodesk 2015)

Water Heating

Water heating is one of the key energy usages in colder regions. The usage can be minimized using high efficient heaters or by using natural resources based on water heaters. Unit to measure amount of energy used for heating required amount of water is Joules (J).

Space Conditioning

Space conditioning is directly related to the occupant thermal comfort. Thermal comfort range of human body is in general constant range. The building envelope acts as transition between the extreme climate outside and controlled environment within the building envelop. Space conditioning systems help us achieve thermal comfort within the envelope about the climate, heating in cold zones, cooling in hot zones, conditioning in moderate zones, etc. There are set benchmarks in different countries to achieve thermal comfort. These parameters are temperature, relative humidity, wind flow, etc. (Fig. 7.4).

With growing climate diversities conditioning has become a major requirement with all buildings across the globe. Thermal comfort can also achieved with different strategies, which can be categorized as follows:

- Active technologies: energy intensive based on efficient systems.
- Passive technologies: systems, which require no energy to run.
- Hybrid technologies: technologies, which require minimum level of energy.

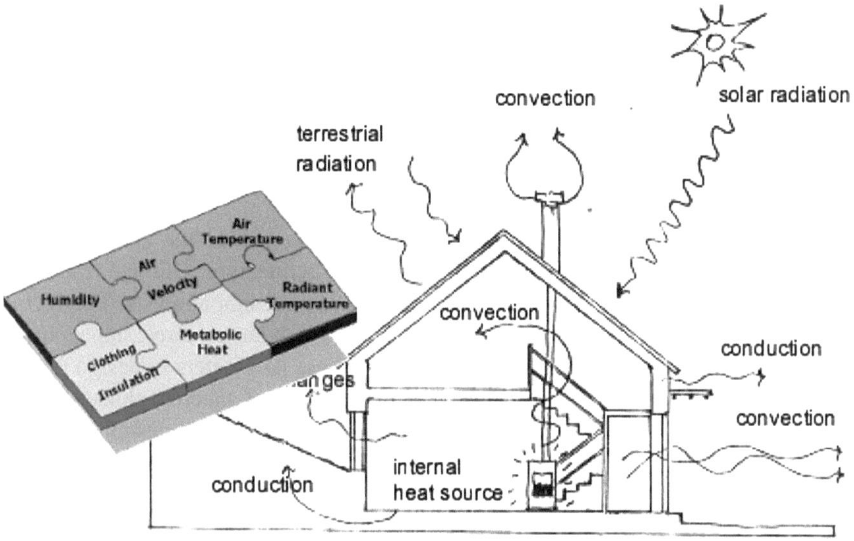

Fig. 7.4 Space conditioning (HSE 2015)

Equipment/Appliances Loads

Households use energy concerning the user behaviour and requirements. Equipment and appliances loads vary in each household depending on the occupant habits. These loads can be categorized into two different sections as per the usage:

Static

Static section covers services, which are generally used by every household, and is constant in usage patterns by different occupants and energy use can be estimated considering a diversity factor into account. These services are refrigerator, geysers, air conditioning (heating/cooling) etc.

Dynamic

Dynamic services majorly cover different appliances used by the occupants as per its respective requirements like washing machine, dishwasher, vacuum cleaner, microwave, computers, laptops, coffee machine etc. These loads cannot be defined and generalized usage. Hence, these loads are considered dynamic in nature.

7.3 Demand Side Management

Energy management is same as for any other commodity we use. There is a demand and supply. For every project the management has to be done from both the sides, i.e. demand side management and supply side management. In this section, we would discuss different techniques and measures in which we can reduce the demand load of a project. Reducing demand load intern reduce grid pressure and the reduced demand has a better option to be offset by renewable resources.

7.3.1 Design Interventions

One of the most basic interventions, which can help reduce the energy demand significantly in any project, is climate responsive architecture. Climate responsive architecture has multiple parameters associated to the design analysis. The section discusses on some of the most crucial parameters, which have a direct impact on the energy demand of the building. Incorporation of these parameters in design process illustrates the building responsiveness to its micro and macro -climate and is designed to incorporate the existing site features in the proposed design (Fig. 7.5).

Orientation

Orientation is one of the essential parameter, which dictates the buildings envelop design. Optimum building orientation is considered to have major facade facing north south as shading design in these directions is easier to handle. Proper orientation also helps channelize wind though the building block.

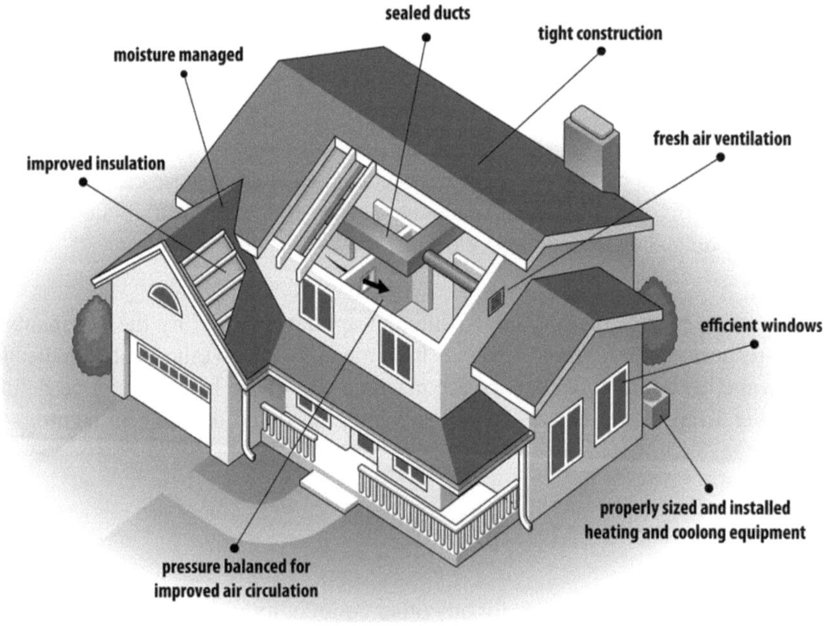

Fig. 7.5 Incorporating the existing site features in the proposed design (TGE 2015)

Massing

Zoning and massing is one of the design parameters, which help channel wind flow and enhance mutual shading between the blocks. Proper massing of blocks channel the summer wind to cross ventilate through the building massing and blocks the winter wind. These design features enhance the project usability of the natural wind for cross ventilation and intern help to reduce HVAC load. Also the mutual shading by building blocks help reduce heat gain in summer and increase heat intake in winters resulting in reduction in cooling load in summers and heating load in winters.

Fenestration Shading

Fenestrations are punctures in the building envelop. Fenestration placement and sizing in the building should be selected according to the orientation of the building. These openings in the building envelop require shading designing to let in glare free day light, reduce direct heat gain through windows. In northern hemisphere north light is considered to be glare free light, while the south light has both high glare and heat; however the glare and heat can be controlled by providing horizontal shading devices. East facing fenestration require vertical shading in late morning to cut glare while the west facing shading is required to cut the maximum heat gain from solar radiation. The trickiest shading design is for the orientation of south-west, as it requires both horizontal and vertical shading devices (Fig. 7.6).

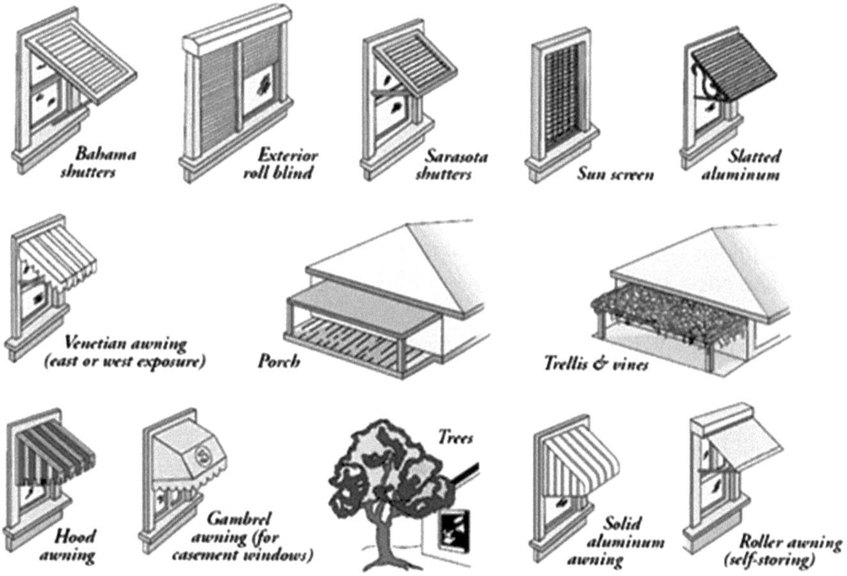

Fig. 7.6 Fenestration shading design (FSEC 2015)

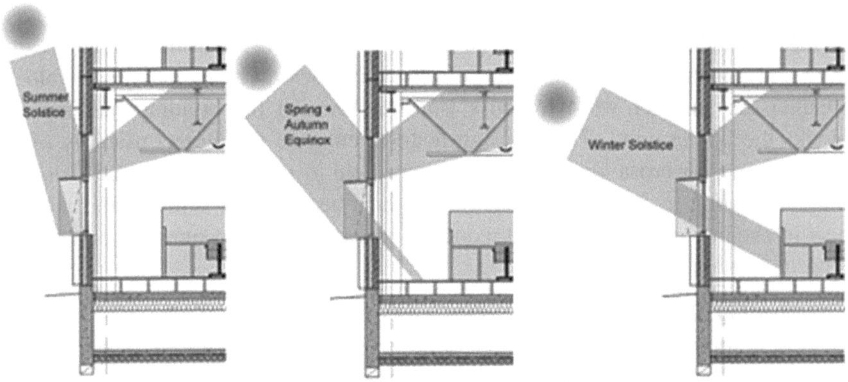

Fig. 7.7 Climate responsiveness while shading design (WBDG 2015)

Another factor, which should be considered during shading design, is the climate responsiveness. The shading should facilitate to cut the summer sun and allow the winter sun to enter the home (Fig. 7.7).

Landscaping

Landscaping is one of the design features which is often neglected and last minute integration with the design, however a proper landscaping design help reduce energy demand in buildings significantly. It helps reduce heat island effect around the building and proper plantation plan can help reduce heat gain by tree shading.

Heat Island Effect (HIE)

Heat island effect is the phenomenon where due to increased hard surfaces and thermal mass the temperature is higher than green areas. This term is generally associated with urban area. The solar incident on thermal mass/hard surfaces absorb maximum heat and releases when the atmospheric temp is less, this intern heats the surrounding air and increases the urban temp around buildings. Strategies to reduce heat island effect includes, increased soft finishes, shading the hard surfaces and by not using thermal mass in finishes.

Plantation of shading tree at appropriate location can help shade the building and reduce heat gain. In addition, water bodies/features incorporated in design can help cool the channelized wind flow, which works as natural cooling mechanism for a building in hot climate.

7.3.2 Efficient Systems

System efficiency is one of the factors, which are user driven and effect directly on the energy demand. Installation of efficient systems result in energy saving;

however, there is a financial upsurge to install a high-end efficient system. The system designing has to balance out financial and efficiency requirement. Some systems may have an incremental upfront cost however; they have a recovery period ranging 3–5 years. System efficiency in a housing project can be increased in lighting fixtures, space conditioning systems, appliances and services.

Efficient Lighting

Using of energy efficient lighting fixture is one of the easiest and most economical ways to reduce energy demand in a housing project. Efficiency of fixtures is measured in efficacy level of fixture and luminaire. Efficacy is measured in lumen per watt i.e. output achieved after ballast losses in a fixture on the designed wattage. Better the efficacy of a luminaire more energy saving can be can be targeted.

Efficacy of a lamp − {Output (Lumen) − Ballast losses (Lumen)}/Wattage (Watts)

When light in incident over any object it can be absorbed, reflected or transmitted.

- Absorption—Transformation of radiant energy to a different form of energy by the intervention of matter.
- Reflection—Process by which radiation is returned by a surface or a medium without change of frequency of its monochromatic components (Fig. 7.8).
- Transmission—Passage of radiation through a medium without change of frequency of its monochromatic components.

Quality of light can be measured in four parameters. Photometry is the science of the measurement of light, in terms of its perceived brightness to the human eye.

- Luminous Intensity (I), measure din candela (cd)—Intensity of black body of 1/60 cm^2, when heated to melting point temperature of platinum.
- Luminous flux (phi), measured in Lumen (lm)—Flux emitted within 1 steradian (sr) by a point source of I = 1 cd, emitting uniformly in all directions.

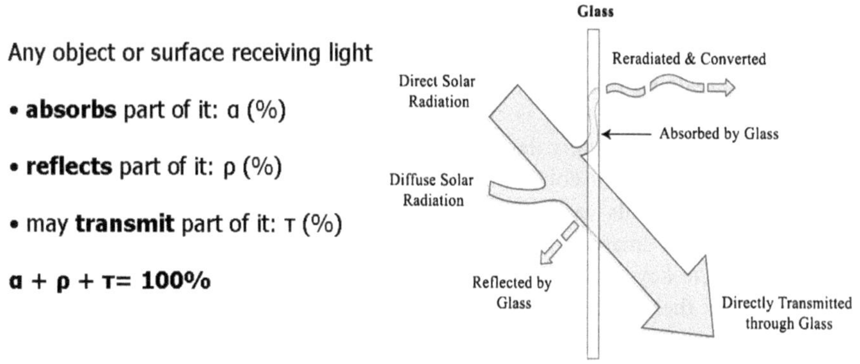

Fig. 7.8 Behaviour of light (Henry et al. 2004)

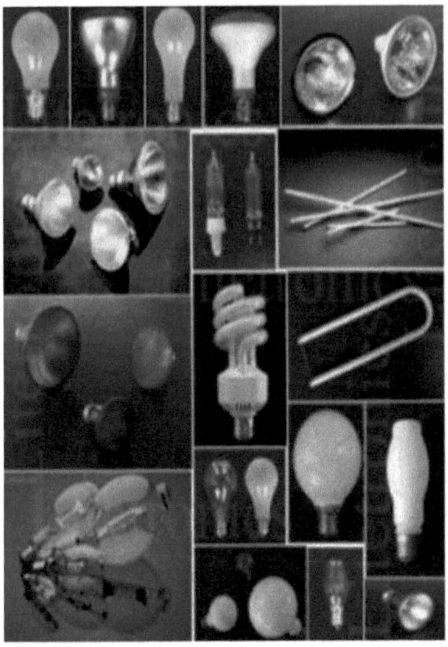

- Incandescent Lamps
- Fluorescent Lamps
- High – Intensity Discharge Lamps
- Mercury Lamps
- Metal Halide Lamps
- High Pressure Sodium Lamps
- Low Pressure Sodium Lamps
- Electrodeless Lamps
- Compact arc xenon &Mercury Lamps
- Electroluminescent Lamps
- Light Emitting Diodes (LED)
- Carbon arc Lamps
- Gaslights

Fig. 7.9 Type of lamps (MES COE 2015)

- Illuminance (E), measured in lux (lx)—measure of illumination of a surface. Density of photons falls in a given surface.
- Luminance, measured in cd/m^2—luminous intensity in a given direction and falls within a given solid angle (Fig. 7.9).

Artificial lighting design is one of the most neglected area in terms of housing lighting design. As per different worldwide codes available, there are recommended lux levels as per the usage of the room. However, we often tend to over design the lighting system, which results in increasing the energy demand drastically. In most of the codes, generally there is a range of lux levels mentioned for respective activity. There are multiple simulation software available in the market to carry out design simulations to access the proposed lux levels at different work plains. Strategies like dedicated lighting design can help reduce excess lighting demand in a building.

Efficient Appliances

All households nowadays have a world of gadgets or appliances to manage the daily chores efficiently. This is one of the loads is residential sector which is dynamic in nature and totally depends on human nature. Factoring appliances load in energy benchmarks for comparing one household to another is one of the major challenges. These loads vary with each house and depend on human habits. With modernization and smart system integration in homes, increasing the efficiency in appliances is the key in demand load reduction.

Worldwide there are multiple star labelling systems available, which rate appliances in terms of energy efficiency. High efficient systems can have an initial installation cost; however, on longer run these help reduce the day to day energy requirement in house folds.

7.3.3 *Building Envelope*

Building envelop design is the key to reduction if energy demand is concerned. In any building envelop, the different components are as follows: Walls, Windows and Roof. These are the components, which interact with the outer climate and provide comfort inside. The designing of these components should to depending on climate and building use.

Selection of Building Material

How does heat Transfer happens in a building? Heat transfer in a building happens through walls, windows and roof from higher temperature outside to the inside of the building through the building envelop. The property of heat transfer through a material happens in three ways:

- Conduction: It is the property of material to transfer heat through direct contact.
- Convection: It is the property of material to transfer heat through movement of air gas or liquid.
- Radiation: It is the property of material to transfer heat by movement of heat through space with relying on air or gas (Fig. 7.10).

To measure the heat transfer in a material the thermal conductivity is the property, which defines the insulation of the building. Conductivity is inversely

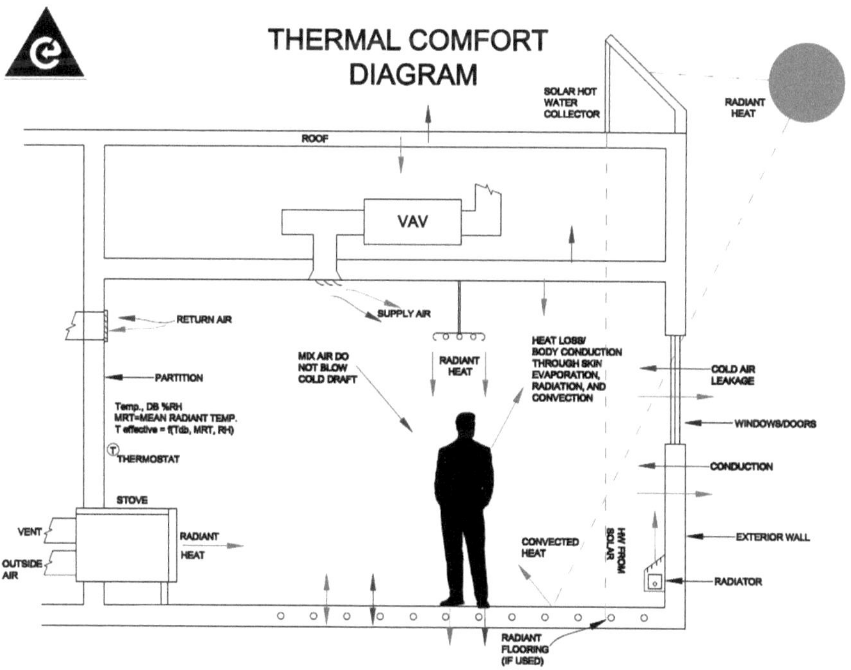

Fig. 7.10 Thermal comfort diagram (TERI 2011)

proportionate to resistivity. The lower the thermal conductivity of a material, the better the thermal performance, i.e. the slower heat will move across a material.

R-value is the measure of resistance to heat flow through a material and its unit is square meter Kelvin per Watt (m^2K/W). The better the R-value of a material is will work as a better insulation.

R = Thickness of material/thermal conductivity

Where,

Thickness of a material is in meter

Thermal conductivity is in Watts per meter kelvin

U value is the measure of heat loss through a material using three ways of heat transfer and its unit is Watt per meter square kelvin (W/m^2K). U value is inversely proportionate to R-value.

U = 1/R+ (heat loss though convection and radiation)

Selection of Glass in a Building

Glass is one of the costliest component is the building material section and it contributes to the maximum heat gain and intern increases energy demand of a building drastically. Use of a glass in a building design is mostly governed to permit day light within the building (reduce artificial lighting) or to provide a connection to the occupant with the surroundings. The building design should address the requirement of glass in a balance state to address both the issues simultaneously. Excess of un-shaded glass only increase glare within the building, which result in installation to, blinds and curtains to reduce glare and intern switch on artificial lighting.

To optimize the use of glass within the building the term WWR (Window to Wall Ratio) is used within the construction industry. WWR is the ratio of glass used with respect to the opaque walling in the building envelop. WWR can be optimized depending on the orientation of the facade and the climate/location of the building.

There are multiple glass sections available in market ranging from cheapest SGU (Single glazed unit) to costliest TGU (Triple glazed unit). The property of glass, which allows to heat exchange, is SHGC (Solar Heart Gain Coefficient). SHGC of glass can be defined as fraction of solar heat incident on window by heat absorbed, transmitted and released inside. SHGC is unit less. Value of SHGC varies from 0 to 1, glass having SHGC 1 being the worst glass used in terms of heat resistance.

Selection of Roofing Material

Roofing is the component of building envelop which is responsible for the maximum heat gain within the building. Reduction of heat gain through roof can be achieved with:

- Shading of roof by pergola, Solar PV systems, roof gardens etc.
- Using high SRI (Solar Reflective Index) paints or reflective surfaces
- Insulating the roof by over deck or under deck insulation (Fig. 7.11).

Shading will reduce incident heat, high SRI paints increase reflectance and reduce absorbed heat and insulation reduce transmittance of heat through roof. Combining the three strategies depending on the climatology helps reduce HVAC load.

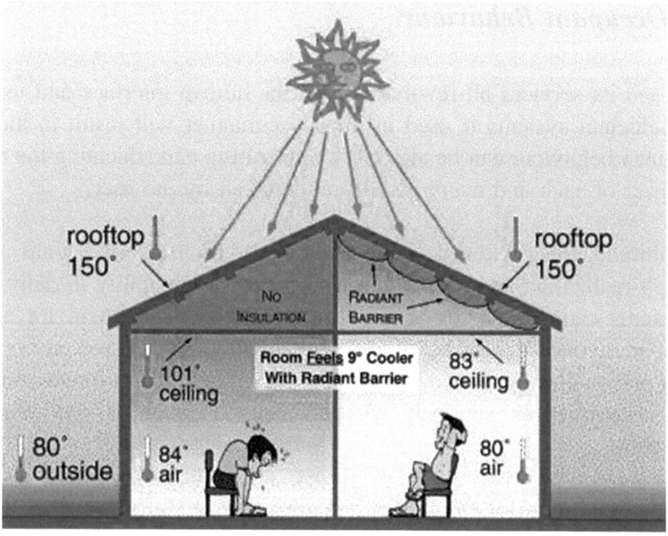

Fig. 7.11 Roofing (Mapawatt 2015)

To increase day lighting in the building skylights are used. These skylights also serve as a major heat-gaining component of a building. To optimize skylight opening in a building is referred as SRR (Skylight to Roof Ratio). Glass used in the skylight should be selected keeping in mind the SHGC criteria of glass selection.

7.3.4 Metering and Submetering

"What gets measured gets managed" (TERI 2011). Metering and sub metering help us monitor our consumption and hence take corrective measures to reduce demand. Nowadays many smart metering systems are available in the market to monitor and control all the system from different portable devices. The metering facility should be provided as per the uses with the project so that relevant strategies can be adopted to reduce demand in respective usage category. Monitoring of usage can be done for following consumption:

- Outdoor lighting.
- Indore lighting and power consumption.
- Common lighting (high rise apartments).

7.3.5 Occupant Behaviour

Building and its services all revolve around the human interface and usage. Even the most efficient systems if used in improper manner will result in huge energy costs. Human behaviour can be affected by informing and educating the user of the direct impact of each and every activity undertaken by the user.

Intent

User attitude governs the decision making for each activity. Until and unless there is self-realization and motivation to achieve sustainability in daily living the smart systems cannot help us achieve sustainability. The intent for sustainable living can be invoked in user by educating them about effects and profits expressed in term of personalized gain and loss. Unless the effects are not direct, user does not take the consequences seriously. A personalized touch is required in mass customization.

Informed Decisions

Decision-making is lot easier if it is informed one. Information is the key for sustainability. Sometimes decision made on perceived information is way off the actual scenario. Smart metering and sub metering is one of the systems which help the user to keep a track on resources being consumed and help them take informed decision during occupancy.

One of the interesting examples of a private developer residential medium rise apartment projects based in India. The project is generating renewable energy (RE) from solar and wind installed within the project facility. Each apartment has been allocated around 200 units of RE for use each month. Now to take informed decision the project has provided a light panel at the entrance of each flat. The light panel has two lights one is red and other is green. Green light in the panel is lighted when the flat user is using the allocated 200 units of energy, which is supplied at a low cost. The moment the user has crossed the 200 units usage in the month the red lights turns on. On which the user is aware he would be charged with higher rate of energy. Such basic systems actually help regulate use of energy in households and help the user to make informed decision during occupancy.

7.4 Energy Use and Fuel Poverty

7.4.1 Definition and Impact

Energy use and fuel poverty are interrelated. Fuel poverty as an issue for sustainable housing goes back to its definition once poor people spending more than 10 % of their income for thermal comfort. Whereas the term has not yet been derived for cooling, in the UK fuel poverty is defined by the "Warm Homes and Energy Conservation Act" as a household which "is to be regarded as living "in fuel poverty" if he is … living on a lower income in a home which cannot be kept warm at reasonable cost" (EU FPN 2013). Hence, a fuel poor household is defined as one, which needs to spend more than 10 % of its income on all fuel use and to heat its home to an adequate standard of warmth. In the UK, according to the ASHRAE-standard this is defined as 21 °C in the living room areas and 18 °C in other occupied rooms. The current definition of fuel poverty states is driven by three key factors: energy efficiency of the home; energy costs and household income.

A new more complex definition of fuel poverty has been used in the UK, based on the Hills review (Hills 2012), resulting in the following definition; when a household's required fuel costs are above the median level; and if they were to spend what is required, then the household would be left with a residual income below the official poverty line. Furthermore, a Fuel Poverty Indicator (FPI) has been created, which shows the degree of poverty, not simply if they are in poverty or not (UK NAREC 2013).

As mentioned, even though the term is mainly used in the UK, Ireland and New Zealand, although discussions on fuel poverty are increasing across Europe (EU FPN 2013), and the concept might also apply everywhere in the world where poverty may be present. In countries where cooling is required it could be applied in an adopted, but yet to be defined manner.

In early 2008, it was estimated by Energywatch that there were around 4.4 million households in fuel poverty in the UK, with just over 3 million in England alone: this was more than double the number in 2003 (Webb 2008). 3 years later, in April 2011 a YouGov survey indicated that the number of households in fuel poverty had risen to 6.3 million households, representing approximately 24 % of all households in the UK (Switch 2011). Research by Confused.com found that 82 % of the UK population had expressed a concern at being able to afford their energy bills throughout winter (VMMN 2011).

In England, applying the original definition of fuel poverty above, 2012 4.82 million households have been in Fuel Poverty compared to those 4.28 million in the previous year, which means an increase of 540,000 (up 13 %). Figure 7.12 also shows the corresponding totals for Northern Ireland, Scotland and Wales. In Northern Ireland, fuel poverty has increased from 380,000 to 425,000 households (up 12 % on 2013). In Scotland, it has increased from 795,000 to 890,000 households (up 12 %) and in Wales, it has increased from 400,000 to 450,000 households (up 13 % on 2013). Across the Devolved Administrations, fuel poverty has increased by 43 % since 2011, the year for which the last complete set of estimations is available (UKACE 2015).

7.4.2 Causes of Fuel Property

Fuel poverty is caused by a convergence of five factors:

- Low income, which is often linked to absolute poverty.

Number in fuel poverty		UK	England	Northern Ireland	Scotland	Wales
All households	2014	6,590,000	4,820,000	425,000	890,000	450,000
	2013	5,855,000	4,280,000	380,000	795,000	400,000
	2011	4,435,000	3,200,000	290,000	580,000	365,000
Families with dependent children	2014	1,440,000	1,070,000	135,000	140,000	95,000
	2013	1,230,000	915,000	115,000	120,000	80,000
	2011	-	590,000	-	-	-
Of which families with under-16s	2014	-	961,000	-	-	-
	2013	-	830,000	-	-	-
	2011	-	530,000	-	-	-
Of which families with under-5s	2014	-	456,000	-	-	-
	2013	-	390,000	-	-	-
	2011	-	245,000	-	-	-
Dependent children	2014	2,570,000	1,940,000	230,000	240,000	160,000
	2013	2,195,000	1,660,000	195,000	205,000	135,000
	2011	-	1,050,000	-	-	-

Fig. 7.12 Number in fuel poverty in the UK (UKACE 2015)

- Corresponding high fuel prices, including the use of relatively expensive fuel sources (such as electricity in the UK, aggravated by higher tariffs for low-volume energy users).
- Poor energy efficiency of a home, e.g. through low levels of insulation and old or inefficient heating systems.
- Under-occupancy: according to UK government statistics, on average those in the most extreme fuel poverty live in larger than average homes.
- Old age of the building.

The sharp rise in fuel prices from 2006 to 2008 has led to an estimated doubling of the numbers in fuel poverty in countries where it is a major problem. A number of illnesses, including cancer can exacerbate the problems associated with fuel poverty (UK NAREC 2000).

7.4.3 Tackling Fuel Poverty

Whilst tackling fuel poverty can be basically considered a legal obligation of any government, energy suppliers recognise their responsibilities, especially for the elderly, and those underprivileged (Energy UK 2015). With this in mind, suppliers have worked closely with social services, citizens' advice bureaus and charitable groups, such as Age UK and Macmillan, to consider the best way to help vulnerable occupants that fall under the Fuel Property line. Thousands of energy customers have been taken out of fuel poverty through the efforts of energy suppliers working with social welfare organisations.

In doing so, the Energy Bill Revolution is an alliance of 170 organisations calling for the Government to make home energy efficiency the UK's priority infrastructure investment (DECC 2013a, b). It desires energy efficiency measures to be provided free for people in fuel poverty, and subsidies for those who practice a certain level of energy efficiency for everyone else. It is proposed that this be paid for by recycling revenues from two carbon taxes that are paid by consumers—the European Emissions Trading Scheme and the Carbon Price Floor. According to recent plans, over the next 15 years, the Government will raise an average of £4 billion every year in carbon taxes; this is enough revenue to insulate a high degree an average of 600,000 homes from being within the Fuel Poverty zone every year. In addition to subsidies, every household could benefit from recycling carbon tax into energy efficiency measures. The Energy Bill Revolution proposes that fuel poor households could be prioritised for assistance. Research conducted by Cambridge Econometrics and Verco10 show that compared to other kinds of public investment and tax breaks, this is the most effective way to promote economic growth and also create employment (CEV 2012). This can be considered the only permanent solution to end Fuel Poverty and bring down energy bills simultaneously.

7.5 Energy Benchmarks

With the massive scale of construction industry, it is hard to compare one of the buildings located in specific geographical location to the other project located to different location. However, the buildings perform with respect to the climate of the location. To compare same typology of building across the globe there has to be set standards, which compare them on equal grounds.

7.5.1 Energy Performance Index (EPI)

Energy Performance Index is the amount of energy utilized per square meter in a year i.e. kWh/m^2/annum. As discussed in Sect. 7.3.1 the different energy uses in a household comprises of lighting, space conditioning and appliances. However, appliances load is a load, which is dynamic in nature, which cannot be standardized across different geographical location. Hence, EPI can be calculated considered in the lighting and space-conditioning load in a building. For instance, EPI is one of the energy comparison benchmark used by GRIHA rating for evaluating green buildings in India (TERI 2011). GRIHA has set benchmark for different type of buildings, location as per climate zones and hours of usage.

7.6 Reduction of Energy Consumption

Various factors play a part in energy consumption of households, for example, home characteristics, components/equipments installed in homes (e.g., devices for heating, cooling, cleaning and lighting) to maintain the comfort of homes, occupants' behaviour (e.g. their consciousness for energy use and occupants' activities in the home), home context (e.g. space, environment, location, time), etc. This section describes different factors that could help reducing the energy consumption including occupant behaviour and energy patterns and energy usage prediction, insulation against energy losses (hot and cold climates), house types and energy consumption/generation, smart metering, energy generation through renewables, Smart home initiatives, and standby energy consumption.

7.6.1 Occupant Behaviour, Energy Usage Patterns and Prediction

Occupant presence and behaviour have large impact on space heating, ventilation, energy consumption of lighting and space appliances (Page et al. 2008). An occupant is present in a room and generates pollutants like CO_2, odour, heat, which can directly change the indoor environment. Because of this change, the occupant will interact with the home environment to maintain the comfort level, for example, by opening the window or turn a ventilator on. This, in turn, can change the energy consumption in homes. Much effort has been made in researching how the users' behaviour can influence energy consumption patterns. The works in Masoso and Grobler (2009), Seryak and Kissock (2000), demonstrated that users' behaviour difference could result in different energy consumption. The occupancy driven control system for HVAC in Erickson and Cerpa (2010), could enable energy savings of 20 %. The researchers Seryak and Kissock (2000) conducted a study on the relationship between electricity consumption and household lifestyle and evaluated the user's awareness of energy and willingness for energy saving. Results showed that energy could be saved through improving users' behaviour by energy-saving education. The light models (Newsham et al. 1995; Reinhart 2004), link the presence of occupants and the use of a lighting appliance. The work in Yamaguchi et al. (2003) proposed a Markov model to simulate the occupants' presence by using weekly profile of the presence probability as input. A stochastic based approach (Page et al. 2008) has been proposed to model and simulate occupant interaction with building such as opening window and blinds. The model also showed the capability to reproduce key properties of occupant presence such as times of arrival and departure, periods of intermediate absence and presence as well as periods of long absence from zone, which could be a direct input for energy consumption. The research in Eriksson and Cerpa (2010), showed that energy consumption during non-working hours is higher than during working hours since the occupants often leave the appliances on. Results demonstrated that the human behaviour is the most important factor that affects energy usage. The work in Dong and Andrews (2009), developed sensor-based modelling and prediction of user behaviour in intelligent buildings and connected the behaviour patterns to building energy management systems. This model only considered single user-behaviour in non-domestic buildings.

At case study (Han et al. 2012), the authors focused on investigating the relationship between human behaviour and energy consumption through the analysis of energy usage. The data was collected from the affordable home typical of Scotland. The study encompassed the data analyses at different time granularities such as hourly and daily energy usage patterns. It is confirmed that domestic energy consumption is affected by the occupants' presences, behaviour and activities. The energy usage patterns at the selected home monitored on weekdays, were prone to differ from those at weekends. In the study, exemplar analysis weekends (saturdays and sundays) are shown in Fig. 7.13.

7.6.2 Insulation Against Energy Losses (Hot and Cold Climates)

Building insulation refers to those systems and materials used to prevent the transmission of heat by conduction through them. It is one of the main passive systems to take into account when designing energy efficient buildings. A good insulator will take care of preventing heat loss in cold weather or climate as well as maintaining indoor environments safe from the amounts of heat outdoors in warmer external conditions. Apart from the benefits derived directly from the energy savings, insulation in homes is also effective to enhance the occupants' thermal and acoustic comfort or to stop fire spreading in case of an emergency.

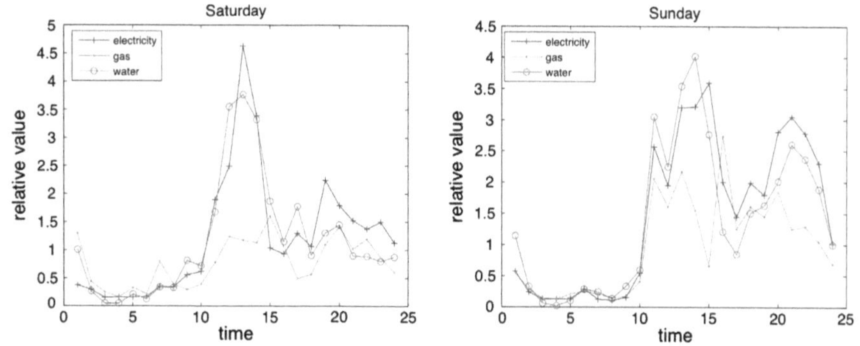

Fig. 7.13 Monitored energy usage patterns of a household from Saturdays and Sundays (Han et al. 2012)

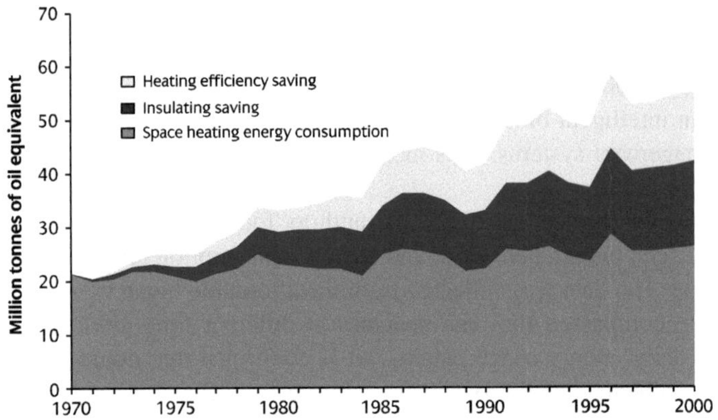

Fig. 7.14 Combined savings from insulation and heating efficiency improvements (DEFRA 2009)

The cost savings from the use of insulation systems combined with improving the energy efficiency of heating equipment are estimated at around 48 % at the beginning of the century (Milsom 2014). That would explain why today many institutions and governments have adopted various strategies and regulations to improve the insulation of old houses as well as to encourage the use of advanced insulation systems in newly constructed homes.

Different measures can be adopted to improve the thermal insulation of a building; their technical characteristics and associated costs will determine their suitability to different scenarios. Below, different measures for this type of isolation are described:

Structural Insulation

Isolation of the structural parts of the housing is carried out by either including insulating materials inside walls, floors and ceilings or by attaching them to the outside in case of not having cavities for the insertion of materials. The structure of the building has implicit heat transfer characteristics, which depend on aspects such as the materials used for its construction, its thickness or the presence of cavities between walls. Insulating materials will be tailored to these in order to improve the heat resistance under the best possible conditions. This entails insulating: Cavity wall insulation, Solid Wall Insulation, Loft or roof insulation, Room in roof insulation, Under-floor insulation and Party wall insulation.

Windows and Doors

An efficient insulation of doors and windows is imperative in order to improve the thermal performance of homes. Recently, new glazing systems have been included, as well as double-glazing, insulating glass and new materials and coatings that enhance the insulating capabilities of traditional doors and windows. These new high-performance doors and windows can become six times more energy efficient than previous models (PCGCC 2009). The new units based on Insulated Glass (IG) achieve optimum performance when used in conjunction with low-E (low transmission of ultra-violet and infra-red light) or reflective coatings, making them ideal for the reduction of heat losses and ensure compliance with local regulations. In addition to windows and doors themselves, the design of the frames and the materials they are made of, would also have to be carefully considered.

Draught Proofing

This is one of the most easy and inexpensive ways of saving energy in homes. Draughts and cracks will allow the flow of indoor-outdoor heat therefore will have to be addressed accordingly. These unwanted gaps need to be filled or covered and they usually appear in windows, doors, loft hatches, electrical fittings on ceilings or walls and in places where construction parts are attached to each other. The combination of the discussed insulation techniques coupled with more efficient heating equipment has allowed significant reductions in the CO_2 emissions impact as shown in Fig. 7.14.

7.6.3 Housing Types and Energy Consumption

Another aspect affecting the energy performance of residential buildings is their shape in terms of geometry, volume and geo-location. It makes sense that in large households, energy consumption should be larger than smaller houses. Furthermore, when a house is of the detached type, it has all its walls in contact with the outside. This means that the risk of heat losses through walls and windows

Fig. 7.15 Detail of the Tabula WebTool showing different types of house based on the year of construction. This tool gives information about the expected energy performance of each type of house included (IWU 2014)

is greater than in the case of semi-attached or attached structures. This is especially true in extreme climates where the difference between indoor-outdoor temperatures will be significantly higher compared to the indoor-indoor in the case of attached homes.

Types of Housing

The most common types of dwelling regarding their exposure to the outdoor conditions could be enumerated as detached (single standalone family building), semi-attached (one side in contact with other houses), attached (other buildings in both sides) and flats (usually surrounded by dwellings in all directions). Depending on the type of building, more or less external surface area will be directly in contact with outdoor weather and climate conditions. Although there may be insulating systems in place (as discussed in previous sections), there is always a minimum heat loss, which will be always directly proportional to the surface exposed as in:

$$Heat\ Loss = \frac{(Area\ of\ Surface) * (Difference\,\frac{indoor}{outdoor}\,Temperature)}{R - value\ of\ Surface}.$$

Due to this, the more open surface the dwelling possess, the more prone to energy inefficiency would be compared to other attached structures. Consequently, for these buildings insulation becomes an even more crucial factor to take into account.

Another important difference between the types of housing is found in the actual volume they occupy. Although initially large homes represent higher consumption than smaller households, there are factors that can make this claim not true for 100 % of cases. For example, in a large home a heating system based on electricity (purely resistive) would incur in a tremendous amount of energy consumption; that is why it is quite difficult to find big houses using this type of heating. However, other smaller houses are better positioned to have this type of electrical equipment (either wall mounted or just portable radiators) because, as they are not so large, utility costs would not be so astronomical.

Energy Certificates and House Types

The information related to the type of residential building can be used for future urban planning, energy trends and shifting or large retrofitting strategies. A European initiative called TABULA (Vimmr 2013) created a framework to study the relationship between the different building typologies and their corresponding energy performance. Thanks to recent European legislations, countries such as Ireland, Denmark, Germany, Italy and many others have compiled large databases containing large amounts of information related to energetic certificates. This data contains information about characteristics of the dwelling and its energy performance. Using this information, a framework can be built to assess the typical performance based on different shapes (see Fig. 7.15 for detail on the TABULA WebTool). Apart from the study of the most common households, each country can assess the real consumption including detailed information about buildings and supply systems (Vimmr 2013).

Quarter	Domestic Smart Meters (Electricity)	Domestic Smart Meters (Gas)	Domestic Smart Meters (All)
Historic installations (1)	59,446e	18,975e	78,421e
Q3 2012	36	32	68
Q4 2012	1,671	1,570	3,241
Q1 2013	12,678	10,963	23,641
Q2 2013	45,456	35,130	80,586
Q3 2013	57,632	35,190	92,822
Q4 2013 (2)	55,603	39,730	95,333
Q1 2014	61,164	37,480	98,644
Q2 2014	60,216	37,113	97,329
Q3 2014	76,227	53,764	129,991
Q4 2014	82,081	60,882	142,963
Q1 2015 (3)	126,515	85,202	211,717
Total	638,725	416,031	1,054,756

(1) Includes historic installations prior to Q3 2012 for the larger 7 suppliers; includes installations prior to Q1 2015 for First Utility and OVO.

(2) Utility Warehouse data included from quarter four 2013.

(3) First Utility and OVO data included from quarter one 2015.

e - Estimated

Fig. 7.16 Number of new smart meters introduced in the UK from 2012. *Source* Smart meters, Great Britain, to end quarterly report, March 2015 (DECC 2015)

Special Housing Types

Other special types of housing deserve particular attention as they differ significantly from the types discussed above in terms of their inherent characteristics. These different types can have special features and requirements in terms of energy efficiency (i.e. special insulation needs, different ventilation conditions or limitations on the orientation or geographic location). Among these types of dwellings, we find Earth-Sheltered, Straw Bale, Log, and Park Homes.

Each of these types needs to be carefully assessed during the designing process to ensure the higher standards of energy efficiency due to their unique nature. More information about these special types of house can be found in the official site for efficient home design at the U.S. Department of Energy webpage (US DOE 2015).

7.7 Smart Metering

The energy meters (for electricity and gas) are devices used by utilities companies to monitor and control the end user consumption. Traditionally, these meters were placed and read mainly for billing purposes. Smart meters are part of a new generation of smart technology, characterized by the ability of transmitting real-time consumption information to the generation/distribution grid. This technology also enables users to connect their smart phones or personal computers to servers allowing monitoring their own consumption.

7.7.1 Smart Meter Initiatives

The number of smart meters installed in households has grown dramatically during the last few years, largely due to government initiatives (i.e. UK or Spain) which require energy companies to bring them to market or provide subsidies to encourage users to use them. On the other hand, intense marketing campaigns have also helped relaunching these devices. The UK government reported a huge increase of the installed meters after starting with their replacing campaigns (see Fig. 7.16).

The increasingly widespread use of smart meters in residential buildings has a number of advantages (DECC 2015):

- Improved efficiency in distribution and grid planning.
- Increased awareness of consumption by the user selection. Use reduction and selection of the best times for consumption.
- Better and more accurate estimate of the actual spending (rather than the traditional estimates).

Technically speaking, the use of these meters does not imply a reduction in consumption directly. However, studies show that users with smart meters in their homes change their consumption patterns so that they tend to get real energy savings compared with households that do not have this technology in place. In Faruqui et al. (2010), it is shown that the use of these devices can lead to savings up to 7 %, even to double this value when used in conjunction with time-of-rates billing tariffs. This indicates that, despite not representing direct savings, smart meters help final users to be more aware of the actual cost (or waste) of energy so they may be able to plan domestic consumption more efficiently. Furthermore, this more accurate monitoring of energy consumption also allows energy companies to be able to make more efficient and stable supply (Faruqui et al. 2010).

7.7.2 Smart Meter Criticism

As discussed above, smart meters do not improve energy efficiency; they simply encourage users to be more careful with their energy usage by giving them more accurate, real-time consumption information. However, some authors argue that some other factors need to be considered in a wider picture. Other factors that may play an important role when modifying user behaviour as discussed in Carroll et al. (2014), where a study of these factors was carried out including the use of different rates, the frequency of billing, or the fact of receiving information regarding the specific consumption of certain appliances.

Because of the difficulty in linking the use of these smart meters with real energy savings, there exist cases in which the various stakeholders wonder whether it is profitable to adopt this technology. Due to this, governmental plans to introduce this technology in the households may become much more challenging. In

Stromback et al. (2011), a thorough investigation of pilot programs was carried out in the Netherlands to identify the most relevant aspects to consider for the success of programs for the introduction of smart meters in homes.

7.7.3 Energy Generation Through Renewables

Home energy generation, also known as micro-generation or urban energy generation consists of autonomous domestic power generation systems mainly based on renewable energy technologies. Since the latest energy crisis and the steady rise in prices, users now consider these alternatives as a real option for energy savings and improved sustainability. Recent advances in renewable generation techniques, which have led to better and more efficient generation systems, are also responsible for the increasing popularity of these technologies.

On the other hand, it has been shown that distributed power generation offers improvements over traditional decentralized generation in terms of efficiency and dynamism (Alanne and Saari 2004) of the supply system. Therefore, domestic generation can have a major role in the generation scheme, especially now that the trend is to implement smart grids that can fully exploit the advantages of decentralization that represents the generation in small houses.

Apart from the stated advantages, these approaches minimize the risk of shortages and are able to leverage in some cases the residual energy from generation (Allen et al. 2008). For example, if water-cooling systems for generating turbines are used, this water can be recirculated through exchangers coupled to deposits that can store surplus heat to convert it into hot water. Although these systems have the potential to reduce energy consumption, users remain reluctant to embark on such projects due to the high costs of installation (Scarpa and Willis 2010). The most common technologies in terms of domestic generation are:

- Solar energy
- Small wind turbines
- Biomass
- Geothermal.

Solar Energy

Solar energy can be harnessed through domestic systems in two ways: using thermal systems and photovoltaic panels (PVs). Solar thermal energy involves capturing solar energy through solar panels containing circulating water inside. The heat collected using these panels is transmitted directly to the water, which will be conducted through a heat exchanger to transfer the heat into a circuit of water and to a storage tank thereafter. The storage tank needs to be particularly well thermally insulated to conserve the high temperature water in times when no sunlight is available.

Meanwhile, solar photovoltaic systems are used to generate electricity from solar radiation. The energy of the photons is used to induce a current in the panels, which is regulated and transformed by power converters and stored in batteries for later use. When generation levels are high, the converters can adapt the characteristics of the electricity generated to those of the main network. Thus, users who generate surplus energy can inject it to the network in exchange for an economic remuneration by the electricity supply companies. The photovoltaic system has the drawback of involving a complex installation and a substantial capital investment. In addition, maintaining the conversion units and accumulators can also result in significant expenditures. However, they have the advantage that can store electricity which can be used at the best convenience (no need to wait for good solar radiation) and even the excess can be sold to power companies; thus improving the investment, which in any case occurs after a few years of savings through this kind of generation.

Solar thermal generation is simpler than PV and therefore less expensive. However, its use is limited to obtaining hot water, and cannot be used to power electric devices like the case of photovoltaic. Moreover, in both cases a significant surface for installation of the panels is needed, although this is improved with the development of integrated panels and more efficient solar cells of smaller area.

Small Wind Turbines

These turbines are the equivalent to a small model of large wind farms installed in large landscapes. Electricity is produced by a wind-driven generator. In this case, similarly to photovoltaic solar energy, electricity is generated directly. The conversion and storage system in this instance is also similar to PV. This technology is still under development and to be able to amortize the design and implementation of such systems, it needs to generate large volumes of energy; unfortunately, small turbines have a moderate regime generation (Bahaj et al. 2007).

Nevertheless, they have the advantage that its installation is somewhat less expensive than PV and wind flow does not have necessarily to stop when night falls. However, these turbines need to be in high places and require some strength in the wind to start generating.

Biomass

These systems are based on boilers that instead of fossil fuels burn biological materials (pellets) for heat production, balancing the emissions produced by burning these materials by saving the emissions from extraction processes and processing required by other fuels. This type of energy is clean and uses relatively cheap materials. However, it presents the problem that the provision of necessary organic matter is still not as representative and usual as the traditional oil and gas companies in addition to fuel prices being more expensive.

Geothermal

This method exploits the advantages of having available water at low temperatures a few meters below the soil. By extracting this cold water, energy can be saved directly in air conditioners and refrigeration (no need to cool this water down). In addition, water remains at a constant temperature of around 10 °C, therefore heat production equipment based on refrigerants can improve the refrigerant fluid evaporation (typically at temperatures about 0–3 °C) thanks to the fixed temperature. In addition, gas-water temperature exchange is more efficient than traditional gas-air exchange typical of climatisation and air conditioning equipment. The obvious disadvantage of these systems is that it requires a place to dig the holes and installation can also be costly and complex. However, this type of energy is clean and efficient once the system is operating.

7.7.4 Smart Home Initiatives

The term Smart Home (or Home Automation) has gained a lot of popularity in recent years. We could define the concept of Smart Homes as those residences where a domestic control is responsible for integrating all the home automation systems (Robles and Kim 2010). The Department for Trade and Industry (DTI) in the UK defined technology as "A dwelling incorporating a communications network that connects the key electrical appliances and services, and allows them to be remotely controlled, monitored or accessed" (DTI 2007).

This means that users can access and monitor information on their house via meters and sensors and can also act directly on the various systems (e.g. lighting, HVAC, windows/blinds, multimedia, appliances…). Although this technology is not new, in recent years it has become particularly relevant because, among others, developments in the Internet Of Things, communication protocols, wireless networks and the development of computing techniques (i.e. agent-based and machine learning) for decision-making and regulation of the different systems.

7.7.5 Smart Home Elements

We could say that a Smart Home consists primarily of the following subsystems (Li et al. 2004).

Smart Home Networks

The technology under these types of networks is based on standardized communication systems that allow all devices on the network to communicate with the central control from which the regulation is done either automatically or directly through orders executed by users. There are several types of network based on the use of electrical circuits in the house for communication; others rely on wireless connections and radio-frequency signals. Some of the most common communication protocols, depending on the means we use to design the network, are KNK, ZigBee, INSTEON or Wavenis (Gomez and Paradells 2010). These protocols differ usually in the physical layer, transmission rates, latency, etc. making them more or less suited depending on the scenario of choice.

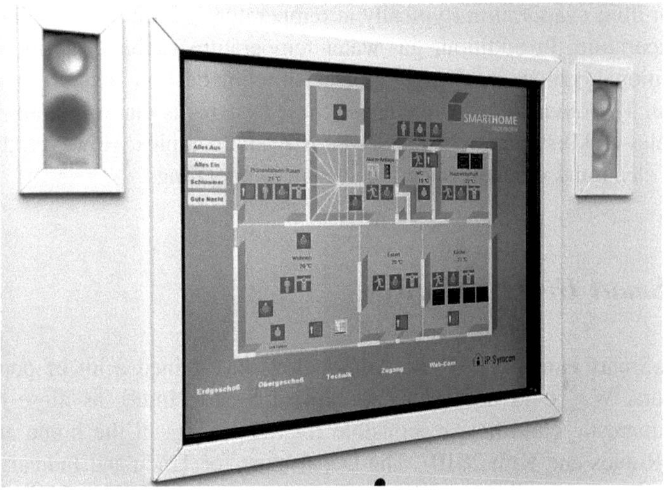

Fig. 7.17 Example of interface (EnergieAgentur.NRW 2014)

Sensors and Actuators

The sensors are directly responsible for monitoring physical aspects that occur in smart environments such as movement, temperature, sound or actual energy consumption (smart meters). Actuators refer to motors or mechanisms that operate certain components of housing that cannot be integrated directly into the system (e.g. windows and doors).

Smart Devices

This includes an increasing number of different devices. Among them are smart homes, smart TVs and other electronic devices that can be equipped with gateways to communicate with the smart network. This is what is known as the Internet of Things (IoT).

Interfaces and Controllers

There are touch screens, thermostats and dedicated controllers, which have been used for controlling smart environments where the information is collected and displayed so as the user is able to decide whether to act on any of the components. Recently, however, devices such as phones or tablets can perform all the functions of home/user interaction without requiring a dedicated exclusive device. Controllers on the other hand, are elements responsible for regulating and managing the system based on user decisions and the state of the different elements of the home network (including the various gateways necessary for the proper functioning both internally and externally through the internet). These devices can be PC computers or devices specifically designed for these tasks. Example in Fig. 7.17.

7.7.6 *Smart Home for Energy Savings*

Although Smart and Pervasive Environments can be used for different purposes such as multimedia control, healthcare or security; energy saving is one of the most attractive features these systems present. Smart homes have always faced the obstacle of being pricey and considered as a dispensable luxury. Due to these considerations, smart environments need to prove themselves as a good investment to increase their chances of success among the most reticent potential customers. Many approaches have been proposed recently to demonstrate how this technology enables dwelling to be more energy efficient, improve user experience and comfort as well as sustainability of these systems in the long run. In the work in Pedrasa et al. (2010) a decision-support was proposed to study the energy consumption in homes through the home automation network to manage the different power sources (also integrated in the home automation) more efficiently. Other example is the patent proposed in Bedros et al. (2011), where they propose a system combining energy-efficient construction and equipment with renewable energy systems to achieve a near-zero-energy NZE home design. In the work in Kailas et al. (2012), there is a discussion on different Home Energy Management (HEM) systems and their role in the final performance of the dwellings and, in turn, of the benefits for smart grids where these homes can be integrated.

7.7.7 *Standby Energy Consumption*

Standby energy consumption is a term that refers to the energy used when determined electric/electronic devices are off, but not disconnected from the mains.

Fig. 7.18 Type of switched extension cord that can be used to save standby energy by totally disconnecting every socket (US DOE 2015)

Certain electronic apparatus and appliances have parts like CPUs and internal clocks that need a small amount of power for its optimal functioning even when in off mode. In addition, equipment containing electrical transformers always has a small consumption as iron and copper loss (about 7 % of the total).

Nevertheless, there are devices that are particularly problematic in this context: network-enabled devices. These devices are constantly monitoring for signals and even though they may be on 'standby' mode, their actual consumption would be close to the nominal value either way. Furthermore, the trend is that the number of such devices will grow dramatically in the coming years. In fact, it has been reported that in 2013 the total estimated number of this type of device was of 14 billion units; and predictions are that this figure will reach 50 billion in 2020 even increasing to 100–500 billion in 2030 and 2050 respectively (Spears 2013). To make things even worse, there is also the problem that certain devices such as computers are left unattended on many occasions (though using low-energy mode or similar), which makes this problem considerably worse. To counter this problem, certain solutions can be adopted. For instance, the US government is conducting campaigns to raise awareness among users that it is appropriate to leave each electronic equipment completely disconnected (see Fig. 7.18). To this, the US bodies recommend the use of switched extensions cords (see figure) that allow total disconnection from all devices connected at that socket/s (Burgett 2015). According to this study, simply by adopting this measure, up to 3.7 % of total energy consumption in the country could be saved up, which is equivalent to about 282 kWh/year of savings.

Another approach proposed in order to tackle this issue is the system described in Byun et al. (2013), where the proposed system studies the consumption profiles of users in conjunction with the standby energy used in the house. This solution is based on the use of the resources of households equipped with home automation systems and intelligent electronic extension cords; enabling to automatically interrupt the power supply when it detects that its use is not necessary.

Although the potential savings regarding energy consumption in standby may seem relatively small, the total amount of devices globally represents a huge quantity of energy, with the potential to not be wasted simply by using practical solutions and changing small daily routines.

7.8 Conclusions

When analysing energy use in housing, it is important to have a good understanding of its breakdown and usage trends, and therefore the relationships between the demand side management and its associated carbon emissions as well as the supply side perspective.

The chapter provided an overview of, and a basic ability to understand the energy use in housing in order to help design and retrofit housing projects toward reduction of energy demand and/or energy end-use efficiency. Moreover, energy benchmarks and best practice guidelines have to be studied for further understanding of the local context and issues like fuel poverty and interventions for adopting energy conscious behaviour. Nowadays, to design a smart home, the integration of renewables and smart technologies is not only essential but also very effective to support demand side management including operation and maintenance.

References

Alanne K, Saari A (2004) Distributed energy generation and sustainable development. Renew Sustain Energy Rev 10(6):539–558

Allen SR, Hammond GP, McManus MC (2008) Prospects for and barriers to domestic micro-generation: a United Kingdom perspective. Appl Energy 85(6):528–544

Autodesk (2015) Sustainability workshop. http://sustainabilityworkshop.autodesk.com/

Bahaj AS, Myers L, James PAB (2007) Urban energy generation: influence of micro-wind turbine output on electricity consumption in buildings. Energy Build 39(2):154–165

Bedros SJ, Markham T, Plocher T, Shetty P, Ekambaram T Mohammed N (2011) Context-aware smart home energy manager. Google Patents. US Patent App 12/852,690

Burgett JM (2015) Fixing the American energy leak: the effectiveness of a whole-house switch for reducing standby power loss in U.S. residences. Energy Res Soc Sci 6:87–94

Byun J, Park S, Kang B, Hong I, Park S (2013) Design and implementation of an intelligent energy saving system based on standby power reduction for a future zero-energy home environment. Consum Electron IEEE Trans 59(3):507–514

Carroll J, Lyons S, Denny E (2014) Reducing household electricity demand through smart metering: the role of improved information about energy saving. Energy Econo 45:234–243

Cambridge Econometrics & Verco (CEV) (2012) Jobs, growth and warmer homes—evaluating the economic stimulus of investing in energy efficiency measures in fuel poor homes. Consumer Focus, London

DEFRA (2009) Adapting to climate change—UK climate projections. Department for Environment, Food and Rural Affairs, London

DECC (2013a) Annual report on fuel poverty statistics 2013. Department of Energy and Climate Change, London

Department of Energy and Climate Change (DECC) (2013b) Fuel poverty report—Updated August 2013. London

Department of Energy and Climate Change (DECC) (2015). Smart meters, Great Britain, quarterly report to end March 2015. United Kingdom Government. Online. https://www.gov.uk/government/uploads/system/uploads/attachment_data/file/433643/Smart_Meters_Quarterly_Statistics_Report_Q1_2015.pdf. Accessed 23 Nov 2015

Dong B, Andrews B (2009) Sensor based occupancy behaviour pattern recognition for energy and comfort management in intelligent buildings. In: 11th international building performance simulation association (IBPSA) conference

DTI (2007) Meeting the energy challenge a white paper on energy. Department of Trade and Industry, Editor 2007, The Stationery Office, Norwich

Energy UK (2015) http://www.energy-uk.org.uk/policy/fuel-poverty.html

EnergieAgentur.NRW (2014) Deutsch: 20. Station der Zukunftsenergientour: Energieeffizienz-Projekte SmartHome in Paderborn. https://commons.wikimedia.org/wiki/File:20._Station_der_Zukunftsenergientour-_Energieeffizienz-Projekte_SmartHome_in_Paderborn_(12100915513).jpg

Erickson V, Cerpa A (2010) Occupancy based demand response HVAC control strategy. In: Proceedings of the 2nd ACM workshop on embedded sensing systems for energy saving

EU Fuel Poverty Network (FPN) (2013) http://www.fuelpoverty.eu

Faruqui A, Sergici S, Sharif A (2010) The impact of informational feedback on energy consumption—a survey of the experimental evidence. Energy 35(4):1598–1608

Florida Solar Energy Center (FSEC) (2015) University of Central Florida. http://www.fsec.ucf.edu/en/

Gomez C, Paradells J (2010) Wireless home automation networks: a survey of architectures and technologies. Commun Magazine IEEE 48(6):92–101

Green Mountain Energy (GME) (2015) New insights on where your home's energy use is going. http://www.greenmountaincleanertimes.com/2013/04/05/new-insights-on-where-your-homes-energy-use-is-going/

Han L, Altan H, Noguchi M (2012) Promoting energy conscious behaviour of occupants through monitoring energy usage patterns. In: Proceedings of 2nd international conference on zero energy mass customized housing (ZEMCH), Glasgow, UK

Henry JC, Peter RS, Chow WK (2004) Dictionary of architectural and building technology. Spon Press, Routledge

Hills J (2012) Getting the measure of fuel poverty—final report of the fuel poverty review (PDF) (Report) (1st edn). Department of Energy and Climate Change. Retrieved 28 Oct 2013

HSE (2015) Health and safety executive. www.hse.gov.uk

IWU (2014) IEE project EPISCOPE. http://www.episcope.eu

Li J, Liu D, Yang B (2004) Smart home research. In: Proceedings of the third conference on machine learning and cybernetics SHANGHAI, pp 659–664

Kailas A, Cecchi V, Mukherjee A (2012) A survey of communications and networking technologies for energy management in buildings and home automation. J Comput Netw Commun 2012

Mapawatt (2015) Outside temperature and energy consumption. http://mapawatt.com/2009/09/13/outside-temperature-and-energy-consumption

Masoso O, Grobler LJ (2009) The darkside of occupants' behaviour on building energy use. Energy Build 42(2):173–177

MES College of Engineering (COE) (2015) Types of lamps. http://www.slideshare.net/

Milsom E (2014) Solid wall heat losses and the potential for energy saving: literature review. Technical report, DECC, Building Research Establishment Ltd., (BRE) UK

Newsham G, Mahdavi A, Beausoleil-Morrison I (1995) Lightswitch: a stochastic model for predicting office lighting energy consumption. In: 3th European conference on energy efficient lighting, pp 60–66

Page J, Robinson D, Morel N, Scartezzini J-L (2008) A generalised stochastic model for the simulation of occupant presence. Energy Build 40:83–98

Pedrasa MAA, Spooner TD, MacGill IF (2010) Coordinated scheduling of residential distributed energy resources to optimize smart home energy services. IEEE Trans Smart Grid 1(2):123–143

Pew Center on Global Climate Change (PCGCC) (2009) Climate techbook: building envelope. Available at http://www.pewclimate.org/technology/factsheet/BuildingEnvelope#8. Accessed 23 Nov 2015

Reinhart C (2004) Lightswitch-2002: a model for manual and automated control of electric lighting and blinds. Sol Energy 77(1):15–28

Robles RJ, Kim T (2010) Applications, systems and methods in smart home technology: a review. Int J Adv Sci Technol

Scarpa R, Willis K (2010) Willingness-to-pay for renewable energy: primary and discretionary choice of British households' for micro-generation technologies. Energy Econ 32(1):129–136

Seryak J, Kissock K (2000) Occupancy and behavioural affects on residential energy use. In: Proceedings of annual conference on American solar energy society

Spears J (2013) Standby energy could power Canada. Proquest technical report 2013

Stromback J, Dromacque C, Yassin MH, Vaasa ETT, Global Energy Think Tank (2011) The potential of smart meter enabled programs to increase energy and systems efficiency: a mass pilot comparison Short name: Empower Demand. Vaasa ETT

Switch (2011) 6.3 Million or almost a quarter of all households are now in fuel poverty. 07 June 2011

Technology Green Energy (TGE) (2015) Energy efficient houses. http://technology-green-energy.blogspot.in/2013/10/energy-efficient-houses.html

TERI (2011) GRIHA manual vol 3: technical Manual for trainers on building and system design optimization renewable energy application. Ministry of New and Renewable Energy, Government of India and the Energy and Resources Institute, New Delhi

UKACE (2015) http://www.ukace.org/wp-content/uploads/2014/02/ACE-and-EBR-fact-file-2014-02-Fuel-Poverty-update-2014.pdf

UK National Renewable Energy Centre (NAREC) (2000). Warm homes and energy conservation act (Report). UK National Renewable Energy Centre (NAREC). Retrieved 28 Oct 2013

UK National Renewable Energy Centre (NAREC) (2013) NAREC distributed energy (2013), "4", ERDF social housing energy management project—final project report (PDF) (report) (1st edn), p 10. Retrieved 28 Oct 2013

US Department of Energy (DOE) (2015) Types of homes | Department of Energy. Energy.gov. N. p. Accessed 23 Nov 2015

Webb T (2008) Fury as fuel poverty soars close to a 10-year record. Observer 20. Retrieved 09 May 2008

Whole Building Design Guide (WBDG) (2015) National Institute of Building Sciences, an authoritative source of innovative solutions for the built environment. http://www.wbdg.org/

Vanquis Money Management News (VMMN) (2011) Bills 'cost 27 % of household income', 16 Nov 2011

Vimmr V (2013) Tabula—residential building typologies in 12 European countries—good practice example from the Czech Republic. Central Europe towards sustainable building 2013

Yamaguchi Y, Shimoda Y, Mizuno M (2003) Development of district energy system simulation model based on detailed energy demand model. In: 8th International IBPSA conference

Chapter 8
Passive Design

Hasim Altan, Mona Hajibandeh, Kheira Anissa Tabet Aoul and Akash Deep

Abstract This chapter reviews passive design strategies and the benefits of using such strategies not only in the building design but also related to the urban context and human factors, which would be linked to urban sustainable design, policies and strategies. The key to designing a passive building is by taking advantage of the local climate (micro-climate) and therefore, climate characteristics and classification can help with identifying approaches as early as site planning and analysis. Therefore, climate and comfort are the two fundamental measures in passive design that require attention. Passive design is a major part of environmental design, and approaches utilising several techniques and strategies that can be employed to the buildings in all types of climates around the world such as orientation, ventilation, shading devices, thermal mass, insulation, daylighting and so on. These techniques and strategies can also be supported by various other parameters such as using technologies (passive and/or active) and customisable controls as well as enhanced by patterns of biophilic design for improving health and well-being in the built environment. There are also passive solar technologies including direct and indirect

H. Altan (✉)
Department of Architectural Engineering, University of Sharjah (UOS), Sharjah, UAE
e-mail: haltan@sharjah.ac.ae; hasimaltan@gmail.com

H. Altan
Welsh School of Architecture (WSA), Cardiff University, Cardiff, UK

M. Hajibandeh
Erasmus University Rotterdam, Rotterdam, The Netherlands
e-mail: monahaji@gmail.com

M. Hajibandeh
Tehran University of Art, Tehran, Iran

K.A. Tabet Aoul
Department of Architectural Engineering, United Arab Emirates University (UAEU), Al Ain, UAE
e-mail: kheira.anissa@uaeu.ac.ae

A. Deep
GRIHA Council, The Energy and Resources Institute (TERI), New Delhi, India
e-mail: akashdeep@grihaindia.org

© Springer International Publishing Switzerland 2016

M. Noguchi (ed.), *ZEMCH: Toward the Delivery of Zero Energy Mass Custom Homes*, Springer Tracts in Civil Engineering,
DOI 10.1007/978-3-319-31967-4_8

solar gains for space heating, solar water heating systems, solar cookers, use of thermal mass and phase-change materials for slowing indoor air temperature swings, solar chimney for enhancing natural ventilation, and earth sheltering that can be considered as part of the actual design. Today, passive design strategies can be easily evaluated with the use of either simple or more sophisticated Building Performance Simulation (BPS) tools such as Ecotect, IES VE, etc.

8.1 Introduction

The design that maintains a comfortable temperature within the building using the climate and natural elements to get the optimum benefit and to reduce or eliminate the independence on mechanical systems for heating, cooling and lighting, is called 'Passive Design'. There are two crucial measures that should be considered for passive design to be beneficial and effective: climate and comfort. The passive design building and urban benefits are explained in the following sections as well as the benefits of passive design on human health and wellbeing are discussed. Many environmental characteristics affect human health, wellbeing and social life.

8.2 Passive Design, Strategies and Site Planning

Passive design is about taking advantage of natural energy flows to maintain thermal comfort. It is about using the appropriate building orientation, building materials and landscaping. The buildings should be properly oriented and the fabric of the building envelope should be specified to prevent or minimize heat gain. Shading also should be provided to minimize solar radiation (Agboola 2011). These techniques and strategies can also be supported by various parameters such as using technologies (passive and/or active) and customisable controls as well as enhanced by patterns of biophilic design for improving health and well-being in the built environment.

When designing a home, the building envelope act as a barrier between natural climate and virtual climate to meet the human comfort level. Some of the major factor, which governs human comfort, is visual, thermal and acoustical comfort. For attaining the desired comfort level, building envelops plays an important role along with technologies used. These technologies which help us achieve the comforts are of active nature, passive nature or hybrid nature i.e. combination of both active and passive.

Passive technologies are systems, which reply on natural resources and help us achieve comfort levels without relying on artificial energy. Choice of passive design techniques is majorly dependent on local climate where the project is located. The techniques are sustainable and use abundantly available natural resources. Integration of such techniques help transform building envelops into living organic creations to sustain human life within. This section will also illustrate components of building envelope, which help us integrate passive designing techniques to attain comfort.

8.2.1 Orientation

Buildings should be planned in such a way that benefit is obtained from shaded indoor and outdoor living areas when the weather is hot and sunny indoor and outdoor areas with wind protection when the weather is cold. Building block orientation governs the passive technologies implementation within the design. It also governs the window sizing and locations, which will affect both lighting space conditioning within a building. Proper orientation can lead to significant reduction in lighting and space conditioning load if coupled with passive design technologies.

Well-designed buildings should be oriented, and the spaces arranged in such a way, that the majority of rooms face towards the equator. In this way, the eastern and western sides are exposed to the low-angle summer sun in the morning and afternoon. The high angle of the sun in the sky in summer makes it easy to shade windows using only a generous roof overhang or horizontal shade. The longer north/south sides of the building benefits from the low angle sun in winter. The roof overhang or shading on the equator side should allow the Sun to shine into the building when its warmth is required in winter, and provide adequate protection from high-angle Sun in summer (Aksoy and Inalli 2006).

If the majority of windows are designed into the equator-facing wall, sun penetration into the building will be maximised. Living areas should be sited to gain maximum benefit from cooling breezes in hot weather and shelter from undesirable winds in winter. This does not mean that the orientation of the building should be varied from north towards prevailing breezes, as it does not have to face directly into the breeze to achieve good cross-ventilation (Mingfang 2002).

8.2.2 Building Shape (Massing)

Well-designed passive building produces less air pollution and greenhouse gasses, and thus it contributes to a more sustainable environment. Good passive buildings not only conserve energy, but also account for hidden environmental benefits. If we are to build a sustainable future for this planet, architects need easy-to-use design tools to help them visualize the many environmental consequences of their design decisions (Littler 1982).

There are a lot of microcomputer design tools; however, a user-friendly one that calculates the environmental costs of each architectural design decision, and then displays an easy-to understand picture of how these costs change from one design to the next is a new version of SOLAR-5 (Milne 1995).

The great advantage of this user-friendly design tool is that the architect can instantly go back and re-design the building and can immediately see if the air pollutant emissions, energy consumption, or operating costs, have been reduced. With this knowledge, architects are in a better position to make informed judgements about the design and operation of their buildings. The increasing complexity in design and performance evaluation of buildings has resulted in the need for the use of computational building performance evaluation and design support tools throughout the process (Hien et al. 2000). It is hope that the development will become an impetus for building designers to utilize the tools for a performance evaluation of their design.

Site planning is an integral part of passive designing. Each components placement result in the governing the microclimate generated around the site. Zoning and massing help to achieve desired microclimate in different climates.

Massing of the building blocks help achieve thermal and visual comfort levels if designed as per the climatological requirements. Building blocks channelize or obstruct the wind flow; they also act as shading devices for surroundings. Building blocks design and geometry can influence the wind flow and velocity. Massing of blocks can help regulate the summer wind and achieve ventilation, and obstruct wind flow in winter season. Figure 8.1 explains some examples of massing to channelize wind flow.

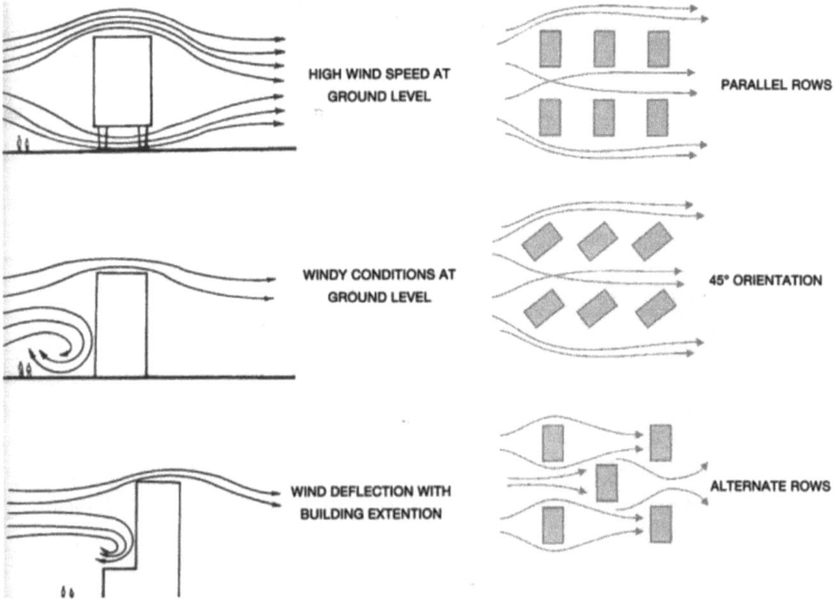

Fig. 8.1 Massing of building blocks influence wind pattern (Pedata 2011)

8.2.3 Choice of Materials

Choice of material depends on the outside climate around the building. However, the properties of material used, which governs their usage, can be noted as three different properties i.e. colour, insulation property and assembly type. Finishes colour will vary the amount of heat and light absorbed and reflected. Lighter the colour greater reflectivity while darker the colour more absorbing property. In addition, the insulation property plays an important role in material selection. Good insulation is required to reduce heat exchange between the internal and external space.

Building components are the key components, which integrate passive design features. Building materials can be classified as two types—visible and invisible. Visible category includes the finishes both internal and external while the invisible include the structural and non-structural. Orientation of a building governs the opening size and locations.

8.2.4 Landscaping (Heat Island Effect)

Heat capacity is a measure of how much heat different types of material can hold. For a given building element, it is found by multiplying the density of its material by its overall thickness, and then by its specific heat. Specific heat is the amount of heat a material can hold per unit of mass. The greater the specific heat, the more energy is required to heat up the material. Thermal mass is basically the ability of a material to store heat (Thani et al. 2012). It can be easily incorporated into a building as part of the walls and floor. Thermal mass affects the temperature within a building by:

- Stabilising internal temperatures by providing heat source and heat sink surfaces for radiative, conductive and convective heat exchange processes.
- Providing a time-lag in the equalisation of external and internal temperatures.
- Providing a reduction in extreme temperature swings between outside and inside (Shashua-Bar et al. 2009).

Material selection to capitalise on thermal mass is an important design consideration. For instance, heavyweight internal construction (high thermal mass) such as brick, solid concrete, stone, or earth can store the Sun's heat during winter days, releasing the warmth to the rooms in the night after it conducts through. Lightweight materials such as plasterboard and wood panelling are relatively low mass materials and will act as insulators to the thermal mass, reducing its effectiveness. Lightweight construction responds to temperature changes more rapidly. It is therefore suitable for rooms that need to heat or cool very quickly (Kleerekoper et al. 2012).

For maximum energy efficiency, thermal mass should be maximised in the equator-facing sides of a building. Any heat gained through the day can be lost through ventilation at night. In using this technique, the thermal mass is often referred to as a 'heat bank' and acts as a heat distributor, delaying the flow of heat out of the building by as much as 10–12 h. Thermal mass design considerations include:

- Where mass is used for warmth, it should be exposed to incident solar radiation.
- Where mass is required for cooling, it is better placed in a shaded zone.
- Buildings may be preheated using electric or hot water tubing embedded in the mass (mostly concrete floors).
- Buildings may be pre-cooled using night-purge ventilation (opening the building up to cool breezes throughout the night), although this requires significant amounts of exposed mass, and may be necessary only at certain times of the year.
- Thermal mass is particularly beneficial where there is a big difference between day and night outdoor temperatures (Kleerekoper et al. 2012).

Both heat capacity and conductivity give rise to what is known as the thermal mass effect. In large heavyweight materials, it can take a significant amount of energy to heat up their surface. This is because much of the energy is actually absorbed deeper into the material, being distributed over a larger volume. With a lot of energy incident on the surface, this absorption can continue until it travels through its entire width, emerging on the inside surface as an increase in temperature. This conduction process can take a significant amount of time, sometimes in the order to 10–12 h with a thick masonry wall. If the energy incident on the outside surface fluctuates, this can set up 'waves' of temperature flowing through the material (Shashua-Bar et al. 2009).

Landscaping elements play critical role in defining microclimate of the site. Elements like amount of hard paving placement of water bodies, placement of shading trees, orientation and location of building blocks etc. Amount of hard paving will affect the heat trapped around the building resulting in heat island effect. In addition, it will increase the run from site thus reducing the dampness around site due to lack of percolation. Water bodies can as natural coolers for hot and dry climate. Where the moisture from the water bond can be used to cool the hot wind blowing above and thus used to cool the building block.

Tree and shrub plantation plays an important role in defining wind flow and also help achieve mutual shading on building and hard paving. They act as sound barrier to obstruct the noise entering in the site when placed around the site boundary and provide shading when placed near the building envelop. However, the height of the tree is a restriction about the number of stories, which can be benefited with this landscaping design feature (Fig. 8.2).

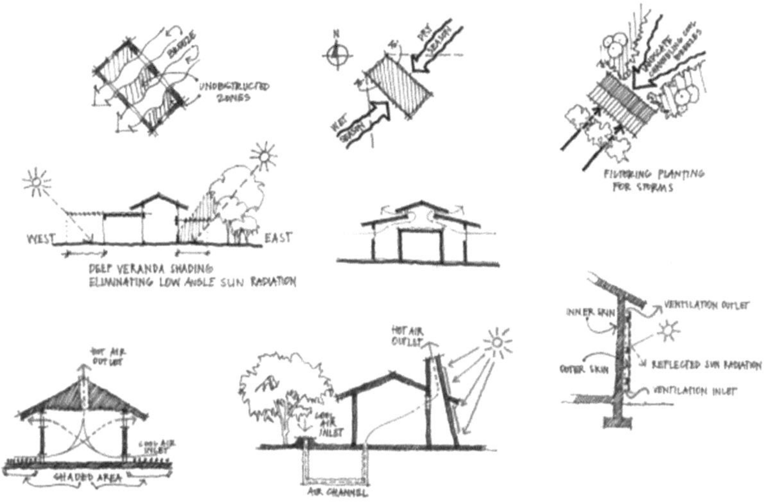

Fig. 8.2 Passive cooling (Amble Resorts 2009)

8.2.5 Water Bodies and Vegetation (Micro-climate)

The key to designing a passive building is by taking advantage of the local climate (micro-climate) and therefore, climate characteristics and classification can help with identifying approaches as early as site planning and analysis.

The modification of urban temperature through landscape approach can be achieved by incorporating sustainable landscape design practices via the interplay of natural vegetation in the hot-humid (Thani et al. 2012).

The knowledge of landscape architecture can be widely utilised to provide design solution to improve many environmental setbacks. Due to its ability to interact between built and natural elements; as well as an interdisciplinary profession, hence, landscape architects have a huge potential to modify the effects of extreme climatic conditions (Brown 2011) through adoption of environmentally-responsive design that incorporate bioclimatic aspects to outdoor environments.

The augmentation of cooling potential in larger scale needed in contributing to the urban thermal modification. Design implication is at early stage of urban and landscape planning where it should consider spatial planning of all natural landscape units and human aspects (Selman 2006) especially for a new development; by linking the landscape and climate of locality. These include the needs to ponder legislative intervention at planning level from relevant authorities (Thani et al. 2012).

In order to design a climate-sensitive city, the development must endeavour to accommodate the adoption of environmentally responsive design that integrates bioclimatic aspects to outdoor environments. The term "bioclimatic" refers to the relations of climate and life, particularly the effects of climate on living things

(Thani et al. 2012). Thus, the bioclimatic aspect in designing an urban landscape is necessary due to its strong relationship with climate, nature and living environment.

To encourage good stewardship and responsible use of natural resources, it requires pertinent execution by policy and legislation. It should be applied as part of the decision-making process in the provision of urban planning stage (Eliasson 2000). In the planning phase, environmentally sensitive area like forest and natural water body as bioclimatic component should be determined. Its development should be conducted as stipulated in the Environmental Quality Act and follow Environmental Impact Assessment (EIA) regulations (Shaharuddin et al. 2009).

The role of vegetation in microclimatic design strategy are varies and can be applied in building-scale, street level or urban and semi-urban level (Thani et al. 2012). By manipulating the benefits from its evapotranspiration process and morphological characteristics, trees whether isolated or planted in-group are always the best urban cooling mechanism. As a great climatic moderator, the use of vegetation can contribute many thermal benefits like providing shade effect, lowering ground and air temperature, reduced solar infiltration, ventilation effect and minimise glare from reflection (Shashua-Bar et al. 2009).

Selection of vegetation should consider more native and ecologically restorative rather than ornamental types and choice of species should reflects the bioclimatic requirements. In addition, the uses of native vegetation will facilitate the adaptation to local climate and reduce the cost of maintenance. However, to keep the aesthetical value of urban beautification purposes, landscape architects can creatively integrate both the ornamental plants and native ones as long as it is undertake the objective to reduce urban temperature.

There are growing interest observed on the design mechanism of water bodies as climatic regulator among urban designers due to its cooling and thermal stabilizer effects in urban environment. The significance of water bodies for urban climate modification includes its influence on transformation of latent and sensible heat fluxes, an "oasis effect" by its evaporation and reduced surface temperature of its surrounding (Thani et al. 2012).

Therefore, the varying ability of cooling effects depends on the proportions of water body area and the threshold with landscape environments. This feature is much more crucial to be put in consideration while incorporating water elements into design.

Seen from landscape architectural view, regardless of whatever climatic region it might, mitigating strategies through design seems to be best option as it is interim adaptive response to accommodate not only climate-related issues but also to many of environmental problems.

8.2.6 Daylighting

Sun as the abundant source of light provide us with daylighting. However, the light received from sun has increased glare and intensity. The building design helps us achieve glare free day light in optimum lux levels. Window orientation and shading

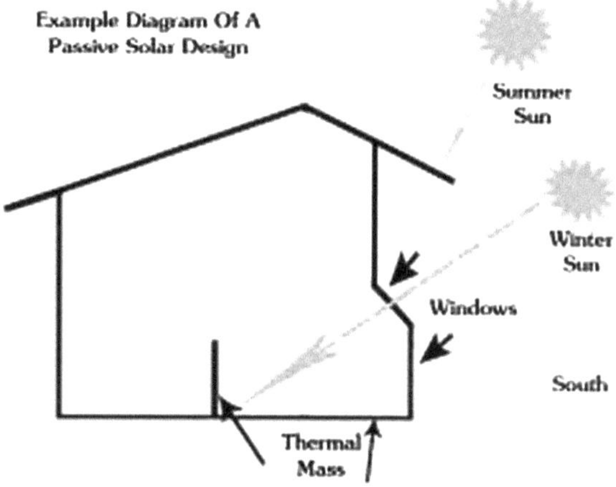

Fig. 8.3 Passive solar design (Flagstaff 2015)

device design are two key passive features, which govern the day lighting integration in the building design. Solar path analysis can be done through varies simulation software to predict the performance of the shading device as per location. Suns' movement is constant i.e. east to west however during the winters the path is generally tilted towards south. Maximum solar glare and heat radiation is accompanied with the south and west sun (Fig. 8.3).

Glare Free Daylight

Glare free day is a diffused form of day light in which the direct incident sunlight is restricted and the reflected light is allowed to enter. The reflected light is diffused and is visually comfortable. North light is generally glare free. Also the morning sun i.e. east sun has early morning glare with low radiation. South and west sun light has the maximum glare.

Types of Shading

Shading design plays a crucial role in decreasing the cost of the building in multiple ways. In addition, it helps cut the direct sun and achieve diffused day light for visual comfort of the occupant. Shading size and type depends on the orientation of the window. Figure 8.4 indicates some types of fenestration designs, which can be adopted depending on the window orientation. In general, horizontal shading element works well in north and south orientation and vertical fenestration are required to cut sun from east and west. However, the trickiest design of shading is concerning the southwest and southeast orientation. As in these directions the altitude of sun is in the midrange. Hence, only horizontal and only vertical shading device will not cut the glare only a combination of both types of shading is required to cut the harsh sun.

8.2.7 Space Conditioning

Space conditioning is one of the most energy consuming services within a household. As the achievable thermal comfort zone varies with different climate

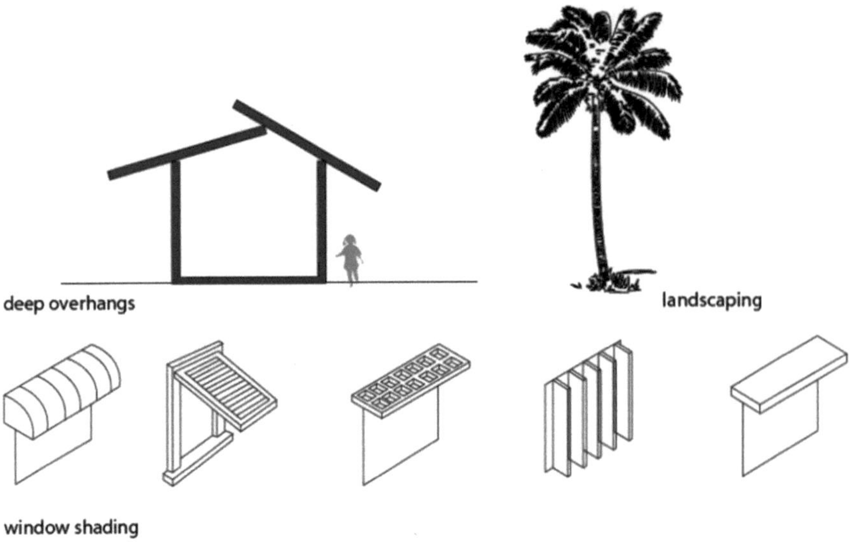

Fig. 8.4 How different openings can affect the design (Yaniv 2012)

zones across the globe i.e. with respect to the outside temperature, relative humidity and air velocity. Different passive designing techniques are associated with different climate zones for achieving respective thermal comfort. This section illustrate some passive cooling and heating techniques, and demonstrated the principle around these techniques.

Passive Cooling. Techniques that focus on restricting heat from entering or removal of heat generated within the building with use of low or no mechanical energy are considered as passive cooling techniques. Following different types of passive techniques can be used to attain thermal comfort.

Preventive Techniques. Preventive techniques combine building design and envelop design. Measures include site planning, building form and orientation, solar access and human behaviour. All of the above measures are controlled at design level and can be incorporated to reduce energy demand significantly in a household.

Natural Ventilation. Natural ventilation is the phenomenon in which the natural wind flow is controlled and circulated through the built environment with in the comfort limits. Massing and site planning influence the wind flow and intensity around the building block. Within a building block placement of opening in envelop and orientation influences the infiltration rate and flow around the occupied space. In humid climate zones, cross ventilation design provides with energy saving techniques.

Shading design also influence the wind flow within the rooms through windows. It helps channelize the wind flow and achieve the desired wind flow intensity as per human comfort. In addition, location of fenestration helps in ventilation through the space. Figure 8.5 shows some strategies with shading design and window location for managing natural ventilation.

Natural ventilation can be achieved in two different ways i.e. cross ventilation ad stack ventilation. Cross ventilation works on the principle of inlet and outlet being controlled with the window location and size, while stack effect use a mechanism of negative and positive pressure. The hot air rises and cold air settles down with the rise of hot air a negative pressure is developed around the area, to neutralize the pressure difference cold air moves towards the negative pressure thus creating a wind flow. One of a strategy to achieve stack effect in the building is solar chimney coupled with courtyard planning.

Radiative. Radiation is the property of object to absorb and emit energy. Roof is the largest surface available to sky in daytime it acts as insulation from harsh sunlight while in night-time long wave infrared radiation emitted from building is more because of the heat absorbed in day time. Designing a roof as a radiator is an effective strategy. Radiative cooling can be achieved either by direct radiant method or by indirect radiant method. In direct radiant technique, the roof is designed as a heat sink, while in indirect radiant cooling building mass acts as a heat sink.

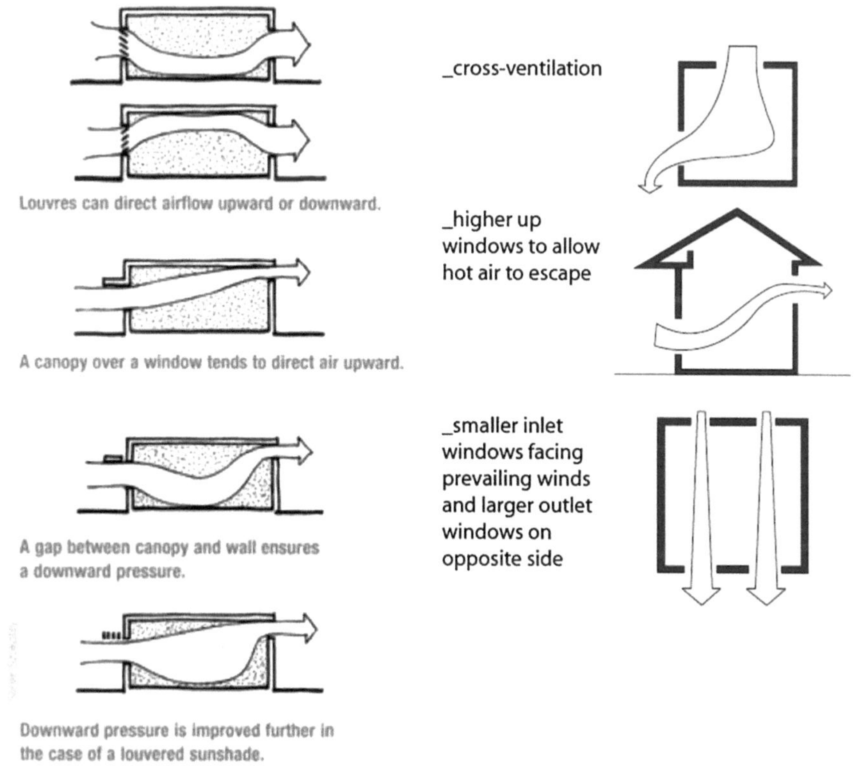

Fig. 8.5 Showing examples of shading design and window locations effecting wind flow (Yaniv 2012)

Evaporative. The systems design relies on evaporative process. Evaporative cooling largely depends on the humidity available in the air. Water in general is used to cool the blowing wind by either passing the wind on water bodies or injecting the wind with mist/water droplets.

Earth Coupling. The technique uses the constant temperature available below the earth surface to reduce the air temp significantly. One of the techniques designed on the earth coupling principle is earth air tunnel where the air is passed through along tunnel buried beneath the earth surface at around 4 m. Earth coupling can be integrated in the design by utilizing earth as sides of building envelop or by using the cooled air through duct work in the ground as cooling source.

Passive Heating. Passive heating utilizes heat from the sun. The use of thermal mass within the design of building is one such technique that can work toward passive heating.

8.3 Other Passive Design Parameters

In addition to the passive design strategies covered in the previous sections, a set of climate responsive passive design parameters can positively contribute to the cooling or heating needs of a building. The next section covers a set of optimal design strategies for cooling or heating purposes, while reviewing their applicability and benefits. In addition, this section highlights the intangible benefits of passive design strategies to human comfort and wellbeing.

8.3.1 Cooling Strategies

Earth Sheltering: Concept, Applications and Benefits

This is a design strategy, which involves covering all or part of the building envelope with earth (Figs. 8.6 and 8.7). It utilizes inherent thermal storage capacity and climate control capabilities of stable subterranean soil temperature to achieve thermal comfort in the building (Kwok and Grondzik 2007). The thermal resistance of the surrounding earth and lowered infiltration of outside air reduce the average thermal load on the building. As outdoor air temperature is significantly higher that ground temperature in summer, less heat is conducted into the house due to reduced temperature differential (Benardos et al. 2014). This design strategy is generally more suitable for temperate climatic regions but may be used in tropical regions with careful design.

Design considerations and guidelines for a successful use of earth sheltering strategy should include (a) An assessment of attitude of the public towards earth-sheltered dwellings. (b) Design of a prototype earth-sheltered house and (c) Simulation of the design for sunlight penetration. (d) An assessment of the subsurface temperature environment. (e) A quantification of the thermal advantages of this subsurface climate by computing the heat fluxes occurring in earth-contact

Fig. 8.6 Earth house. Eastate, Dietikon Switzerland (Roland 2010)

Fig. 8.7 Earth covered homes, Keldur Iceland (Chris 73 2006)

building envelopes. (f) Whole building energy consumption assessment of the designed prototype earth-sheltered house. (g) Cost implications of earth-sheltered buildings (Al-Temeemi and Harris 2004). The advantages of earth sheltering range from human to building matters. These include:

1. Energy conservation.
2. Protecting aesthetic and historic qualities of the area.
3. Achieving low visual impact on the landscape as well as freeing or "recycling" surface space.
4. Little disturbance of the surrounding environment and limited visual pollution.
5. Lower building maintenance costs due to limited exposed surfaces.
6. Better protection against weather conditions.
7. Better noise and vibration damping.
8. Underground space is the preferential setting for hosting uses and infrastructures while attaining a small footprint in terms of environmental impacts (Benardos et al. 2014).

Evaporative Cooling Systems; Concept, Applications and Benefits

This traditional passive cooling technique is a process that relies on the effect of evaporation as a natural heat sink. Operationally, when hot and dry air comes in contact with water, it begins to evaporate with the help of latent energy taken from the air. Thus, the air becomes cooler whereas its relative humidity ratio increases (Cuce and Riffat 2016). Climatic applicability of this strategy depends on the temperatures and humidity levels of the considered region (Givoni 1994).

Evaporative cooling can be classified as direct and indirect systems. Direct evaporative cooling systems involve direct evaporative cooling and cool-air downdraft to passively cool hot dry outdoor air. In addition, direct evaporative coolers are commonly used for residential buildings. In this type of evaporative cooling, the reduction of temperature is followed by an increase of moisture content. It is the simplest and oldest form of air conditioning. Typical commercial evaporative coolers have an effectiveness of 50–70 % (Santamouris and Kolokotsa 2013).

An ingenious wet porous cooling plate as a building wall is proposed in which cooling is performed via evaporation of the porous material. Outdoor climatic conditions facilitate the cooling effectiveness as well as the plate's thickness. This system leads to an average temperature of about 5–8 °C below the ambient (He and Hoyano 2010). Several designs of exist of DECs, which include: Drip Type, Spray-type, Rotary Pad and Textile-mill evaporative coolers (Minke et al. 1994).

Indirect Evaporative Cooling (IEC) Systems

These can be an alternative option in hot humid climates where the humidity is considerably high throughout the whole day. The system uses coolers where the evaporative cooling is delivered across a heat exchanger, which keeps the cool moist air separated from the room. Unlike the direct system, this does not cause an increase of the air humidity. The hot outside air is passed through a series of horizontal tubes that are wetted on the outside. As secondary air stream blows over the outside of the coils and exhausts the warm, moist air to the outdoors. Indirect evaporative cooling typically has an effectiveness of almost 75 % (Santamouris and Kolokotsa 2013). Examples of IECs include: Simple-dry and regenerative dry surfaces, Plate-type heat exchanger, Pennington heat-storage wheel, Gafford "run around", Indirect evaporative radiant and two-stage evaporative (Minke et al. 1994). The benefits of evaporative cooling includes:

1. Low-energy and cheap alternative to active cooling.
2. Can serve as a stack ventilator at night.
3. Provides a significant approach carbon emissions and energy consumption.

Night Ventilation of Thermal Mass

This covers the use of night ventilation to purge the building structure of the heat accumulated during the daytime (Beggs 2010). This refers firstly, to the thermal capacity of the materials i.e. the ability to absorb, store and release heat. However, cooling of buildings can be achieved via ventilated air in the night period. As outdoor night temperatures are usually lower than indoor temperatures in summer, cooling is possible in the building by natural night ventilation. Ventilated air enhances convective heat losses from mass elements and dissipates the released heat to the lower temperature outdoor (Balaras 1996).

It works on the principle that heavy external walls delay the heat transfer from the outside into the inside spaces in summer. In addition, higher internal mass of materials implies that the rise in air temperature is slow. The resulting gradual heating of the building is thus achieved and maximum indoor temperature is reached only during the late hours when the outside air temperature is already low. Heat originating from high thermal capacity (mass) walls inside is "removed" by ventilation in the evening and night (Shaviv et al. 2001).

Applications of Night Ventilation of Thermal Mass

In terms of heat removal capability, ventilation is at its least effective during the daytime, as temperature differential between the interior and exterior of a building is small. Night ventilation is shows more practical benefit in heavyweight buildings since the temperature differentials are much greater than during the daytime. As cool outside moves air over the surface of high thermal mass structures good heat transfer occurs. This can be done either by natural or by mechanical means. In some cases, floor voids can be used as an additional driver for air current. The inflowing air is exposed to the cooled slab and pre-cooled before its introduction to the building. Floor void allows displacement ventilation to thus be exploited for cooling (Beggs 2010).

Benefits of Night Ventilation of Thermal Mass

The advantages of this strategy may be summarised as the following:

1. In naturally ventilated buildings, thermal mass is effective for reducing the air temperature fluctuation
2. Reduction of peak heating or cooling load, and subsequently building energy consumption
3. Yields gradual build-up of heat in the interior and passive cooling at night (Yang and Li 2008).

Fig. 8.8 Earth cooling tubes, one angel square, Manchester, UK (Dixon 2012)

Earth Cooling Tubes

These are underground tubes, which bring cooled outdoor air into an interior space. Cooling and sometimes dehumidification occurs as air travels below the ground. The cooling effect is advanced when there is a significant between the temperature of the outdoor air and the soil at the tube depth (Kwok and Grondzik 2007) (Fig. 8.8).

Applications of Earth Cooling Tubes

Closed Loop System. Outdoor air is -sometimes assisted by fans- is introduced directly into the interior after passing through the tubes. Cross ventilation to draw the cooled air from the tube In and out of the indoor space. Cooled room air is circulated through the tubes back into the occupied space.

Open System. Unlike the closed system, cross ventilation to draw the cooled air from the tube in and out of the indoor space (Kwok and Grondzik 2007).

Benefits of Earth Cooling Tubes

1. Provides passive space cooling
2. Requires simple materials and skills
3. Easy to install in new or existing buildings

8.3.2 Heating Strategies

Active Solar Systems

Active Solar Thermal systems utilize a solar thermal collecting device, which may be built into the building envelope element (e.g. wall, window, shading, or roof). When integrated into the façade design, they can be categorized as Wall-based, Window-based, Balcony/sunshield based or Roof-based (Zhang et al. 2015).

Active solar space systems use collectors to heat a fluid, storage units to store solar energy until needed, and distribution equipment to provide the solar energy to the heated spaces in a controlled manner (Kalogirou 2013). They can also be integrated as part of the roof (solar integrated roof collectors). They can be used for water heating for domestic use e.g. bathing, swimming pool heating (Kwok and Grondzik 2007). They are generally more efficient than passive systems and often easier to retrofit. These systems are capable of continued operation even during a power outage, thus providing sustained comfort for users. There are three main types of active solar thermal systems.

(i) **Air systems and space heating**

In these systems, sun heats intake air in an air solar collector built into a south facing roof or wall; fans move the warm air down a gap in the outer wall or floor structure. They can be integrated into the wall system or prefabricated as Closed Air Systems or Open Air Systems.

(ii) **Ground source heat pumps**

These are essentially made up of a heat pump and a system for exchanging heat with the ground. It includes various systems that use the ground, ground water and surface water as a heat source. They utilize the energy stored in the earth's crust, transferred to the earth's surface by solar radiation. The two main factors affecting heat transfer from the ground to the collector are the collector's surface area and the thermal properties of the ground at that location. Another determining factor is the working fluid as it significantly determines the thermal characteristics of the heat pipes. It defines the temperature and transferable heat flux that are achievable during at operation points. The most important characteristics of a working fluid are its latent heat, melting and boiling point, vapour pressure, heat conductivity, wettability of the inner tube material wall and the corrosion resistance of the tube material against the working fluid (Meisel et al. 2015) (Fig. 8.9).

In terms of application of ground source heat pumps, the pipes are laid horizontally 200–400 m long about 1 m below the ground; placed to and from at about 1 m or apart, or in coils. Other forms include, groundwater wells ("open" systems), borehole heat exchangers (BHE), and "geostructures" (foundation piles equipped with heat exchangers) (Sanner et al. 2003). Benefits of ground source heat pumps include the following:

1. They provide balanced heating.
2. Appropriate for small installations, particularly for new construction.
3. Vertical systems can be used where land area is limited and require less pipe and pumping energy.

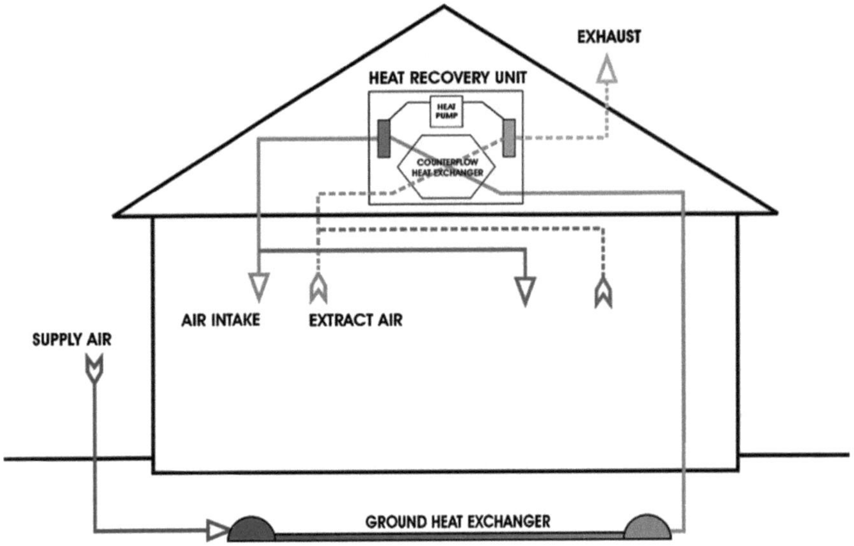

Fig. 8.9 Ventilation unit with heat pump and ground heat exchanger (Kobraklb 2012)

4. "Geostructures" can be very cost-effective as the additional cost of integrating plastic heat exchanger tubes is small.
5. When combined with the building thermal mass, can be used as a flexible load to balance supply and demand, allow the reduction of operation costs at the consumer side (Carvalho et al. 2015; Sanner et al. 2003).

Hot Water Heating Systems. A solar water heater is a combination of a solar collector array, an energy transfer system, and a storage tank. The solar collector array, which absorbs solar radiation and is responsible for conversion into heat, is the most important part. A heat transfer fluid (water, non-freezing liquid, or air), which is sent through the collector then absorbs this heat. This heat can then be stored or used directly (Kalogirou 2013) (Fig. 8.10).

Active Open-Loop Systems. These operate water for use to heat inside the collector and its circulation is performed by using the stack effect. The difference in water density due to the higher temperature of the collectors compared to the tank causes a momentum that transfers hot water up into the tank (Bessa and Prado 2015). It is also called drain-down systems and they can operate in either manual or automatic mode. It relies on two solenoid valves to drain water, two temperature sensors, a timer and a standard controller.

Closed-Loop or Active Indirect Systems. These pump a heat-transfer fluid, usually water or a glycol–water antifreeze mixture, through the solar water heater. This heats the water that then flows into the home. They are popular in climates prone to freezing temperatures. In Swimming Pools; require radiation loss covers to reduce heat loss at night. Benefits of hot water heating systems are as follows:

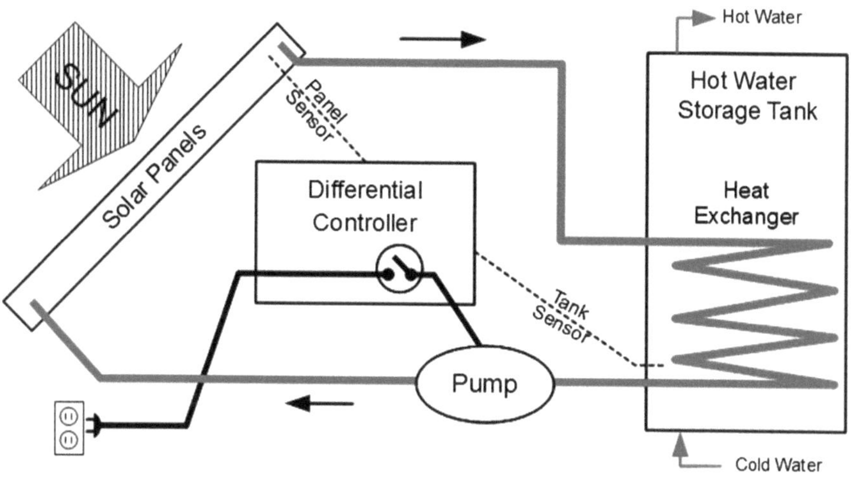

Fig. 8.10 Solar hot water basic diagram (Wikimedia 1965)

1. The solar collectors require simple construction.
2. Construction is basically inexpensive and is cost effective.
3. Characteristically, they do not require anti-freeze.

8.3.3 Passive Design and Human Comfort, Health and Wellbeing

The recourse to passive design strategies is considered fundamental to achieve a high performing building. Beyond the tangible resources' savings, there is a widespread acknowledgement that passive design contribute to human comfort, satisfaction and wellbeing. Furthermore, the benefits basis of to achieve a high performing building accepted as the Passive design The recourse to passive design strategies has been, not only because of the benefits on the building and urban scale but also because it benefits the occupants health and well-being as well. Air, water and light are the main elements that the human being depends on, and since we spend the major of our time indoors, passive design helps improve the quality of these elements to keep the building occupants healthy.

Biophilic design is one of the new design strategies that could contribute to passive design and green walls are considered to be one of the biophilic design strategies that also have a passive design effect on the building, which will help to protect the human health and improve his well-being as well as satisfy his needs to connect with the nature.

Indoor Air Quality (IAQ)

Indoor air quality could be affected by the harmful chemicals (organic compounds (VOCs) and heavy metals), gas emissions (Carbon monoxide (CO), sulphur dioxide (SO_2) and nitrogen oxides (NO X)), dust and smoke from either natural or man-made reasons such as: burning fossil fuels, agriculture activities, manufacturing industries and cleaning products (Kampa and Castanas 2008). Indoor and outdoor air pollution is one of the world's biggest killers according to the World Health Organization (World Health Organization 2014); it can cause upper chronic respiratory and heart problems along with other illnesses caused by the building air quality that were identified as Sick building syndrome (Joshi 2008), as well as deadly diseases for adults such as lung cancer asthma, and chronic bronchitis (Kampa and Castanas 2008). Add to that bad indoor air quality can affect the children too by causing acute respiratory infections, pulmonary tuberculosis and eye irritation among many other diseases (Bruce et al. 2000).

Access to Daylight and View

Beyond the provision of adequate lighting to perform the task, people consider natural light as one of the main amenities of windows and believe that it is a superior illuminant to electric lighting (Heerwagen and Heerwagen 1986). Although both daylight and sunlight provide illumination, investigation of human reactions has considered the two aspects from different viewpoints. Literature on daylight seems more concerned with the amount and quality of the light provided, whereas, the literature on sunshine concentrated on its psychological benefits.

Daylight contributes to the quality of the interior by providing a directional component as well as models objects by the horizontal direction of daylight coming from sidewall windows (Phililips 2004). Further, daylight provides variety and interest to an interior through the short-term variation of natural light as compared to the continuous and uniform artificial lighting (Heerwagen and Heerwagen 1984). It changes in intensity, direction and colour characteristics as opposed to the uniform lighting that lacks visual stimulus and variability generating boredom (Raiford 1999; Heerwagen and Wise 1998). Further, the unvarying electric light can lead to low level sensory deprivation manifested by "impairment of organized thinking, oppression and depression, confusion, suggestibility and general irritability (Aldworth and Bridgers 1971). Under a daylight-stimulating skylight, shift workers showed an improved performance of cognitively complex tasks when lighting conditions decreased (Boyce et al. 1993).

Sunlight preference on the other hand, was dependent on its general availability, the occupant's activity and the possibility to control it. The main benefits related to sunshine were expressed as a pleasurable atmosphere due to its brightness and warmth (Collins 1975; Boubekri et al. 1991), the belief that it has biological benefits which extend to the maintenance of health and wellbeing (Ott 2000; Kuller 2001). Although sunshine has been found desirable in interiors by 93 % of interviewed households, 91 % of hospital patients and 73 % of office workers (Ne'eman 1974), the importance of access to sunlight is not universal and may be influenced by the climatic and cultural context (Tabet Aoul 1989; Terman 1986; Kuller and Wetterberg 1993; Heerwagen 1998; Kim et al. 2012; MacGowan 2010).

Beyond the mere human preference of natural illumination, natural light has a number of non-visual related effects such as the synthesis of vitamin D and more importantly through the regulation of man's chronobiological system. Moderate disruption of the chronobiology, which regulates the sleep-wake up cycle, can lead to jet lag (Kuller and Wetterberg 1993). A severe disruption may result on the Seasonal Affective Disorder (SAD), a recurrent annual depression with an onset in winter and an elated mood in spring (Terman 1986; Heerwagen 1998; Stone 1998; Kuller 2001). The lack of exposure to light affects the secretion of melatonin and may require light therapy.

Natural light is also in a large number of studies claimed to improve productivity (Tabet Aoul and Shelley 1993; Stone 1998; Heerwagen 1998; Kuller and Lindsten 1992; Veitch and Newsham 1998). Although, the attempt to link daylight and productivity is claimed to be quixotic as so many factors interact in the environment (Erhardt 1993), there are evidences that a more satisfied worker translate into a more "productive" worker in the form of reduced absenteeism, improved concentration and greater output (Erhardt 1993; Heerwagen and Wise 1998; Erwine and Heschong 2002). Further, recent research indicates that students showed a significant improvement in learning rates and performed significantly better on standardized tests in classrooms where windows and skylights let more daylight in the classroom (Erwine and Heschong 2002). Similarly, day lit retail spaces were found to account for 40 % more sales then non-daylit ones (Heschong 2002).

Abundant literature exists in terms of window design for daylight. However, it is likely that the sole presence of window increases the perception of the contribution of daylight to a space. Well's study showed that beyond 6 m from the window the occupants overestimate the proportion of daylight in the overall illumination (Wells 1965). This illusion of a larger amount of daylight at the distant workplace indicates that having access to daylight, regardless of the distance, might be enough to satisfy the need for daylighting.

Access to View Out

The view out and the contact with the outside world it offers is the most cited benefit of the window. A view to the outside is believed to be a good 'visual rest centre', which permits the eye to re-focus at distant scenes in contrast with the typical close work found in offices. It also provides a sense of orientation allowing assessing the time of the day and keeping track of the weather as well as feeling in contact with the outside world. Therefore, avoiding feelings of being cooped up, isolated and claustrophobic. Further, evidences point out to a relationship between the lack of windows in the workplace and job dissatisfaction, feelings of isolation and depression (Finnegan and Solomon 1998). There is ample evidence from the reviewed literature that "although a view out is generally regarded as desirable, in some restricted and monotonous situations, it becomes almost a necessity" (Collins 1975; p. 76).

Access to nature directly or visually has been identified as a positive contributor to human health, comfort and wellbeing. The adequate usage of vegetation through biophilic design strategies could affect achieve the human comfort goals passively. Biophilic design strategies, benefits, and patterns and the implementation of green living walls are reviewed for their environmental and human psychological well-being.

Biophilic Design

Biophilic design is designing the built environment using vegetation and other natural systems and their processes. Wilson defined Biophilia as "the study of the human response to the natural environment and the relationship between humans and natural systems while the Biophilia hypothesis is the innate tendency to focus on life and life like processes" (Wilson 1984). Kellert relates the strong need for connection to nature to "a set of complex learning rules that have been ingrained in our genetic history and that the need to relate to natural processes is biological and is essential to our physical and mental well-being" (Kellert 1993). The key features, attributes and qualities of a Biophilic Building are: prospect, refuge, water, biodiversity, sensory variability, biomimicry, playfulness, enticement (Heerwagen and Hase 2001). Biophilic design patterns Fourteen are divided in three categories (Browning et al. 2014).

Nature in space: that is mainly contributed to the direct physical presence of nature within the built environment. It includes visual connection with nature, nonvisual connection with nature, non-rhythmic sensory stimuli, thermal and airflow variability, presence of water, dynamic and diffuse light, connection with natural systems. Some examples are: potted plants, flower beds, courtyard gardens, green roofs, green and green living walls.

Natural analogues: that is mainly related to integrating natural objects, materials, colours, shapes, sequences and patterns in the built environment. In addition, it encompasses patterns such as: biomorphic forms and patterns, material connection with nature, complexity and order. Such category could be implemented through the furniture and décor elements.

Nature of the space: this category addresses the nature's spatial configurations. The biophilic design patterns that are encompassed within the nature of the space are prospect, refuge, mystery, risk and peril. These configurations are best experienced when they are integrated with the nature in space and natural analogues.

One of the widely constructed, advanced technologies to implement the above mentioned patterns while integrating to a building and achieve the best biophilic experience are Green Walls.

Green Walls

There has been a rising interest by designers concerned about climatic changes, to adjust the vertical spaces of the built environment by integrating green living walls to increase green areas. Green Living Walls, also known as vertical walls are basically vertical walls, covered with vegetation and natural habitat as well as some water delivery systems, in the interior or exterior of the building or could be standing alone within the built environment and they differ in their types and sizes, and each type can encompasses different kind of vegetation, materials and technologies (Beatley 2009). However, the two main categories are Green Facades and Living Walls. The table below explains briefly the systems, type of plants, which will affect their behaviour as passive energy systems (Pérez et al. 2011a, b) (Fig. 8.11).

Classification of green vertical systems of buildings.

	Extensive systems		Intensive systems
Green facades	Traditional green facades Double-skin green facade or green curtain	Modular trellis Wired Mesh	
			Perimeter flowerpots
Living walls			Panels Geotextile felt

Fig. 8.11 Classification of green vertical systems of buildings (Pérez et al. 2011a, b)

Green Facades: such as the traditional and double skin green façade or green curtain and the perimeter flower pots (Pérez et al. 2011a, b) are made from climbing plants that uses soil at the base of the wall that grows on it either with the use of plant supports such as wood and meshwork or could grow without the need for additional infrastructure (Fig. 8.12).

On the other hand, living walls are completely artificial systems that use a continuous or modular planted-up units, panels and/or geotextile felts (Pérez et al. 2011a, b). The plants are either rooted in the structure directly or use growth medium that could be loose; where the soil is in packages installed on the wall; or the medium is coir fibre or felt mat that's thin and of multiple layers (Fig. 8.13).

Benefits of the Green Walls

Additionally, other systems such as: green hedges, green screen walls and live curtains may act similarly to a green wall. Green hedges are interchangeable with green façades or living walls for some of their features and ecosystem services. While green screens are made of a climbing plants pre-grown on a freestanding, galvanized steel framework, and established as an instant hedge. On the other hand, live curtains combine the features of green facades and living walls (Fig. 8.14).

Fig. 8.12 Traditional and double skin green façade or green curtain (Pérez et al. 2011a)

Fig. 8.13 Section through a typical green wall (*left*), loose medium (*middle*), mat medium (*right*) (Pérez et al. 2011b)

Fig. 8.14 Green hedges (*left*), green screen (*middle*), live curtains (*right*) (Pérez et al. 2011b)

Implementing Green Walls within the built environment has several objectives such as providing shade, reducing energy consumption from the wall façade (Pérez et al. 2011b), cleaning the air by removing carbon, enhance the microclimate and minimize The UHI effect, retain storm water and convey a sense of time and weather, reduce noise pollution and increase the biodiversity and greenery within the built environment (Loh 2008).

However, several other objectives have an effect on the human. Green Walls affect the human cognitive functioning, psychology, physiologically health and well-being through reducing stress, lowering blood pressure and the heart rate. As well as it can improve the metal attentiveness and temper cognitive fatigue add to that it helps in calming, providing physical comfort and shifting the individual focus to relax the eye muscles (Browning et al. 2014).

8.4 Conclusions

The chapter provided a review of, and a basic ability to understand passive design strategies in general including the approaches and benefits of using such strategies not only in the building design but also related to the urban context and human factors. The key to designing a passive building is to take advantage of the local climate (microclimate) and therefore, climate characteristics and classification can help with identifying approaches as early as site planning and analysis. Therefore, climate and comfort are the two fundamental measures in passive design that require attention. Passive design is a major part of environmental design and its approaches utilising several techniques and strategies that can be employed to achieve sustainable design also enhanced by patterns of biophilic design for improving health and well-being in the built environment.

References

Agboola OP (2011) Importance of climate to architectural designs in Nigeria. Environ Issues Agric Developing Countries 3(1):15–28

Aksoy UT, Inalli M (2006) Impacts of some building passive design parameters on heating demand for a cold region. Build Environ 14(12):1742–1754

Aldworth R, Bridgers D (1971) Design variety in lighting. Lighting Res Technol 3(1):8–23

Al-Temeemi AA, Harris DJ (2004) A guideline for assessing the suitability of earth-sheltered mass-housing in hot-arid climates. Energy Build 36(3):251–260

Amble Resorts (2009) New Panama resort, the resort at Isla Palenque, Panama, taps trend in passive cooling, drawings by 4240 architecture, Inc. Available at http://amble.com/ambler/2009/08/press-release-ambles-new-panama-resort-taps-trend-in-passive-cooling/

Balaras CA (1996) The role of thermal mass on the cooling load of buildings. An overview of computational methods. Energy Build 24(1):1–10

Beatley T (2009) Biophilic urbanism: inviting nature back to our communities and into our lives. Wm. Mary Envtl. L. Pol'y Rev 34:209

Beggs C (2010) Passive solar and low energy building design. In: Energy: management, supply and conservation, 2nd edn. Routledge, United Kingdom, pp 315–342

Benardos A, Athanasiadis I, Katsoulakos N (2014) Modern earth sheltered constructions: a paradigm of green engineering. Tunn Undergr Space Technol 41:46–52

Bessa VM, Prado RT (2015) Reduction of carbon dioxide emissions by solar water heating systems and passive technologies in social housing. Energy Policy 83:138–150

Boubekri M, Hulley RB, Boyer LL (1991) Impact of window size and sunlight penetration on office workers' mood and satisfaction: a novel way of assessing sunlight. Environ Behav 23 (4):474–493

Boyce P, Beckstead J, Eklund N, STROBEL R, Rea M (1993) Lighting the graveyard shift: The influence of a daylight-stimulating skylight on task performance and mood of night-shift workers. Lighting Res Technol 29(3):105–134

Brown RD (2011) Ameliorating the effects of climate change: modifying microclimates through design. Landscape Urban Plan. doi:10.1016/j.landurbplan.2011.01.010

Browning WD, Ryan CO, Clancy JO (2014) 14 Patterns of biophilic design. Terrapin Bright Green LLC, New York

Bruce N, Perez-Padilla R, Albalak R (2000) Indoor air pollution in developing countries: a major environmental and public health challenge. Bull World Health Organ 78(9):1078–1092

Carvalho AD, Moura P, Vaz GC, de Almeida AT (2015) Ground source heat pumps as high efficient solutions for building space conditioning and for integration in smart grids. Energy Convers Manag 103:991–1007

Chris 73 (2006) Iceland Keldur earth covered homes.JPG—Own work. Licensed under CC BY-SA 3.0 via Wikimedia Commons. Available via: https://commons.wikimedia.org/wiki/File:Iceland_Keldur_Earth_covered_homes.JPG. Accessed on 28 Nov 2015

Collins B (1975) Windows and people: a literature survey—psychological reaction to environments with and without windows. Natural Bureau of Standards, Gaithersburg, MD
Cuce PM, Riffat S (2016) A state of the art review of evaporative cooling systems for building applications. Renew Sustain Energy Rev 54:1240–1249
Dixon D (2012) One angel square earth tube. Jpg—own work. Licensed under CC BY-SA 3.0 via Wikimedia Commons. Available via: https://commons.wikimedia.org/wiki/File:One_Angel_Square_earth_tube.jpg. Accessed on 28 Nov 2015
Eliasson I (2000) The use of climate knowledge in urban planning. Landscape Urban Plann J 48:31–44
Erhardt L (1993) Views on the visual environment. Lighting Des Appl 24(4):16–17
Erwine B, Heschong L (2002) Lighting for learning. Lighting Des Appl 76–78
Finnegan MC, Solomon LZ (1998) Work attitudes in windowed and windowless environments. J Soc Psychol 115:291–292
Flagstaff (2015) Example of a passive solar design. Village land shoppe. Available via: http://www.greenflagstaffrealestate.com/passive_solar.html
Givoni B (1994) Passive low energy cooling of buildings. Wiley, New York
He J, Hoyano A (2010) Experimental study of cooling effects of a passive evaporative cooling wall constructed of porous ceramics with high water soaking-up ability. Build Environ 45(2): 461–472
Heerwagen J (1998) Design, productivity and well being: What are the Links? In: American Institute of Architects conference on highly effective facilities, Cincinnati, Ohio, 12–14 March
Heerwagen JH, Heerwagen DR (1984) Heat reflecting glazing: effects on comfort and functioning at work. In: Proceedings of 9th solar conference, Columbus, Ohio, pp 390–395
Heerwagen J, Heerwagen D (1986) Lighting and psychological comfort. Lighting Des Appl 47–51
Heerwagen JH, Hase B (2001) Building biophilia: connecting people to nature in building design. Environ Design Constr 3:30–36
Heerwagen J, Wise J (1998) Green building benefits: differences in perceptions and experiences across manufacturing shifts. HPAC Magazine 57–63
Heschong L (2002) Daylighting and human performance. ASHRAE J 44(6):65–68
Hien WN, Poh LK, Feriadi H (2000) The use of performance-based simulation tools for building design and evaluation—a Singapore perspective. Build Environ 35(8):709–736
Joshi SM (2008) The sick building syndrome. Indian J Occup Environ Med 12(2):61–64
Kampa M, Castanas E (2008) Human health effects of air pollution. Environ Pollut 151(2): 362–367
Kalogirou SA (2013) Solar energy engineering: processes and systems. Academic Press, Cambridge
Kellert S, Wilson EO (eds) (1993) The Biophilia Hypothesis. Island press, Washington, DC
Kobraklb (2012) Ventilation unit with heat pump and ground heat exchanger.png 6128—own work. Licensed under CC BY-SA 3.0 via Wikimedia Commons. Available via: https://commons.wikimedia.org/wiki/File:Ventilation_unit_with_heat_pump_%26_ground_heat_exchanger.png
Kim JT, Shin JY, Yun GY (2012) Prediction of discomfort glares from windows: influence of the subjective evaluation of window views. Indoor Built Environ 21(1):92–97
Kleerekoper L, Escha MV, Salcedob TB (2012) How to make a city climate-proof, addressing the urban heat island effect. Resour Conservation Recycl 64:30–38
Kwok AG, Grondzik WT (2007) The green studio handbook environmental; strategies for schematic design, 2nd edn. Architectural Press, Oxford
Kuller R (2001) The influence of daylight and artificial light on diurnal and seasonal variations in humans—a bibliography. Publication CIE 139–2001
Kuller R, Lindsten C (1992) Health and behavior of children in classrooms with and without windows. J Environ Psychol 12:305–317

Kuller R, Wetterberg L (1993) Melatonin, cortisol, EEG, ECG and subjective comfort in healthy humans: impact of two fluorescent lamp types and two light intensities. Lighting Res Technol 25(2):71–81

Littler JGF (1982) Overview of some available models for passive solar design. Comput-Aided Design 14(1):15–18, Available online 27 Feb 2003

Loh S (2008) Living walls-a way to green the built environment. BEDP Environ Design Guide 1 (TEC 26):1–7

MacGowan D (2010) Discomfort glare criteria. Lighting Res Technol 42(1):121–122

Meisel P, Jobst M, Lippmann W, Hurtado A (2015) Design and manufacture of ceramic heat pipes for high temperature applications. Appl Therm Eng 75:692–699

Milne M (1995) SOLAR-5.4 project. Department of Architecture and Urban Design University of California-Los Angeles Box 951467 Los Angeles, California 90095-1467, Available via online http://www.aud.ucla.edu/energy-design-tools

Mingfang T (2002) Solar control for buildings. Build Environ 37:659–664

Minke G, Hauser G, Bansal NK (1994) Passive building design: a handbook of natural climatic control. Elsevier, Amsterdam

Ne'eman E (1974) Visual aspects of sunlight in buildings. Lighting Res Technol 6(3):159–164

Ott JN (2000) Health and light: the extraordinary study that shows how light affects your health and emotional well-being. Ariel Press

Pedata, L. (2011). Environmental design IED Master. http://www.laurapedata.com/

Pérez G, Rincón L, Vila A, González JM, Cabeza LF (2011a) Behaviour of green facades in mediterranean continental climate. Energy Convers Manag 52(4):1861–1867

Pérez G, Rincón L, Vila A, González JM, Cabeza LF (2011b) Green vertical systems for buildings as passive systems for energy savings. Appl Energy 88(12):4854–4859

Phililips D (2004) Daylighting: natural light in architecture. Architectural Press, New York

Raiford R (1999) Blue light specials. Buildings 3(2):28–29

Roland ZH (2010) Dietikon - Lättenstrasse Erdhaus Peter Vetsch IMG 6128 - Own work. Licensed under CC BY-SA 3.0 via Wikimedia Commons. Available via: https://commons.wikimedia.org/wiki/File:Dietikon_-_L%C3%A4ttenstrasse_Erdhaus_Peter_Vetsch_IMG_6128.JPG. Accessed on 28 Nov 2015

Sanner B, Karytsas C, Mendrinos D, Rybach L (2003) Current status of ground source heat pumps and underground thermal energy storage in Europe. Geothermics 32:579–658

Santamouris M, Kolokotsa D (2013) Passive cooling dissipation techniques for buildings and other structures: the state of the art. Energy Build 57:74–94

Selman P (2006) Planning at the Landscape scale. Routledge, Taylors and Francis Group, London, New York

Shaharuddin A, Noorazuan MH, Yaakob MJ (2009) Fenomena Pulau Haba Bandar dan isu alam sekitar di Bandaraya Kuala Lumpur. In Malay. GEOGRAFIA OnlineTM Malaysian. J Soc Space 5(3):57–67

Shashua-Bar L, Pearlmutter D, Erell E (2009) The cooling efficiency of urban landscape strategies in a hot dry climate. Landscape Urban Plann 92(3–4):179–186

Shaviv E, Yezioro A, Capeluto IG (2001) Thermal mass and night ventilation as passive cooling design strategy. Renew Energy 24(3):445–452

Stone NJ (1998) Windows and environmental cues on performance and mood. Environ Behav 30 (3):306–321

Tabet Aoul K (1989) Functions of window and fenestration design in a hot climate. Proc ISES, Kobe, Japan, pp 192–197

Tabet Aoul K, Shelley G (1993) An investigation on the effect of access to windows on visual comfort and productivity. In: Proceedings of 3rd European conference on architecture, Florence, Italy, 17–21 May, pp 157–160

Terman M (1986) The photic environment and physiological time-keeping. In: International daylighting conference proceedings I, pp 356–359

Thani SKSO, Mohamad NHN, Idilfitri S (2012) Modification of urban temperature in hot-humid climate through landscape design approach: a review. Soc Behav Sci 68:439–450

Veitch J, Newsham G (1998) Lighting quality and energy-efficiency effects on task performance, mood, health, satisfaction and comfort. J Illum Eng Soc 107–129

Wells BWP (1965) Subjective responses to the lighting installation in a modern office building and their design implications. Build Sci 1:57–68

Wikimedia (1965) https://upload.wikimedia.org/wikipedia/commons/1/1e/SolarHotWaterBasicDiagram.png

Wilson, EO (1984) Biophilia, the human bond with other species. Harvard University Press, Cambridge (Massachusetts), p 157

World Health Organization (WHO) (2014) Public Health, Environmental and Social Determinants of Health, Geneva

Yang L, Li Y (2008) Cooling load reduction by using thermal mass and night ventilation. Energy Build 40(11):2052–2058

Yaniv R (2012) Building beyond: a trade school in Swaziland, Africa, passive design techniques. https://swazischool.wordpress.com/2012/09/11/passive-design-techniques/

Zhang X, Shen J, Lu Y, He W, Xu P, Zhao X, Dong X (2015) Active solar thermal facades (ASTFs): from concept, application to research questions. Renew Sustain Energy Rev 50:32–63

Chapter 9
Active Systems

Jun-Tae Kim, Jin-Hee Kim, Ahmed Hassan, Caroline Hachem and Fred Edmond Boafo

Abstract The 21st Resolutions of the recent United Nations Framework Convention on Climate Change (UNFCCC) Conference of the Parties (COP) outlined, promotion of sustainable energy, sustainable management of natural resources, and maintaining environmental integrity through mitigation of greenhouse gas emissions; towards sustainable development. Renewable energy is sustainable and natural resource is the source of clean energy. Solar energy is the most abundant and inexhaustible renewable energy resource; the sun's daily irradiation to the earth reaches approximately 10,000 times more energy than the average daily energy consumption of Earth. However, harnessing the sun's energy efficiently at a reasonable cost comes as a challenge. Building-integrated, building-attached and standalone active systems that convert solar energy into useable forms of heat and electricity as well as wind power and geothermal systems are reliant on solar energy as identified and discussed in this chapter. General planning and methods of integration of various active systems are described in this section.

J.-T. Kim (✉) · J.-H. Kim · F.E. Boafo
Kongju National University, Gongju, South Korea
e-mail: jtkim@kongju.ac.kr

J.-H. Kim
e-mail: jiny@kongju.ac.kr

F.E. Boafo
e-mail: febs@kongju.ac.kr

A. Hassan
United Arab Emirates University, Al Ain, United Arab Emirates
e-mail: Ahmed.Hassan@uaeu.ac.ae

C. Hachem
University of Calgary, Calgary, Canada
e-mail: caroline.hachem@ucalgary.ca

© Springer International Publishing Switzerland 2016

M. Noguchi (ed.), *ZEMCH: Toward the Delivery of Zero Energy Mass Custom Homes*, Springer Tracts in Civil Engineering,
DOI 10.1007/978-3-319-31967-4_9

9.1 Introduction

Broadly, the term active solar systems refer to systems that use mechanical and electrical equipment to harness and convert solar energy into heat and electric power. Active systems include: solar thermal collectors that can be used for domestic hot water (DHW) and space heating; photovoltaic (PV) that can be used to produce electricity; hybrid photovoltaic/thermal (PV/T) systems that can generate thermal and electrical energy simultaneously; building-integrated photovoltaic (BIPV) which is an integrated and functional building envelope with electricity producing PV cladding; building-integrated photovoltaic thermal (BIPVT) which is a hybrid system that combines PV and solar thermal collector technologies, as well as other building envelope components such as insulation, the entire BIPVT system is a functional building envelope component, as well as simultaneously produce thermal and electrical energy; wind power systems that derive its source from solar energy which creates variation in air flow pressure, able to be converted into usable energy; as well as in geothermal systems where the ground acts as a heat source in winter and a heat sink in summer, also stipulated by solar conditions. Common to all these active solar systems is a heat extraction medium; usually being air or water.

This chapter will seek to define and describe the basic working principles of active systems, as well as consider some case studies showing proven integration and feasibility of the various systems.

9.2 Solar Thermal (ST) Systems

Heating demand in house hold comes mainly from producing hot water and space heating, the proportion being dependent on the climatic conditions of the site under consideration. Solar energy available naturally and abundantly subject to certain variation depending on climatic conditions can be utilized to meet such demands efficiently and cost effectively. Solar thermal collector works on intercepting solar radiation through a certain area (aperture), absorbing solar radiation on a surface (solar collector) which convert and retains it into thermal energy. The thermal energy is then transferred to a heat transfer fluid (HTF) and can be either directly used for space or water heating or stored in a container. In most cases the solar thermal system produces the temperature required for domestic applications however in some climatic conditions, an additional heating set up may be needed as auxiliary or a backup system (Fig. 9.1). Based on the working principle, the solar thermal system has the following components:

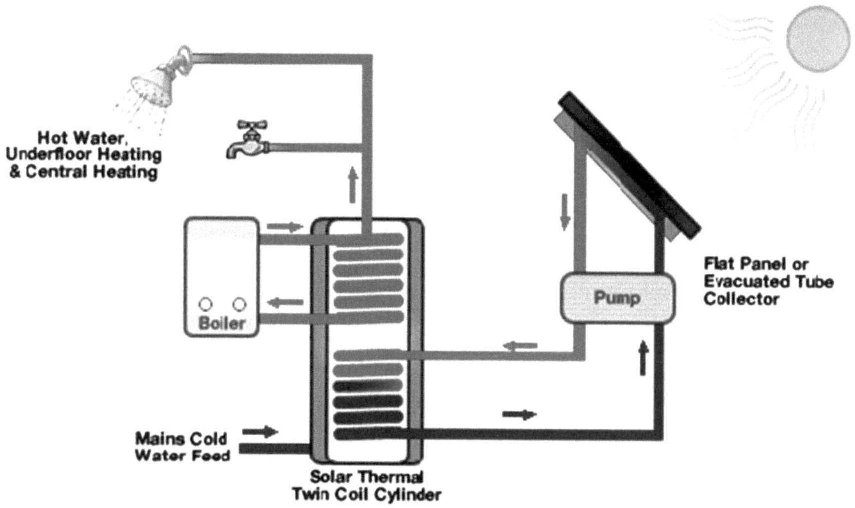

Fig. 9.1 Schematics of a solar water heating system with and auxiliary boiler (Solar Utilities 2013)

9.2.1 Solar Collector

Solar collectors (SC) absorb solar radiation, convert it into thermal energy, retain it from flowing back to ambient and transfer it to a circulating HTF. SC is the basic part of solar thermal energy set up and also major cost contributor. The HTF is air, water, oil or a refrigerant which transfers heat from solar collector to hot water tank while circulating in a loop between the collector and the storage. The SC has two basic designs in terms of intercepting the incoming solar radiation, namely; concentrated and non-concentrated SC. In non-concentrated design solar radiation hit the surface as they are coming from sun, while in concentrated solar collector some mechanism is used to converge the radiation in order to increase the solar radiation intensity on the smaller target surface. In this case, size of the collector surface reduces depending on whether the SC is being tracked along the position of the sun or is static. Heat losses are observed at the surface of solar collector due to convection and radiation which can be minimized by applying glazing top cover. It allows the coming radiations (near infrared) from sun to pass through; the glazing however hinders the thermal losses by stopping far-infrared radiations. Absorber is the core part of the SC which converts the solar radiation into thermal energy and to a large extent affects the output efficiency of the collector (Fig. 9.2). The absorber has different layers depending on design and geometry considerations. A metal sheet is used as substrate material due to higher thermal conductivity for enhanced heat transfer, having an infrared red reflective layer on top which is connected to the absorber layer through a bonding layer by adhesive which restricts the metal atoms from the substrate to enter the absorber when exposed to very high temperatures. In

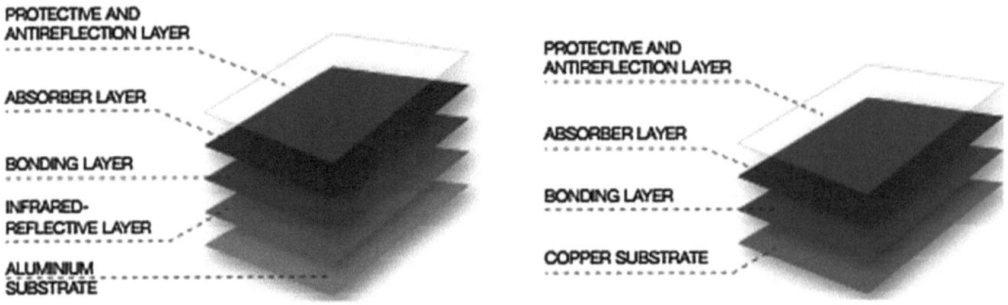

Fig. 9.2 Construction layers of a solar thermal collector system

this way, optical characteristics of the collector are secured while a multilayer cermet structure is used as absorber. The outermost layer is kept anti-reflective, mechanically strong, and scratch resistant to avoid radiation losses, environmental and seismic impacts, and damage while transportation usually employing fused quartz which has higher mechanical strength and is resistant to scratch (Nabila et al. 2012; Green Systems 2015; Tinox Energy 2015).

The SC in domestic applications is generally categorized into two types based on the collector design, namely; flat plate collectors and evacuated tube collectors. Flat Plate Collectors are used mostly for domestic hot water applications having an insulated box with absorber. Absorber sheet is welded with a highly conductive pipe heat exchanger for efficient heat transfer. Evacuated Tube Collectors are comprised of an array of single or twin wall glass tubes with a vacuum that provides excellent insulation against heat loss. Single wall evacuated tubes normally have a fin that has the absorber coating, similar to that used in the flat plate collector. Twin wall evacuated tubes have the absorber coating on the inner tube and the space between the two tubes is "evacuated" to form vacuum Parabolic Trough Collectors (Fig. 9.3); which reflect solar radiation through a dish like parabolic surface back to a small target to increase the solar radiation intensity by 15–20 times for domestic solar cooling applications thereby achieving temperatures of 100–250 °C which can in turn run the absorption chillers to produce solar driven absorption cooling (Fernández-Garcia et al. 2010; Apricus 2015; Price et al. 2002).

Fig. 9.3 Flat plate solar collectors (*left*), evacuated tube solar collectors (*middle*) and parabolic trough collectors (*right*)

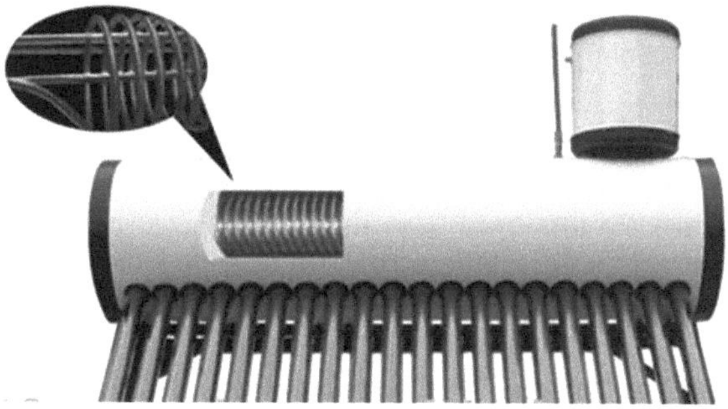

Fig. 9.4 Heat exchanger system for an evacuated tube collector employing copper as heat exchanger material

9.2.2 Heat Exchanger

Heat exchanger fabricated from highly conductive aluminum or copper based alloy is employed to transfer heat from the solar collector to thermal storage (Fig. 9.4). In solar thermal systems, generally three types of heat exchangers are used, namely; liquid to liquid, air to liquid and liquid to air. Heat exchanger employs a heat transfer fluid (HTF) to act as medium that carry heat from SC and transfer it to storage tanks through heat exchangers generally being water, air, glycol, hydrocarbon oil, refrigerants/phase change fluids and Silicones. Several parameters affect the performance of HTF such as latent heat, coefficient of expansion, viscosity, specific heat capacity and phase change characteristics of boiling, freezing and flash points. Application of heat transfer mostly depends on the weather conditions of the regions thereby being for colder and warmer climates. A good fluid should have less viscosity and higher latent heat to flow with less power inputs and carry maximum heat respectively (Energy 2010).

9.2.3 Planning of Solar Thermal Technology for ZEMCH

The applications of solar thermal energy encompass water to so-called, space-heating, desiccant based dehumidification, and absorption chiller based solar cooling and desalination. Small, medium and large-scale systems in solar thermal energy are still being developed in different ways to be integrated into housing at a single house scale, housing cluster scale and community based district heating and cooling system. Although solar thermal heating systems for use in heating, desalination and dehumidification are well developed, solar thermal cooling systems still need further research and development to be widely adopted for ZEMCH (Renewables 2015). Solar thermal systems are easy to integrate into ZEMCH as they provide basic necessary services of water heating, space heating and space cooling. Depending on the climatic conditions and heating/cooling demands, the ST systems can either be installed as standalone providing themselves all the heating/cooling demand or can be combined with the existing traditional heating and cooling systems to save energy inputs. In terms of organizing spaces for the ST systems, water and space heating systems can be installed on the roof or wall. However the solar cooling system are subject to solar radiation concentration and solar tracking and would need dedicated space away from the building as the collectors produce higher temperatures. Design of the collector system depends on the weather conditions, the collector type and the temperature requirements. (Patel et al. 2012).

Solar thermal may also be used for cooling buildings of ZEMCH in order to reduce their energy consumption, especially in warm climates. The systems can either be used as standalone in moderate cooling demand climatic conditions or can be integrated into the traditional mechanical cooling systems in the hot climate with high cooling demands. The cooling process involves evaporation and condensation in two configurations namely open system utilizing solid and liquid sorption method and a closed system utilizing adsorption as well as absorption principle. Detailed energy modelling soft-wares are available such as TRNSYS, to develop a particular design for configuration of interest and analyze the design and performance of a solar thermal collector system taking statistical weather data of the site. In order to estimate the capacity and performance of the solar thermal collector, collector characteristic curve reported in any text can be used with the weather data. The performance of the solar thermal system depends on the climatic conditions of the site, type of SC, type of HTF, tilt of the collector and the nature of application in the building. For an optimized performance, the SC is always recommended to be tilted at latitude of the place facing the respective hemisphere.

Solar thermal systems are well established for water heating where they can easily be designed and integrated with the plumbing systems. However the solar space heating and desalination systems can be challenging to integrate. For solar thermal to be effective for space heating, a much larger surface area of collector is required; approximated at 0.5–1 m^2 of collector area per 10 m^2 of floor area to be heated. The large collector area also needs larger thermal storage capacity (up to 1000 L) which generally remains redundant in summer months create a major hurdle for adaptation of ST space. Thermal desalination processes by using solar energy is one of the promising yet challenging areas in terms of system integration and space requirements (Fig. 9.5). Solar desalination is classified into direct and indirect, where solar energy through SC distillate directly in the solar collector while indirect desalination combines conventional desalination techniques such as multistage flash desalination (MSF), vapor compression (VC), reverse osmosis (RO), membrane distillation (MD) and electro dialysis, with solar collectors for heat generation. Direct solar desalination is simpler in operation and less costly compared with the indirect technologies; however the former requires large land area and has a relatively low productivity (Qiblawey and Banat 2008).

Fig. 9.5 Solar thermal desalination system (Woody 2015)

9.3 Photovoltaic (PV) Systems

Photovoltaic (PV) systems are emerging as an important part of the trend towards energy source diversification (Wiginton et al. 2010; Neuhoff 2005; Pearce 2002). PV technology implementation is still limited however; constituting less than 1 % of global energy production (Wiginton et al. 2010).

The electricity generated by a PV system constitutes only a fraction of the solar radiation absorbed by the system surface, referred to as the electrical conversion efficiency of the PV modules. The remaining energy is partly converted to heat (Poissant and Kherani 2008). Existing electrical efficiency of some of the commonly used PV modules such as polycrystalline and monocrystalline silicon ranges currently between 20 and 25 % while the electrical efficiency of amorphous silicon (a-Si) PV reaches 10 % (Green et al. 2011). Thin film silicon modules are being developed with increasing efficiency, currently reaching some 16 %. The electrical conversion efficiency is measured under standard conditions (solar irradiation of 1000 W/m^2 and cell temperature of 25 °C).

The performance of a PV system depends mainly on the tilt angle and azimuth of the collectors, local climatic conditions, the collector efficiency, and the operating temperature of the cells. During the winter months, the insolation can be maximized by using a surface tilt angle that exceeds the latitude of the location by 10–15°. In summer an inclination of 10–15° less than the site latitude maximizes the insolation (Duffie and Beckman 2006). The PV system is commonly mounted at an angle equal to the latitude of the location in order to reach a balance between winter and summer production (Kemp 2006). In locations where snow accumulation is an issue, the tilt angle should be selected to take into account this factor.

The orientation of the PV panels affects both the electricity generation and the time of peak generation. PV system orientation can be selected to better match the grid peak load (Holbert 2009). This can affect the annual return value of the produced electricity, especially in locations where electricity value changes with time of use (Borenstein 2008).

9.4 Hybrid Photovoltaic/Thermal Systems (PV/T)

Hybrid photovoltaic/thermal systems (PV/T) combine PV modules and heat extraction devices to produce simultaneously power and heat (Tripanagnostopoulos et al. 2002). Heat extraction from the PV rear surface is usually achieved using the circulation of a fluid (air or water) with low inlet temperature. The extraction of thermal energy serves two main functions. It is exploited for space heating and solar hot water applications, and it serves for cooling the PV modules, thus increasing the total energy output of the system (Charron and Athienitis 2006).

The total electrical and thermal energy output of the PV/T systems depends on several factors including solar energy input, ambient temperature, wind speed, and heat extraction mode. For locations with large space heating requirements, air based PV/T systems can be particularly advantageous and cost effective (Tripanagnostopoulos et al. 2002).

9.5 Wind Power Systems

Similar to the solar radiation universally available which can be exploited to produce solar energy, wind is another universally available natural source of energy production. Wind energy is considered to be a potential candidate to manage energy demand in houses due to its abundant supply, relatively lower cost, and environmental friendly energy conversion process. Through decades of research and development, tremendous improvements are made in the design of wind turbine resulting in steady efficiency improvements along-with effectiveness for building integration (Fig. 9.6). To make them market competitive however wind turbines still have to go through further improvement to compete with the cost, reliability and robustness of existing fossil fuels based energy systems (Manwell et al. 2010).

9.5.1 Working Principle

Due to variation in altitude, presence of coastline and vegetation inequality, Earth absorbs different amount of solar radiation and eventually results in different ambient temperatures that cause low atmospheric pressure at hotter terrain and

Fig. 9.6 An innovative design of wind turbine suitable for building integration (West Devon Borough Council 2013)

relatively high pressure on the colder part. Air being a fluid blows from high to low pressure regions. As altitude rises, speed of wind increases and it becomes more laminar approximately 100 ft above the ground. Wind turbine exploits the pressure available in the wind flow to convert into electricity through shaft power. Flowing wind rotates the blades mounted on the rotor which in return rotates the shaft that eventually spins the electric generator resulting in electricity production. In simple terms, working principle of wind turbine is opposite to ventilation fan because it takes energy from wind, rotates the wings or blades of fan causing the rotation of motor to generate electricity.

In one way, wind energy derives its source from the solar energy which creates variation in pressure that causes air to flow which can be converted to usable energy. In each energy conversion process, there are always energy losses which are accumulated to result in system losses which eventually reduces the energy yield of the system and the actual available energy is quite lower compared to available potential which highlight the potential for improvement in each conversion process to come up with more future efficient wind energy conversion technology (Manwell et al. 2010; Paul 2004). The main components of the wind turbine system (Fig. 9.7) contribute to the energy conversion process and can be a focus for improvements in future designs.

Wind turbine blades are the first interaction between wind pressure and energy conversion process. The blades are made of certain geometry with small orifices at the surface. The orifices act to optimize the relationship between wind velocity and wind pressure through valves that can adjust pressure and velocity by varying the diameter of orifice present at the exit in case the pressure or velocity exceeds the specific limit in stormy conditions. The speed of air at the blade tip is also kept within a range to yield an optimum blade spin of 20 rotations per minute for optimized energy production and turbine stability (Parera 2008; Energy Center of

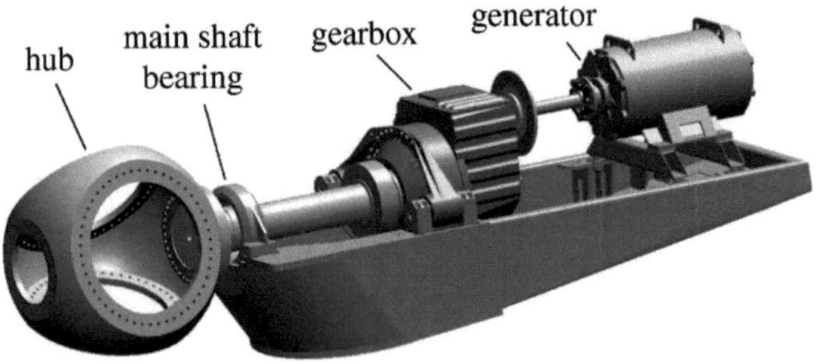

Fig. 9.7 Components of a wind turbine system (Kotzalas and Doll 2010)

Wisconsin 2010). The following structural and housing components of the wind turbines system are employed.

The nacelle is the housing of gear assembly and generator behind the rotors, the gear assembly controls the rotational speed of the rotor fed into the generator. The Tower Prime purpose of the tower is to mount rotor at certain height to benefit from lamina and fast flow of wind. The height of this cylindrical shape tower varies from 150 to 200 ft with its diameter almost 10 ft in non-building integrated applications in order to get smooth air. In building integrated applications, it can exploit the building height and reduce the cost dramatically which otherwise can reach as high as half of the total system cost. The Base is the essential part for erection of the whole assembly and for its firm sustainability. In general, concrete reinforced with steel bars are used. However in building integrated applications, it can benefit from existing building roofs as the base and can further reduce the cost of the system (Manwell et al. 2010; Paul 2004).

9.5.2 *Types of Wind Energy Systems*

Wind turbines are categorized depending on the axis of spin of rotor into horizontal wind turbine and vertical axis wind turbine.

9.5.2.1 Horizontal Axis Wind Turbine—HAWT

In this design, blades rotate around the horizontal axis and shape of blades looks like propeller of the plane. In small turbines, gear box and generator are fixed at the top of tower and their direction is normal to the wind flow. While servo motors are used along with wind sensors that adjusts the direction of rotor depending upon the direction of air flow. Safe distance is kept between tower and blades to avoid any inconvenience in the case of very fast winds. In some cases blades are slightly tilted to overcome the future risks. The horizontal wind turbine (Fig. 9.8) has advantage of laminar air flow, increased wind velocity due to tall structure and higher conversion

Fig. 9.8 Horizantal axis wind turbine installed on a building in urban settings with gear box assembly and generator fixed on the rotor (Atmospheric Water Systems, Inc. 2010)

efficiency due to blades direction normal to the wind flow. However it has disadvantages of higher cost, additional set up required to keep the blades normal to wind flow and needs brakes in case of very high wind speeds to avoid damage.

9.5.2.2 Vertical Axis Wind Turbine—VAWT

In vertical axis wind turbines (Fig. 9.9), the rotors are designed in vertical direction thereby eliminating the need for adjustment of direction of rotors which makes it suitable to continuous varying winds directions and turbulent wind flow conditions. The gear box assembly and generator are fixed on ground instead of mounting on the tower which reduces the cost of the systems and makes it easily accessible for maintenance. The vertical axis wind turbines are further divided into two types, namely; Darrieus Wind Turbines also known as "eggbeater" due to their shape have higher efficiency however less reliability due to higher fatigue on its blades and need a separate arrangement to start the turbine, and Savonius Wind Turbine which have lower efficiency however are more reliable due to drag type turbine, effectiveness in turbulent conditions and absence of additional mechanism to start the turbine (Manwell et al. 2010; Paul 2004).

9.5.3 Design of the Wind Energy Systems

Selection of material depends upon the required characteristics. For example, in the case of blade design, material should be stiff and have reasonable strength to bear

Fig. 9.9 Vertical axis wind turbine installed on the roof (Keough et al. 2014)

wind loads besides being lightweight to minimize the inertial and gyroscopic resistances. Initially steel was preferred due to its stiffness and aluminum due to its lightweight, nevertheless composite materials are employed currently due to their superior lightweight and stiffness characteristics compared to steel and aluminum therefore most of the turbine blades are currently being manufactured from fiber-glass due to its highest strength—to—weight ratio however it renders higher cost. In addition to fiber glass, turbine blades are also manufactured from epoxy to its better fatigue properties and being less prone to being brittle with age. Steel is still used for erection of tower while steel reinforced concrete is used for secure base (Tangler 2000; Negm and Maalawi 2000).

Small wind turbines for household are cost competitive compared to the larger industrial and turbines where they can easily be installed at roof top. The energy produced from wind turbines in not stable due to its dependence on the availability and magnitude of wind speed which is highly variable. The problem of fluctuating energy production is being solved in three ways. Firstly, by integrating the wind turbine with grid and drawing power from grid when under wind turbine under-produces and supply to the grid when the wind turbine overproduces than the house demand. Secondly, installing electrical energy storage battery pack to store the excess energy produced and draws it when needed. Lastly, designing a combination of different renewables, such as photovoltaic, fuel cells, biodiesel, and biomass with wind turbine to overcome supply and demand mismatch in terms of magnitude and time of production and utilization.

While integrating to houses (Fig. 9.10), the major barrier had been noise from the rotors and blades, however innovations in the contemporary small scale house integrated deigns, and noise has been reduced to affordable limits. From architectural perspective, these turbines are available in different shapes and various colors

Fig. 9.10 Wind Turbine installed on Croydon City House, UK (*left*) and a community house in Kenya (*right*)

that can well integrate with relevant designing requirements and constraints with working life being over 20 years make this option more attractive for building integration (Paul 2004).

9.5.4 Performance of Wind Turbine System

Selection of wind turbines starts with analysis of diurnal and seasonal wind data at the site of installation. The regions with long coastline are rich in wind energy due to diurnal changes of pressure and temperature over the ocean and earth. After confirming the abundant and regular winds through weather data, demand of energy should be calculated in order to select the appropriate turbine for particular energy demand (West Devon Borough Council 2013). Performance of wind turbine

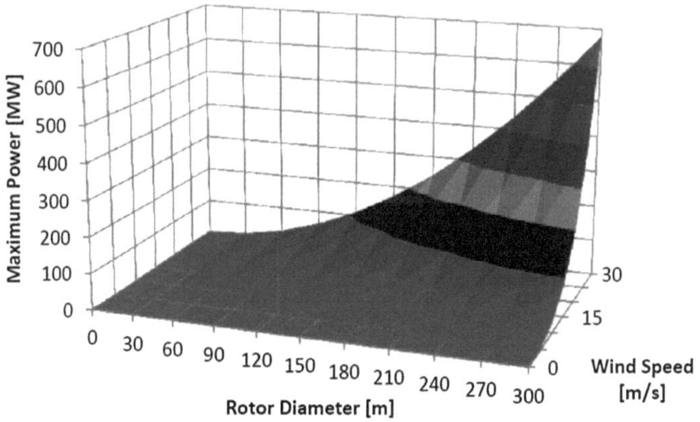

Fig. 9.11 Maximum power output of wind turbine as a function of rotor speed and diameter (Koroneos et al. 2003)

depends on wind speed, rotor diameter and type of turbine (horizontal or vertical). The relation between the wind speed, rotor diameter and power output from the turbine can be used to predict the performance of wind turbine for a particular place for designing the wind turbine system (Fig. 9.11). The power increases as a square of rotor speed and cube of the wind speed.

9.6 Geothermal Systems

Literally, geothermal means earth (Geo–) + heat (Thermal). Geothermal technology relies on the fact that the earth (beneath the surface) remains at a relatively constant temperature throughout the year, warmer than the air above it during the winter and cooler in the summer, very much like a cave. Geothermal system consists of three basic parts, namely; ground heat exchanger, heat pump, and delivery system (ductwork) (Fig. 9.12). The geothermal heat pump takes advantage of this by transferring heat stored in the earth or in ground water into a building during the winter, and transferring it out of the building and back into the ground during the summer. The ground, in other words, acts as a heat source in winter and a heat sink in summer. Geothermal Heat Pumps (GHP) are mainly used in buildings for space heating, cooling and sometimes domestic hot water supply. These systems transfer heat by pumping water or a refrigerant (a special type of fluid) through pipes just below the Earth's surface, where the temperature is a constant 50 to 60 °F. Heat pumps allow transformation of heat from a lower temperature level to a higher one by using external energy (e.g. to drive a compressor). The amount of this external energy input, be it electric power or heat, has to be kept as low as possible to make the heat pump environmentally and economically desirable.

In contrast to other heat pumps, such as air-to-air heat pumps, ground source heat pumps can store extracted heat in summer and make this heat useful again in the heating mode in winter. During the winter, the water or refrigerant absorbs

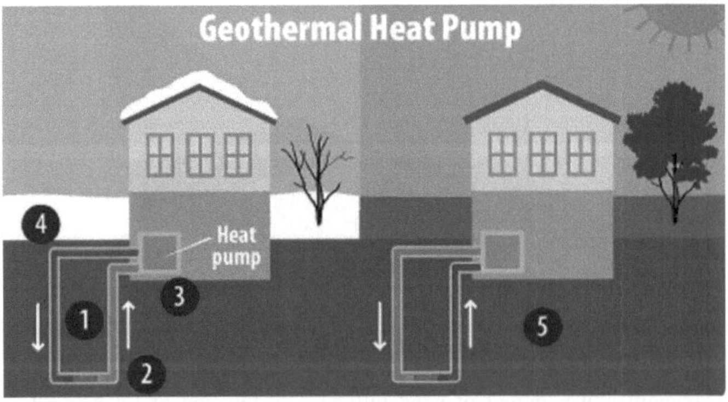

Fig. 9.12 Principle of the GHP system (EPA)

warmth from the Earth, and the pump brings this heat to the building above. In the summer, some heat pumps can run in reverse and help cool buildings.

The principle of the geothermal system with heat pump is as follows:

1. Water or a refrigerant moves through a loop of pipes.
2. When the weather is cold, the water or refrigerant heats up as it travels through the part of the loop that's buried underground.
3. Once it gets back above ground, the warmed water or refrigerant transfers heat into the building.
4. The water or refrigerant cools down after its heat is transferred. It is pumped back underground where it heats up once more, starting the process again.
5. On a hot day, the system can run in reverse. The water or refrigerant cools the building and then is pumped underground where extra heat is transferred to the ground around the pipes.

Positive characteristics of geothermal include: capability to provide base load power; no seasonal variation; immunity from weather effects and climate change impacts; compatibility with both centralized and distributed energy generation; re-source availability in all world regions, particularly for direct use. Barriers to usage include high capital cost, resource development risk, lack of awareness about geothermal energy and perceived environmental issues.

9.6.1 Geothermal System Building Cases

1. *Landhut Ludmilla*

Landhut Ludmilla residential estate is located in Stadt Landshut, 84034 Landshut, Niederbayern, Bayern. It is a 65.79 km^2 residential complex (Fig. 9.13a) comprising apartment buildings and single family homes. Different types of ground collectors were installed to tap geothermal energy (Fig. 9.13b). The building complex is a new building and currently being evaluated, however that very good temperature conditions on the heat sources and thus desirable effect especially during heating season is envisioned (Landhut 2015).

2. *Leaf House*

In the Leaf House, the heat and cold generation is carried out by the geothermal heat pump exchanging with the ground through three vertical probes of 100 m each (Fig. 9.14). This solution is used both for cooling and for heating thus avoiding the use of boiler and air conditioner. For most summer days it is not even necessary to have any heat pump, since the water automatically cools by passing through the underground probes. The heat pump efficiency provided is high (Coefficient of performance (COP) of 4.6) thanks to two factors: on one side the nature of the ground, rich of water, enables the thermal exchange and on the other the adoption of a floor radiating heating and conditioning system to maintain relatively low temperatures of the heat carrier in winter (less than 40 °C) and relatively high in

(a)
(b)

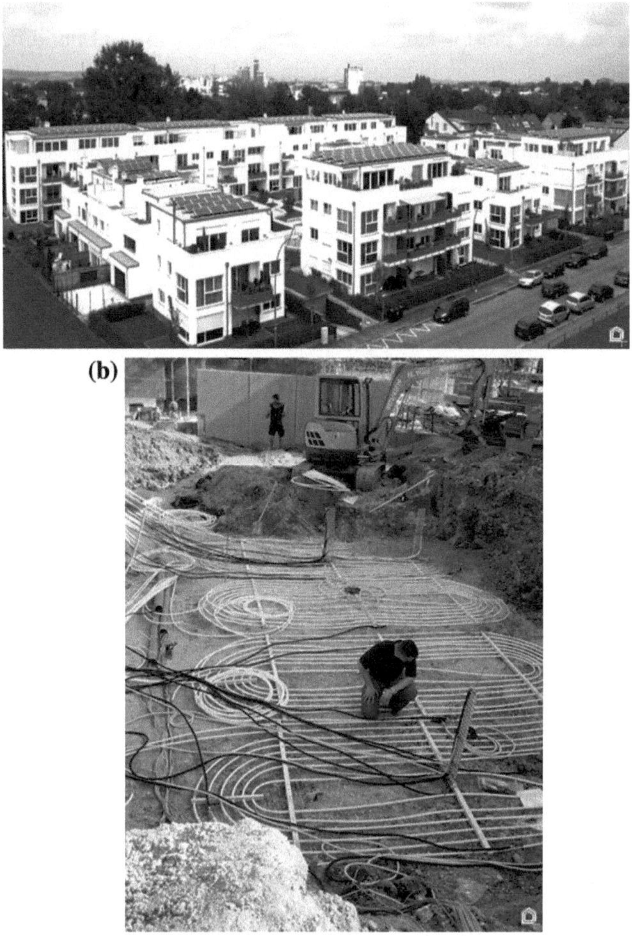

Fig. 9.13 **a** View of Landhut Ludmilla residential site. **b** Laying the measurement technology for the ground collectors (Landhut 2015)

Fig. 9.14 (*Left*) Leaf House side view and (*right*) geothermal system fittings (Leaf 2015)

summer (more than 17 °C). GSHP systems provided higher source temperatures in heating mode, and lower heat sink temperatures in cooling mode; thus reducing the source-sink temperature differential and increased the system's efficiency (COP) in the Leaf House.

9.7 Heat Recovery Systems (HRS)

Heat recovery is the collection and re-use of heat arising from any process that would otherwise be lost. The process might be inherent to a building, such as space heating, ventilation, among others, or it could be carried out as part of process activity, such as the use of ovens, furnaces, boilers, among others. Heat recovery can help to reduce the overall energy consumption of the process itself, or provide useful heat for other purposes. Some common sources of waste heat that present opportunities for cost-effective heat recovery include: ventilation system, boilers, air compressors, refrigeration plant, and high temperature exhaust gas streams from furnaces, kilns, ovens and dryers, among others. Recovered heat can be used for pre-heating fresh air used to ventilate the building, pre-heating combustion air for boilers, ovens, furnaces, etc.; space heating, drying, among others. In most cases, heat recovery is far more efficient when the heat source and heat sink are coincident—meaning they are physically close together and occur at the same time.

Heat recovery ventilation (HRV) exchanger is a ventilation facility that is able to remove stale air and bring fresh air into the building, while at the same time using the heat from the air being removed to either heat or cool the incoming fresh air. These systems are useful in homes with a tight building envelope because of the degree of control they allow over both ventilation and heating systems. Heat recovery works by continuously replacing contaminated stale air with warmed, fresh, filtered air. A heat exchanger is used to transfer the stale air's heat to the incoming air, both sensible and latent heat may be transferred. Stale, moist air is extracted from rooms of a home. The moist air is then ducted to a central unit located normally in the loft space in a house, or in a utility room or cupboard (Fig. 9.15).

Fig. 9.15 Principle of HRV exchanger (Coastal ventilation 2015)

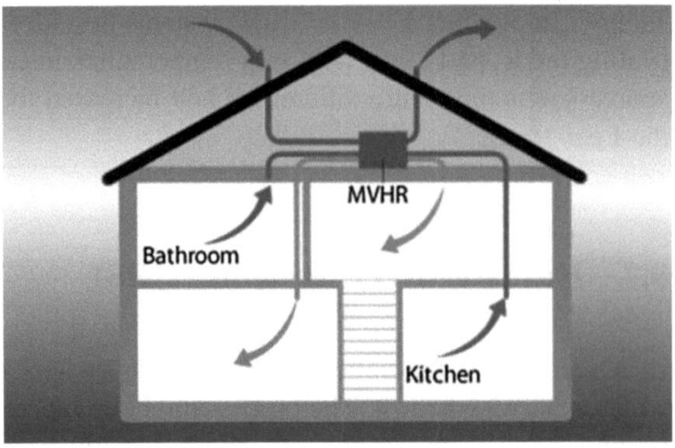

Fig. 9.16 Principle of MVHR system (Renew green energy 2015)

This extracted air passes over a heat exchanger before being ducted to outside. Simultaneously, fresh air is drawn into the unit from outside via a pollen filter, and is warmed by the high efficiency heat exchange cell. This tempered, fresh air is then delivered through supply vents into the living spaces.

A mechanical ventilation with heat recovery (MVHR) system removes moist stale air and delivers fresh air into a building, in an energy efficient way. It works by extracting warm, moist air from around the home, putting it through a heat exchanger to extract the heat and then expelling the air to the outside. At the same time it takes fresh generally colder air from the outside and uses the heat from the heat exchanger to warm the incoming fresh air before delivering it around the house (Fig. 9.16). Typically stale air would be removed from kitchens, bathrooms and toilets and fresh air would be introduced into living and sleeping areas such as lounges and bedrooms. The MVHR system allows occupants to control the speed with which air is changed in a house, without relying on extractor fans or opening and closing windows. The system can be managed using a control panel possibly linked to humidity or carbon dioxide sensors or boost switches.

9.8 Active Daylighting Systems

Daylighting reduces energy used for mechanical lighting, offers psychological benefits, as well as having a sustainable effect on the environment, among others (Kim and Kim 2010; Sapia 2013). Conventional glazing systems and passive daylighting techniques such as roof skylights and clerestories can have problems of glare, overheating, restricted depth of living space penetration, ability to collect direct sunlight and diffused skylight only, substantially reduced illuminance levels when the sun is low (during the mornings and late afternoons), among others. On the other hand, active daylighting systems, contrary to passive daylighting systems, use movable devices to mechanically track the sun path; increasing the yield and efficiency of daylight collection at an intensity that allows mechanical lighting to be turned off. Over the years, developments in computer-controlled mechanisms and high-reflectivity optical materials have advanced the frontier of active daylight systems (Kim and Kim 2010; Oh et al. 2013; Sapia 2013; Mayhoub 2014). Solar tracking in active daylighting systems is generally achieved by using a set of lens or sensors, or by employing electronic logic based on mathematical formula derived from pre-programmed solar path chat detailing the sun's positioning per a given time of day, which focuses light through a parabolic mirror to send it deep into the interior. Active daylighting systems improve the penetration of daylight, thus increase illuminance levels in deep building spaces.

9.8.1 How It Works

The functional components of the active daylighting system include mirrors/lenses, sensors and diffusers (Fig. 9.17). The control box regulates the active daylighting system to track the sun path; for instance from an hour after sunrise to an hour before sunset allowing the end user/sensor to turn mechanical lights off during daylight hours. Mirrors or lenses direct natural sunlight through a highly reflective light duct and diffuser lens into the interior. A number of active daylighting systems are categorized under commercially available systems, demonstration systems, prototype stage systems, and theoretical ideas (Mayhoub 2014).

9.8.2 Commercially Available Systems

Sunlight redirecting method. These systems generally use GPS to track sun path and redirect sunlight into a light well using mirrors. Some cases of sunlight redirecting method are as follows:

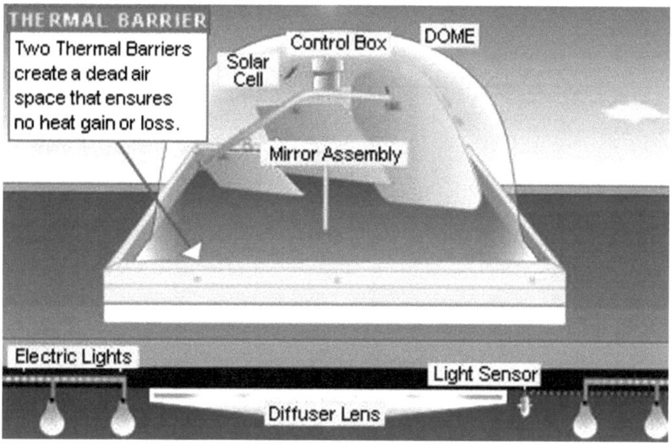

Fig. 9.17 Functional components of active daylighting system

1. *Ciralight SunTrackers*

Ciralight SunTrackers (Fig. 9.18) uses a solar powered GPS motor under a dome that keeps single or triple dynamic and highly reflective aluminum mirrors in perfect alignment with the sun. Daylight is reflected by the mirror controlled by GPS. Meanwhile a clear, thermally formed, highly impact resistant acrylic or polycarbonate dome provides superior strength and UV resistance.

Figure 9.19 shows illuminance levels in a retail depot with installed Ciralight SunTrackers projecting daylight into the interior space. It was predicted that

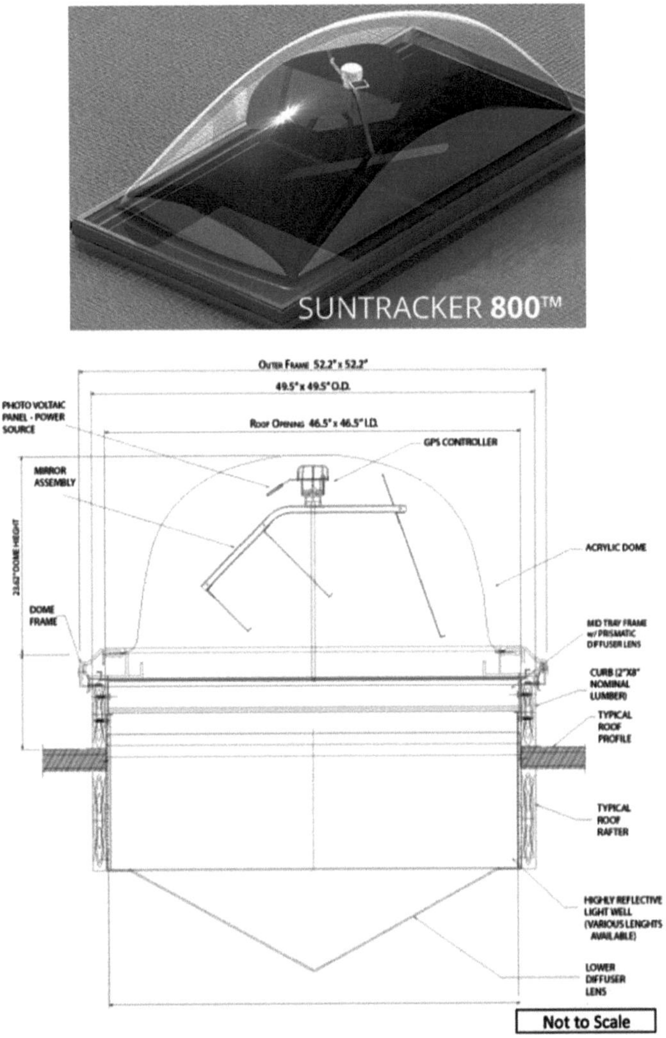

Fig. 9.18 Ciralight SunTracker: (*top*) image and (*bottom*) schematic view (Ciralight 2015)

Fig. 9.19 Office Retail Depot, Greensboro, North Carolina (electrical lighting is off) (Ciralight 2015)

Table 9.1 Comparison analysis

Product	Technology	Roof area required	VLT	SHGC	U-value	Payback (years)
Ciralight SunTracker	Active	1–2 % of roof	0.91	0.32	0.40	2–4
Solar tube	Passive	5–10 %	0.65	0.49	0.82	16
Normal skylights	Passive	7–10 %	0.63	0.55	0.76	N/A

reduction in energy use alone will provide % a 15–35 % return on investment for the retail depot. From Table 9.1, Ciralight SunTrackers require less roof area, provide more light for longer hours with superior thermal comfort, and has the shortest payback period as compared to some passive daylighting techniques (Ciralight 2015).

2. *Sundolier System*

Sundolier uses a computer controlled two-axis tracking system (with mounted mirrors) and a luminaire daylight harvester to collect maximum available daylight (Figs. 9.20 and 9.21).

Fig. 9.20 Sundolier system: (*left*) macrograph and (*right*) building scale sectional view (Sundolier 2015)

Fig. 9.21 Warehouse daylighting with Sundolier systems Loveland, Colorado (Sundolier 2015)

Concentrating Method. These systems use a sun tracking sensor to track the sun path and collected sunlight is concentrated through highly efficient lenses; connected optical fibers transmit sunlight into living spaces. Some cases of sunlight concentrating method are as follows:

1. *Himawari Solar Lighting System*

The Himawari daylighting system is said to be the first commercial optic daylighting system (Mayhoub 2014). It has an efficient collection of sunlight throughout the day using a combination of lenses and optical fiber cable (Fig. 9.22). The outdoor collector catches the maximum quantity of sunlight throughout daytime. The collector consists of two parts, namely: a set of highly efficient lenses and a sun-tracking sensor driven by twin-axle motors and covered by an acrylic dome.

The other is a controlling unit housed in a supporting base. The lenses automatically track the sun (together with an internal clock mechanism, a sun-sensor, and a microprocessor to calculate the position of the sun) throughout the day to

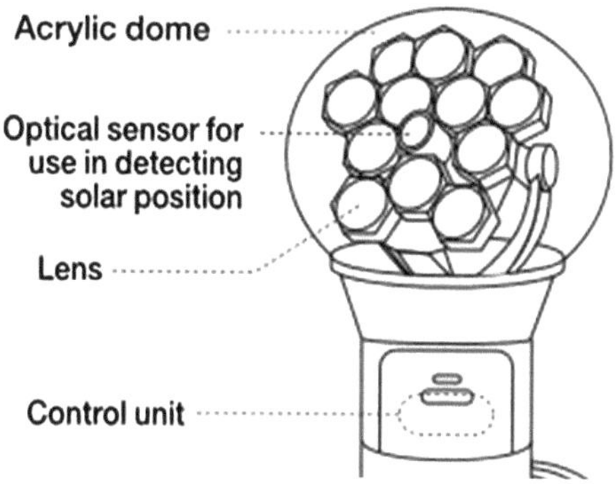

Fig. 9.22 Schematic view of Himawari Solar Lighting System (Himawari 2015)

position themselves at a right angle to the sunlight. In fine weather, the exact location of the sun is ascertained by the sensor. When the sun is hidden behind clouds, the system switches to its internal clock mechanism. Collected sunlight is condensed about 10,000 times through a highly efficient lens. The inlet-end of an optical fiber cable is positioned at the focal point of the lens. The sunlight enters the cable through the inlet and proceeds through the fiber cable, repeating its full reflection, and is emitted from the outlet of the fiber cable. A highly pure quartz glass fiber cable transmits the visible rays with very little attenuation. A cable made of flexible optical fibers can be installed in any new or existing building to transmit sunlight to wherever it is needed (Fig. 9.23). The entire system shuts off after sunset and positions itself so that it's ready for the next sunrise (Himawari 2015).

2. *Parans Solar Receiver System*

Similarly, the Parans system uses an array of lenses to collect sunlight and optical fibers transfer the sunlight through the building structure into the indoor environment (Figs. 9.24 and 9.25).

Fig. 9.23 Applications of Himawari daylighting system: external installations

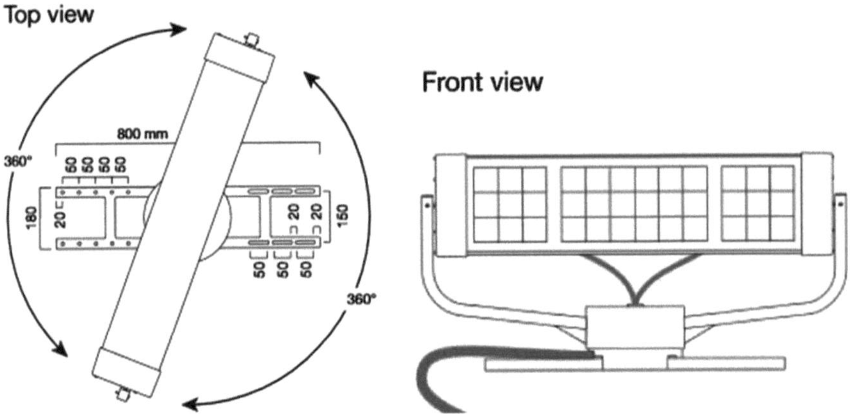

Fig. 9.24 Schematic view of Parans solar receiver (Parans 2015)

Fig. 9.25 Application of Parans daylighting system for a private residential home, UK

9.9 Building Integration of Solar Components

9.9.1 Building Integrated Solar Thermal

Solar thermal systems can be integrated into ZEMCH to either supply the heating demand independently or combined with the existing traditional heating system (gas boiler or electrical heating) for improved energy efficiency. The solar thermal has three main components; a solar collector (SC), energy pack and a hot water storage tank. The additional components depend on the design of integrating the solar thermal system to building and the operational mode of the system.

9.9.1.1 Standalone Solar Water Heating System

The solar collectors can be mounted on the roof, walls or ground to absorb solar radiation to convert into heat which is removed by heat exchanger pipes connected beneath the collector which carry HTF. The HTF enter into an energy pack consisting of a pump and a controller to maintain and regulate HTF flow in the heat exchange through the SC. After releasing heat to the water inside the tank, HTF is pumped back to the solar collectors to be remove the heat and this process keeps on as long as there is enough solar radiation incident on the solar collector (Fig. 9.26). The controller inside the energy pack ensures the heat transfer fluid is directed inside the hot water tank only when the required temperature is reached (Garg 1985).

9.9.1.2 Hybrid Solar and Traditional Water Heating System

Depending on the climatic conditions, weather fluctuation and heating demand of a house, the solar thermal collector may not supply all the heating demand and need an auxiliary heating system through oil/gas fired boiler or electrical heater. The hybrid system is little more complex and bulky compared to stand alone system employing an additional boiler, pump and heat exchanger between boiler and storage tank and a controller between the boiler and the collector. The controller interfaced with boiler and pump regulates the HTF flow rate depending on the collector outlet temperature. The controller interfaced with the boiler and water storage tank controls the turning on and off of the auxiliary boiler based on water temperature inside the water storage tank. In case the water temperature at the outlet of the storage tank being supplied for domestic heating drops below the desired temperature, the controllers enforces the auxiliary heating set up to turn on and

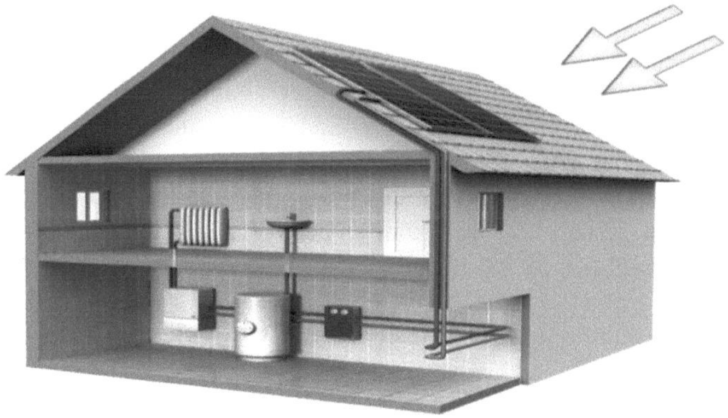

Fig. 9.26 Roof applied solar hot water system integrated into the existing plumbing and water supply network in a house for water and space heating applications without auxiliary heating (Vantage 2013)

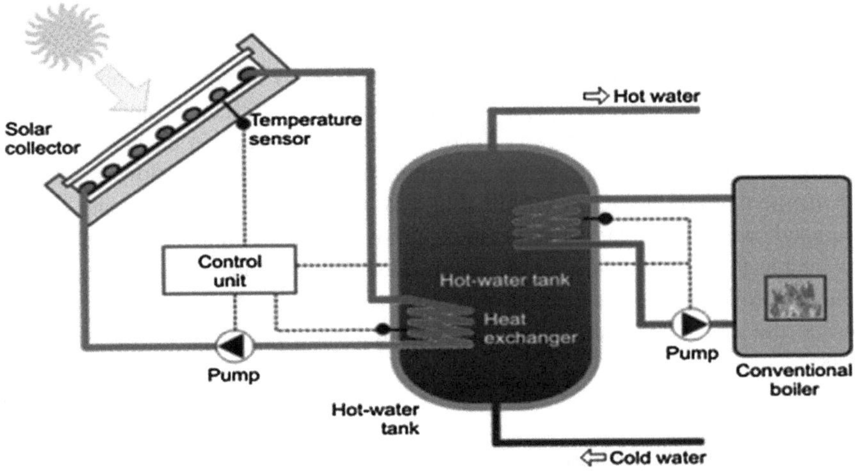

Fig. 9.27 Hybrid water heating system employing a solar thermal system coupled with the conventional oil/gas fired boiler to guarantee a stable temperature from the system (Pradhan 2015)

supply the additional thermal energy required to maintain a stable temperature at the outlet of the storage tank (Fig. 9.27).

9.9.1.3 Solar Assisted Desiccant Dehumidification System

A desiccant dehumidification system is an alternative means of removing humidity from air by moisture absorption in a desiccant material contained in a rotating wheel between incoming outdoors air and return air. The technique is very energy efficient compared to condensation coil in a traditional cooling system. The system consists of a rotating wheel containing desiccant material which removes humidity from the incoming hot air and supplies the air after cooling through cooling coil. The desiccant on absorbing moisture becomes saturated in humidity and needs to discharge the moisture to regenerate to be able to remove humidity on the next run. The cooled less humid return air from indoors can remove the humidity from the saturated wheel however it needs to be heated up to 80 °C in order to dry the desiccant of moisture. The energy required for this heating is supplied by integrating a solar thermal collector system heat the return air which in turn removes humidity from the desiccant wheel (Fig. 9.28) (Chan 2015).

9.9.2 Building Integrated Photovoltaic (BIPV) Systems

PV systems can be used as an add-on over the building envelope (building add–on photovoltaic system (BAPV)), or integrated into the envelope system (BIPV).

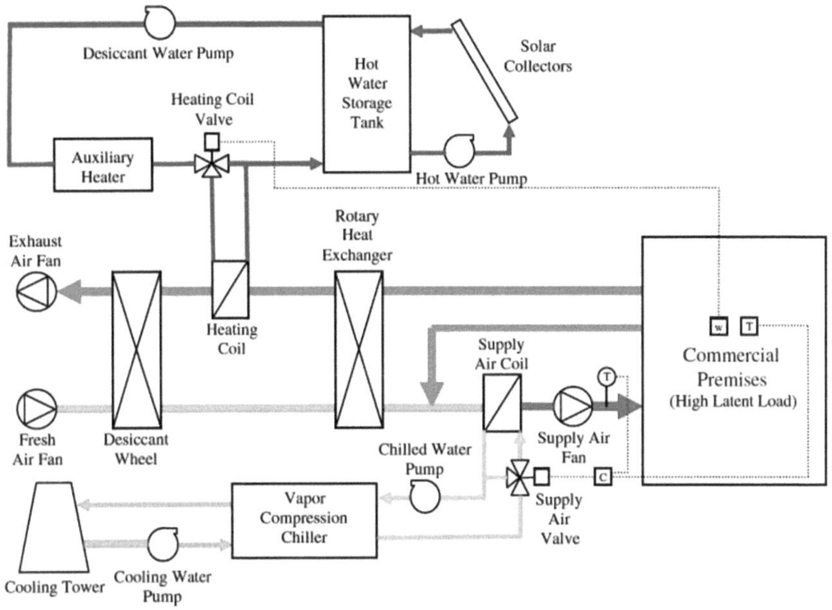

Fig. 9.28 Desiccant dehumidification system employing integrated solar thermal collector system for heating of the return air required for desiccant regeneration in desiccant wheel (Chan 2015)

BAPV requires additional mounting systems while the BIPV system is an integral part of the building envelope construction and has therefore the potential to meet all its requirements (such as mechanical resistance, weather protection, etc.). For instance in Canada, BIPV systems are estimated to have the potential of providing 46 % of the total residential energy needs (Pelland and Poissant 2006). This figure is determined based on a conservative methodology which estimates the available area of roofs and facades for integration of grid connected PV systems, while accounting for architectural and solar constraints (Technical Report IEA-PVPS T7-4 2002).

9.9.2.1 Integration of PV in the Building Envelope

Integration of PV panels into the building design as BIPV is gaining much attention. For instance, the International Agency of Energy (IEA) has launched IEA Task 41—Solar Energy and Architecture (IEA Task 41 2009) to investigate the architectural integration of solar collectors in buildings. The mission of this task includes the identification of barriers for integration of solar collectors, providing guidelines for integration and identification of successful examples of architectural integration of PV systems and solar thermal collectors, around the world. Advantages of building-integrated photovoltaic systems include architectural, technical and financial aspects. Some of these advantages are summarized in the following:

- The electricity is produced on site, thus reducing the cost and impact of transport and distribution (Müller 2005).
- Elimination of the structural framework required to support free standing solar collectors. This can help in offsetting the cost associated with the additional support structure, as well as the cost of multiple roof penetrations for the supports (Pearsall and Hill 2001).
- BIPV panels are designed to substitute the external skin of the building envelope (i.e. PV as a cladding), or to substitute the whole technological sandwich (e.g. semitransparent glass-glass modules as skylights), and therefore it can counterbalance the price of the building materials and systems it replaces (Pearsall and Hill 2001).
- No additional land area is required, since the building surfaces are used to mount the system, thus allowing its application in dense urban areas (Pearsall and Hill 2001).
- PV systems have generally a long life span and require no maintenance (Müller 2005).
- PV systems offer a multitude of architectural design solutions, ranging from urban planning scale to specific building components (e.g. shading devices, spandrels, etc.) (Kaan and Reijenga 2004).

BIPV systems have few disadvantages as compared to add on PV modules (BAPV), the most significant is its higher cost (Pearsall and Hill 2001). This cost however is continuously decreasing (see below). The application of BIPV systems is more suitable for new buildings than to retrofitted buildings.

9.9.2.2 Aspects of Integration

A multidisciplinary approach is required to achieve a successful integration of BIPV systems. Several aspects should be considered including architectural, functional and technical aspects. A summary of some of these considerations is presented below.

- Architectural/aesthetic integration: Several ways of architectural integration have been identified (IEA Task 7 2000). These include neutral integration, where the system does not contribute to the appearance of the building, or prominent integration, where the BIPV system is distinguished from the total building design. An important criterion of a good architectural integration is the overall coordination with the design of the building.
- Functional integration: Solar collectors can be engineered to serve multiple functions. Examples include passive solar design elements (awnings, light shelves, etc., see below) and as roof and façade cladding materials (Keoleian and Lewis 2003).

- Technical integration: This refers to the integration with the building systems, such as the structural, mechanical and electrical systems. Electrical integration includes voltage and current requirements, wiring methods, in addition to the utility integration.

Building envelope incorporating BIPV systems must be designed to resist water infiltration that may penetrate the framework into the BIPV interlayers, must provide a weather seal and control thermal transfer. In addition, the BIPV systems must be able to withstand the stresses that a building envelope is subjected to, including thermal expansion.

9.9.2.3 Methods of Integration of PV in the Building Envelope

BIPV systems can be designed to cover a part or total area of roofs or facades, or as added components on these surfaces.

Roofs:

- There is an intense interest to integrate PV systems in the roofs especially in residential or low rise buildings, since it can provide an ideal exposure to solar radiation. BIPV products are becoming commercially available, that can substitute some types of traditional roof claddings such as tiles, shingles and slates (Fig. 9.29). These BIPV products are developed to match existing building products and are therefore compatible with their mounting systems.
- Prefabricated roofing systems (insulated panels) with integrated thin film laminates (Fig. 9.30) are starting to penetrate the market as well. These PV "sandwiches" constitute complete PV systems that comprise PV modules with mounting and interface components. Such products often include dummy elements to facilitate the aesthetical integration.

Semi-transparent PV systems can be used in skylights, where semi-transparent crystalline or translucent thin film panels are most commonly employed.

Façades:

- A PV system can substitute the external layer of façades, as a cladding component, or it can substitute the whole façade system (e.g. curtain walls—opaque or translucent). In the case of PV as external cladding, the back is usually ventilated, to avoid overheating of the panels and lowering the electrical efficiency of the system. The heated air can be employed for space or water heating. Different curtain wall structures can offer a multitude of architectural appearances (Fig. 9.31).

External Components:

PV panels can be employed as external components to serve various functions such as shading devices, spandrels or balcony parapets, as well as window shutters and awnings (Fig. 9.32).

Fig. 9.29 (*Top left*) Shingle PV, (*top right*) PV Tiles, (*bottom left*) PV slates (Uni-Solar 2011), (*bottom right*) PV laminates (Solar Power Panels 2011)

Fig. 9.30 (*Left*) Solar sandwich (Best Solar Energy 2011), (*right*) solar roof system (Systaic 2011)

9.9.2.4 Cost of PV Systems

The average price of PV is currently around 3$ per Watt peak (Wp) (Solarbuzz 2011). Due to market extension and the increased production volume, prices are dropping steadily (Fig. 9.33). A feasible long term cost potential of PV module that

Fig. 9.31 (*Left*) GreenPix Media Wall, (Beijing, China (© Simone Giostra & Partners/Arup); (*right*) Solar decathlon (façade from Onyx)

Fig. 9.32 (*Left*) PV panels as window shutters, (Colt international 2011); (*right*) Solar awnings, (Solar Awning inbalance-energy.co.uk)

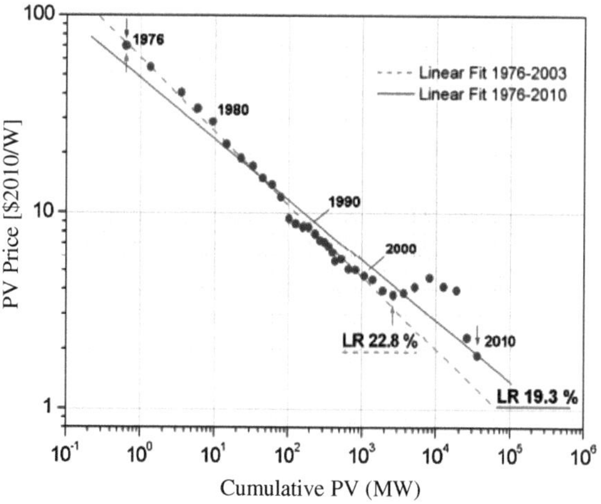

Fig. 9.33 PV cost index per cumulative production (Breyer and Gerlach 2010)

ranges between 0.3 USD/Wp and 0.6 USD/Wp is estimated (Curtright et al. 2008; Pietzcker et al. 2009). At this price, it could be economical to integrate PV systems not only on the equatorial facing roofs and/or facades, but also on west and east sides of the building.

PV price depends directly on the Watt-peak capacity of a panel. *Watt Peak* is defined as the power a module can deliver under standard test conditions (solar irradiation of 1000 W/m^2, and temperature of cells at 25 °C). Consequently, it is possible, given a predetermined budget, to cover different surface areas of roofs or façades by using different available technologies. For instance the decision maker can opt for a small PV area (used as overhangs or other similar functions), where the PV system has high electrical efficiency and high price, or to a large surface area, with low electrical efficiency and low cost.

The financial return of PV electricity generation can be broken into two main parts: (1) Initial immediate return through subsidies (see below), and (2) Energy saving and selling to the utility (Holbert 2009). Immediate return on the cost of PV systems is usually obtained through utility and governmental incentives that provide investment subsidies. In these cases the authorities refund part of the cost of installation of the system, as well as offering the owners (in commercial and residential projects) a premium price for all the renewable power produced at their site.

Moreover, selling the excess electricity to the grid can reduce the payback period for the original cost of PV systems. Currently, there are several incentives that buy the electricity at a price that can significantly reduce the payback period. These incentives include feed-in tariffs (FIT), and time of use (TOU). These are summarized below.

- Feed-in tariffs/net metering: the electricity utility buys PV electricity from the producer under a multiyear contract at a guaranteed rate. The solar buyback rate can be large enough to cover the cost for the remaining electricity need at the going residential rate. Currently in Canada, only Ontario offers significant incentive through the feed-in tariff program for renewable energy employed to encourage installation of PV systems for renewable energy generation (CMHC 1998).
- Time of use plans: According to this plan, the cost of electricity varies as a function of time and day (due to demand variations). When demand is high the electricity price is high and vice versa. In locations where prices of electricity vary with TOU, annual return on energy produced may be a more important object than the amount of energy produced. Favorable timing of the PV electricity generation can increase its value by up to 20 %. This premium value of PV can be improved by 30–50 % when price responsive demand and peaking prices strategies are used (Borenstein 2008).

9.9.3 *Building Integrated Photovoltaic Thermal (BIPVT) Systems*

One major drawback of PVs is low efficiency, relative to the amount of incident solar radiation (Anderson et al. 2009). Aside that, the temperature of PV modules reduces the efficiency of PV systems (Brinkworth et al. 1997). More so for BIPV systems where the PV modules are integrated together with other materials, as functional building envelope components. The extraction of heat from the space between PV modules and other building envelope components will lower the PV module temperature and thus a desirable effect on increasing the efficiency of the entire system (Kim et al. 2014). The extracted heat can be used for space heating and service hot water systems. The end product is hybrid building-integrated photovoltaic-thermal (BIPVT) system which integrates PV and solar thermal collector technologies, as well as other building envelope components such as insulation; the entire system utilizes the sun's energy to simultaneously produce electrical and heat energies (Fig. 9.34). Heat extraction can be achieved by a fluid stream like air or water (Kim and Kim 2012). Due to sharing of resources like materials and functions in the integration, BIPVT system becomes cheaper than separate products (Agrawal and Tiwari 2010).

BIPVT Case I. The Solar XXI building was constructed in 2006 at LNEG campus in Lisbon, Portugal. It has been integrated into the IEA database comprising about 200 nearly, net zero or positive energy buildings. The building integrates passive solar designs augmented with BIPVT. The BIPVT is installed in the south façade as an integral form, creating an air cavity was created between PV back sheet and the building envelope. As such the system is constituted of an outer layer (PV modules) and an inner layer (building envelope). Fitted vents are mechanically operated depending on occupant comfort needs and weather conditions to admit or dismiss warmed air (Fig. 9.35). During high solar radiation winter days, air heated by BIPVT can reach temperatures of 30 °C (Aelenei and Gonçalves 2014).

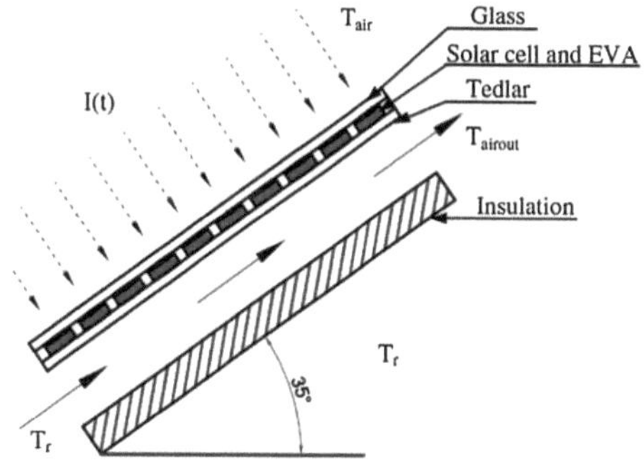

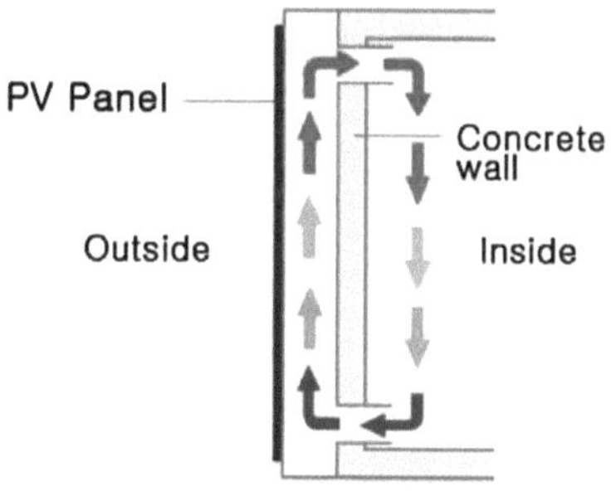

Fig. 9.34 Schematic of: (*Left*) roof top BIPVT (Agrawal and Tiwari 2010) and (*right*) façade BIPVT (Kim and Kim 2012)

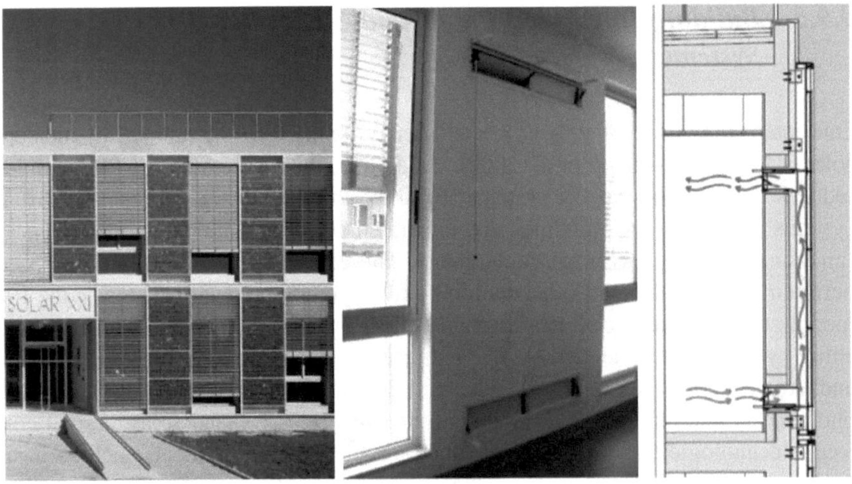

Fig. 9.35 Solar XXI BIPVT system: (*Left*) façade, (*middle*) operable vents behind BIPVT, and (*right*) Schematic diagram showing flow direction of air (Aelenei and Gonçalves 2014)

BIPVT Case II. The John Molson School of Business (JMSB) building solar façade in Concordia University was constructed in 2008. On sunny days during heating season, large amounts of fresh air are heated with a temperature increase of about 20 °C. The warmed air is then delivered to the building HVAC system where it is further heated if required (NSERC 2012) (Figs. 9.36 and 9.37).

Fig. 9.36 BIPVT façade (NSERC 2012)

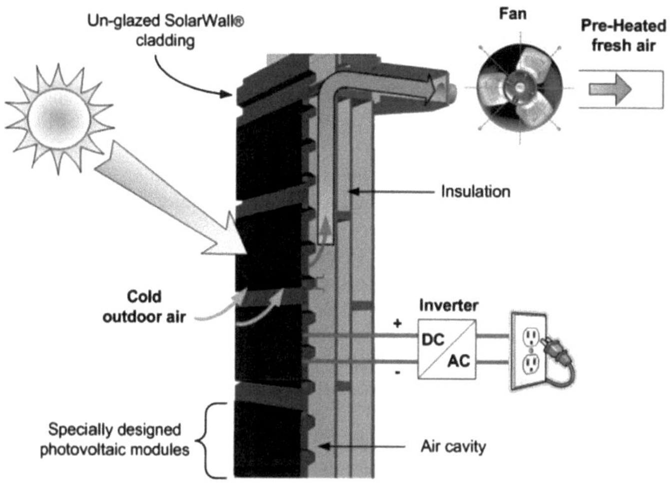

Fig. 9.37 Process flow diagram (NSERC 2012)

References

Aelenei L, Gonçalves H (2014) From solar building design to net zero energy buildings: performance insights of an office building. Energy Procedia 48:1236–1243

Agrawal B, Tiwari G (2010) Life cycle cost assessment of building integrated photovoltaic thermal (BIPVT) systems. Energy Build 42:1472–1481

Anderson TN, Duke M, Morrison G, Carson JK (2009) Performance of a building integrated photovoltaic/thermal (BIPVT) solar collector. Sol Energy 83:445–455

Apricus (2015) Solar thermal collectors. Available via DIALOG. http://www.apricus.com/solar-thermal-collectors.html#.VlYQltIrLIU. Accessed 25 Nov 2015

Atmospheric Water Systems, Inc. (2010) What does AWS atmospheric water really cost? San Luis Obispo CA

Borenstein S (2008) The market value and cost of solar photovoltaic electricity production. Working paper, Center for the Study of Energy Markets (CSEM) University of California Energy Institute, Berkeley campus

Breyer C, Gerlach A (2010) Global overview on grid-parity event dynamics, 25th EUPVSEC/WCPEC-5, Valencia, 6–10 Sept

Brinkworth B, Cross B, Marshall R, Yang H (1997) Thermal regulation of photovoltaic cladding. Sol Energy 61:169–178

Chan D (2015) Solar energized liquid desiccant air conditioning—a review (Part II). Available via DIALOG. http://marine-services-singapore.blogspot.ae/2015/02/solar-energized-liquid-desiccant-air.html

Charron R, Athienitis A (2006) Design and optimization of net zero energy solar homes. ASHRAE Trans 112:286–295

Ciralight Global Inc. (2015) Ciralight's Suntracker daylighting devices. Available via DIALOG. http://www.ciralight.com/. Accessed 25 Nov 2015

CMHC (1998) Tap the Sun: passive solar techniques and home designs. Canada Mortgage and Housing Corporation, Ottawa

Coastal ventilation (2015) Heat recovery ventilator system. Available via DIALOG. www.coastalventilation.co.uk. Accessed 25 Nov 2015

Curtright AE, Morgan MG, Keith DW (2008) Expert assessments of future photovoltaic technologies. Environ Sci Technol 42:9031–9038
Duffie JA, Beckman WA (2006) Solar engineering of thermal processes. Wiley, Hoboken
Energy Center of Wisconsin (2010) Wind power in Wisconsin. In: Energy Center of Wisconsin. Available via DIALOG. http://www.ecw.org/windpower/web/index.html of Energy Center of Wisconsin
Energy.gov (2010) Heat exchangers for solar water heating systems. Available via DIALOG. http://energy.gov/energysaver/heat-exchangers-solar-water-heating-systems. Accessed 24 Nov 2015
Fernández-García A, Zarza E, Valenzuela L, Pérez M (2010) Parabolic-trough solar collectors and their applications. Renew Sustain Energy Rev 14:1695–1721
Garg H (1985) Solar water heating systems. D.Reide Publishing Company, New Delhi
Green Systems UK (2015) Solar thermal glossary. Available via DIALOG. http://greensystemsuk.com/solar-thermal-glossary/ Accessed 23 Nov 2015
Green MA, Emery K, Hishikawa Y, Warta W (2011) Solar cell efficiency tables (version 37). Prog Photovoltaics Res Appl 19:84–92
Himawari (2015) Solar lighting system. Available via DIALOG. http://www.himawari-net.co.jp/e_page-index01.html. Acesssed 25 Nov 2015
Holbert KE (2009) An analysis of utility incentives for residential photovoltaic installations in Phoenix, Arizona, 39th North American Power Symposium (NAPS 2007)Washington State University, Downloaded from IEEE Xplore on 10 Jan 2010
IEA Task 7 (2000) Photovoltaic power systems in the built environment, Report IEA-PVPS 7-03: 2000, Photovoltaic building integration concepts. In: Proceedings of the IEA PVPS Task VII Workshop. EPFL, Lausanne, Switzerland, 11–12 Feb 1999
IEA Task 41 (2009) Solar energy and architecture, solar heating and cooling programme, International Energy Agency. Available via DIALOG. http://www.iea-shc.org/task41/. Accessed 14 Apr 2011
Kaan H, Reijenga T (2004) Photovoltaics in an architectural context. Prog Photovoltaics Res Appl 12:395–408
Kemp WH (2006) The renewable energy handbook: a guide to rural energy independence, off-grid and sustainable living. Aztech Press, Tamworth, ON
Keoleian GA, Lewis GMD (2003) Modeling the life cycle energy and environmental performance of amorphous silicon BIPV roofing in the US. Renew Energy 28:271–293
Keough R, Mullaley V, Sinclair H, Walsh G (2014) Design, fabrication and testing of a water current energy device. Project report, Memorial University of Newfoundland
Kim J-T, Kim G (2010) Overview and new developments in optical daylighting systems for building a healthy indoor environment. Build Environ 45:256–269
Kim J-H, Kim J-T (2012) A simulation study of air-type building-integrated photovoltaic-thermal system. Energy Procedia 30:1016–1024
Kim J-H, Park S-H, Kang J-G, Kim J-T (2014) Experimental performance of heating system with building-integrated PVT (BIPVT) collector. Energy Procedia 48:1374–1384
Koroneos C, Spachos T, Moussiopoulos N (2003) Exergy analysis of renewable energy sources. Renew Energy 28:295–310
Kotzalas MN, Doll GL (2010) Tribological advancements for reliable wind turbine performance. Philos Trans R Soc Lond A Math Phys Eng Sci 368:4829–4850
Leaf community (2015) Leaf House. Available via DIALOG. http://energy.loccioni.com/wp-content/uploads/2009/10/Brochure_Leaf_House1.pdf
Ludmilla energy-plus housing estate (2015) Landhut Ludmilla. Available via DIALOG. http://www.eneff-stadt.info/en/pilot-projects/project/details/ludmilla-energy-plus-housing-estate-landshut/
Manwell JF, McGowan JG, Rogers AL (2010) Wind energy explained: theory, design and application. Wiley, New Jersey
Mayhoub M (2014) Innovative daylighting systems' challenges: a critical study. Energy Build 80:394–405

Müller HFO (2005) Integration of photovoltaic in low energy buildings, University of Sharjah. J Pure Appl Sci 2
Nabila I, Razika I, Azzeddine M (2012) Effect of glazing number on the performance of a solar thermal collector. Int J Sci Res 2319–7064
Negm HM, Maalawi KY (2000) Structural design optimization of wind turbine towers. Comput Struct 74:649–666
Neuhoff K (2005) Large-scale deployment of renewables for electricity generation. Oxford Rev Econ Policy 21:88–110
NSERC (2012) Building integrated combined solar thermal and electric generation demonstration project at Concordia university. Available via DIALOG. https://www.concordia.ca/content/dam/concordia/docs/quartier-concordia/Solar_PanelsENG.pdf. Accessed 25 Nov 2015
Oh SJ, Chun W, Riffat SB, Jeon YI et al (2013) Computational analysis on the enhancement of daylight penetration into dimly lit spaces: light tube vs. fiber optic dish concentrator. Build Environ 59:261–274
Parans (2015) Product information. Available via DIALOG. http://www.parans.com/eng/sp3/. Accessed 25 Nov 2015
Parera J (2008) Blade for wind turbine. United States of America Patent US 7,435,057 B2, 14 October 2008
Patel K, Patel MP, Patel MJ (2012) Review of solar water heating systems. Int J Adv Eng Technol 146–149
Paul G (2004) Wind power: renewable energy for home, farm, and business, 2nd edn. Chelsea Green Publishing Company, White River Junction, Vermont
Pearce JM (2002) Photovoltaics: a path to sustainable futures. Futures 34:663–674
Pearsall NM, Hill LR (2001) Photovoltaic modules, systems and applications, edMA3.doc, ch. 15
Pelland S, Poissant Y (2006) An evaluation of the potential of building integrated photovoltaics. In: 31st annual conference of the solar energy society of Canada (SESCI). Montréal, Canada, 20–24 August
Pietzcker R, Manger S, Bauer N, Luderer G, Bruckner T (2009) The role of concentrating solar power and photovoltaics for climate protection. In: Proceedings 10th IAEE European energy conference. Vienna, Austria, Sept 7–10
Poissant Y, Kherani NP (2008) PV systems and components. In: 4th Canadian solar buildings conference on SBRN BIPV Workshop, Powerpoint Presentation. Toronto, Canada, 23 June
Pradhan B (2015) Solar thermal technology for rural India. SlideShare
Price H, Lupfert E, Kearney D et al (2002) Advances in parabolic trough solar power technology. J Solar Energy Eng 124:109–125
Qiblawey HM, Banat F (2008) Solar thermal desalination technologies. Desalination 220:633–644
Renew green energy (2015) Mechanical ventilation heat recovery system. Available via DIALOG. www.renewgreenenergy.co.uk. Accessed 25 Nov 2015
Renewables-made-in-Germany (2015) Solar thermal energy. Available via DIALOG. http://www.renewables-made-in-germany.com/en/renewables-made-in-germany/technologies/solar-thermal-energy/solar-thermal-energy/technologies-and-applications.html. Accessed 26 Nov 2015
Sapia C (2013) Daylighting in buildings: developments of sunlight addressing by optical fiber. Sol Energy 89:113–121
Solar Utilities Ltd. (2013) Solar thermal. http://www.solarutilities.co.uk/how-solar-thermal-works.php. Accessed 08 Nov 2015
Sundolier (2015) Lighting system product overview. Available via DIALOG. http://www.sundolier.com/. Accessed 25 November 2015
Tangler JL (2000) The evolution of rotor and blade design. National Renewable Energy Laboratory, NREL/CP-500-28410
Tinox Energy (2015) Solar absorber coatings. Available via DIALOG. http://www.almecogroup.com/uploads/1074-ALMECO_TinoxEnergy_ENG_S402_05_2013_mail.pdf. Accessed 24 Nov 2015

Tripanagnostopoulos Y, Nousia Th, Souliotis M, Yianoulis P (2002) Hybrid photovoltaic/thermal solar systems. Sol Energy 72:217–234

Vantage Energy (2013) Solar energy. Available via DIALOG. http://www.venergy.ca/solarenergy.html# Accessed 26 Nov 2015

West Devon Borough Council (2013) Micro scale wind turbines and permitted development. Available via DIALOG. http://www.westdevon.gov.uk/CHttpHandler.ashx?id=5557&p=0. Accessed 25 Nov 2015

Wiginton LK, Nguyen HT, Pearce JM (2010) Quantifying roof top solar photovoltaic potential for regional renewable energy policy. Comput Environ Urban Syst 34:345–357

Woody T (2015) Water-cleaning technology could help farmers. Available via DIALOG. http://www.nytimes.com/2014/02/17/technology/water-cleaning-technology-could-help-farmers.html?_r=0. Accessed 26 Nov 2015

Chapter 10
Zero Energy Homes

Laura Aelenei, Antonio Frattari, Laurent Riscala, Haşim Altan, Arman Hashemi, Kheira Anissa Tabet Aoul and Masa Noguchi

Abstract In the last 50 years a fifth of the planet's inhabitants had a strong development that deeply changed their habits and their life quality. For this enhancement, the people of the developed areas paid a high price. A large use of energy, produced from non-renewable sources as fossil fuels, increased the carbon dioxide (CO_2) emissions into the atmosphere with several problems and a huge impact on the nature. As a consequence, there is a need to rethink the design of buildings, cities and their organizations. The challenge for the new sustainable cities is to grow according to the lifestyles of today and tomorrow, while implementing a better relation between the nature and the mankind and restoring the lost human contacts. An option for doing this is to design and develop Zero Energy Homes (ZEH) reducing to the minimum the impact of pollution and the exploitation of non-renewable sources. In particular, the following aspects should be considered: to use of renewable and recycled materials; to improve the energy efficiency of buildings; to introduce more efficient energy systems that use alternative and clean sources; and to introduce building automation systems

L. Aelenei (✉)
Laboratório Nacional de Energia e Geologia (LNEG), Paço do Lumiar, 22, Lisbon 1649-038, Portugal
e-mail: laura.aelenei@lneg.pt

A. Frattari
University of Trento, Trento, Italy
e-mail: antonio.frattari@unitn.it

L. Riscala
ECOxIA, Yerres, France
e-mail: laurent.riscala@ecoxia.com

H. Altan
British University in Dubai (BUiD), Dubai, UAE
e-mail: hasim.altan@buid.ac.ae; hasimaltan@gmail.com

H. Altan
Canadian University Dubai (CUD), Dubai, UAE

H. Altan
Cardiff University (Welsh School of Architecture, WSA), Cardiff, UK

© Springer International Publishing Switzerland 2016

M. Noguchi (ed.), *ZEMCH: Toward the Delivery of Zero Energy Mass Custom Homes*, Springer Tracts in Civil Engineering,
DOI 10.1007/978-3-319-31967-4_10

(to optimize the energy consumption). In this lecture the following topics will be presented: definition of ZEH, including a review of definitions, parameters influencing the definition and examples, criteria to build or refurbish to a ZEH standard and some questions and examples related with the design, construction and operation of new Zero Carbon Homes.

10.1 Definition of the ZEH

10.1.1 Introduction—Holistic Approach and Definitions Review

Zero-energy buildings have gained more attention since the publication in 2010 by the European Union Council of the Energy Performance Building Directive (EPBD) recast (EPBD 2010). According to Directive, by 31 December 2020, all new buildings should meet higher levels of performance than before by exploring more the alternative energy supply systems available locally on a cost-efficiency basis and without compromising the comfort in order to ensure that they are nearly zero-energy buildings. A "nearly zero-energy building" refers to a high energy performance building of which annual primary energy consumption is covered to a very significant extent by energy from renewable sources, including energy from renewable sources produced on-site or nearby. Since the Directive does not specify minimum or maximum harmonized requirements as well as details of energy performance calculation framework, it is up to the Member States to define the exact meaning of "high energy performance" and "amount of energy from renewable sources" according to their own local conditions and strategic interests. Although the European Directive refers to "nearly" zero-energy building, the terminology used for this building performance can also be referred as "Zero-Energy Buildings", "Net Zero-Energy Buildings", "NZEB", "NetZEB", "nZEB".

Zero-Energy Buildings have been the object of various studies in the recent years as various countries have set this performance as a long-term goal of their energy policies (Torcellini et al. 2006; Ayoub 2009; Aelenei et al. 2011; Sartori et al. 2012; Marszal et al. 2011).

A. Hashemi
University of Brighton, Eastbourne, UK
e-mail: A.Hashemi@brighton.ac.uk

K.A.T. Aoul
United Arab Emirates University, Abu Dhabi, UAE
e-mail: Kheira.Anissa@uaeu.ac.ae

M. Noguchi
The University of Melbourne, Melbourne, Australia
e-mail: masa.noguchi@unimelb.edu.au

Zero energy housing falls into two categories: self-sustainable or net (Noguchi 2008). The former type is a standalone house whose operational energy relies solely on its own power generation and storage so that it is disconnected from a commercial grid or disuses the power from the outside sources. The latter is the one whose energy 'use' becomes net zero over a fixed period of time. In addition, a house whose energy 'bill' becomes net zero under the same conditions is termed net zero-energy-cost housing. The notion of zero carbon housing today is from time to time likened to that of the above-mentioned homes; perchance, the performance may entail the further steps to cover CO_2 emissions that derive from not only the operation but also the construction and demolition—i.e. over the house's full life cycle.

Regarding Net Zero Energy Buildings (NZEB) performance, four main types of NZEBs can be identified: Net Zero Site Energy, Net Zero Source Energy, Net Zero Energy Cost and Net Zero Energy Emissions (Marszal et al. 2011; Aelenei et al. 2013). Net Zero Site Energy means that the annual balance is based on the grid interaction at the boundary of the building site, i.e. the overall energy delivered to the building from the utility grid has to be offset by the overall energy feed into the grid. In the Net Zero Source Energy definition, which is the one that matches the currently used by EPBD recast in a nearly zero-energy context (EPBD 2010), the energy (delivered from and feed into the grid) has to take into account primary energy conversion factors. Net Zero Energy Cost buildings definition is based on an economic balance (the energy bills of a building are equivalent to the amount of money the utility pays the owner for renewable energy the building feeds into the grid) whereas in the Net Zero Energy Emissions case, buildings produce and export at least as much emissions-free renewable energy as they import and use from emission-producing sources on an annual basis (Torcellini et al. 2006). The same author identified the following main definitions of a Zero Energy Building (Torcellini et al. 2006): Net Zero Site Energy: a site ZEB produces at least as much energy as it uses in a year, when accounted for at the site; Net Zero Source Energy: a source ZEB produces at least as much energy as it uses in a year, when accounted for at the source. Source energy refers to the primary energy used to generate and deliver the energy to the site; Net Zero Energy Costs: in a cost ZEB, the amount of money the utility pays the building owner for the energy the building exports to the grid is at least equal to the amount the owner pays the utility for the energy services and energy used over the year; Net Zero Energy Emissions: a Net Zero Emissions Building (NZEB) produces at least as much emissions-free renewable energy as it uses from emissions-producing energy sources (in other words, a Zero Carbon Building).

Although there is no standard approach for designing and realizing a Net Zero Energy Building (there are many different possible combinations of building envelope, utility equipment and on-site energy production equipment able to achieve net-zero energy performance and also the balance boundary, which defines which consumers are included in the balance differs in known approaches) there is some consensus that zero energy buildings (ZEB) design should start from passive sustainable design as this level of performance is achieved as a result of executing two fundamental steps: (a) reduce building energy demand and (b) generating electricity

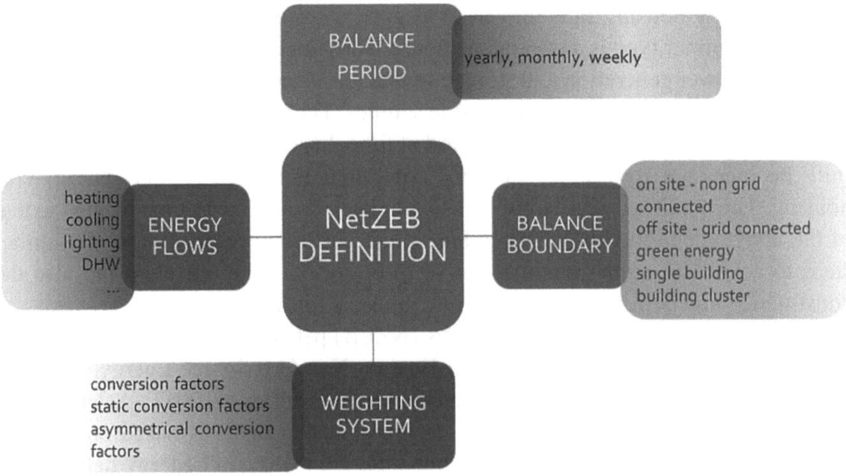

Fig. 10.1 ZEH parameters (Aelenei 2012)

or other energy sources to get enough off-sets to achieve the desired energy balance from renewable energy systems (RES). As one can easily imagine passive approaches play a crucial role in addressing NZEB design as they directly affect the heating, cooling, ventilation and lighting loads of the building's mechanical and electrical systems and, indirectly, the strive for renewable energy generation. The combination of design measures and strategies together with other energy balance parameters (Fig. 10.1) should be considered for a consistent zero energy balance definition.

10.1.2 ZEH Parameters Influencing Definition

10.1.2.1 Energy Balance and Boundary

If one draws an imaginary boundary in the nearby of a building (to account for renewable energy produced on-site and/or nearby), the energy balance may be schematically represented (Fig. 10.2). Accordingly, zero-energy buildings exchange energy with the grids (electricity, heating or cooling, gas or biomass) in the form of energy carriers that is converted from or on to primary sources using credits. Accordingly, the Energy Balance (EB), for different energy carriers, is, between the energy delivered (ED) to building and the energy exported (EE) into the public grids, writes:

$$EB = \sum_{i} EE_i \times f_{e,i} - ED_i \times f_{d,i} \qquad (10.1)$$

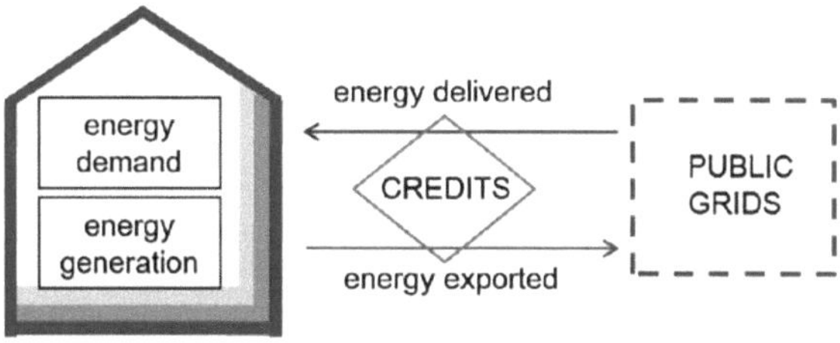

Fig. 10.2 Schematic representation of energy balance of zero energy buildings (Aelenei et al. 2013)

where f are factors which are used to convert the physical units into other metrics, such as primary energy or equivalent carbon emission.

In view of the abovementioned analysis model (Eq. 10.1), one can draw the conclusion that three different scenarios are possible, depending on the value of the energy balance. In the case of a neutral annual energy balance (i.e. the building use no more energy than it produce), the building is commonly referred as a Zero-Energy House. If the building falls short of the neutral balance then it can be referred to as a "nearly Zero Energy House". In the scenario where the balance is positive (when the building produces more energy than it consumes) the building is referred as a Plus Energy Building. The simple balance approach described so far becomes rapidly complex if one considers other features. For instance, if the boundary is drawn around a group of buildings instead (zero energy community), additional concerns regarding grids and conversion factors together with community-based infrastructure and industry need to be considered as well. In such cases, it is possible that plus-energy buildings may provide the additional amount of energy to nearly zero energy buildings from the same community and contribute in this way to the zero balance target of the entire community.

10.1.2.2 Weighting System

Primary energy indicator sums up all delivered and exported energy into a single indicator using primary energy factors. Therefore, the metric of the energy balance should allow comparison of different forms of energy (electricity, natural gas, biomass and solid fuels). Using primary energy as an indicator raises a question concerning the conversion factors that should be applied (Voss et al. 2011). The averaged conversion factors may be either derived from actual national statistics or from European similar figures and they are usually strategically determined in order to give priority to a particular category of energy fuel. A good example is the case of the asymmetrical weighting factors where the primary energy conversion factor for energy delivered by the grid is different from the factor for energy exported into the grid to encourage on-site generation. In cases where carbon dioxide is considered appropriate, conversion factors from primary energy to carbon dioxide can also be considered. This approach provides additional information about the consequences of energy use, in the terms of CO_2 emitted to the atmosphere. However, due to the fact that carbon cycle has a strong dynamic character, accounting for emissions in the same context can be a tricky business (Black et al. 2010).

10.1.2.3 Balance Period

The standard energy calculation procedure is annual due to the need of accounting the whole range of operating energy of a building typical for a complete meteorological cycle. Climate plays a dual role in ZEB (houses mainly) as it is a driver for space heating and cooling and a driver for supplying renewable energy resources at the same time. Using time intervals shorter than one year for calculus of the energy balance (seasonally, monthly or daily) is useful for the analysis of the interaction of the building with the electricity grid and other energy grids (Hermelink 2013). According to the same study, a yearly energy balance is not capable to provide the complete interaction with the grid as this procedure assumes the grid as an infinite storage. Buildings incorporating renewable energy systems are often characterized by a mismatch between the energy need and the energy generated on site. For instance, a seasonal calculus of the energy balance may result positive in summer (due to higher solar potential and lower energy needs) and negative in winter. As the consequences of mismatch are a matter under investigation, perhaps the best strategy to adopt in this respect is to reduce the absolute value of the potential mismatch between demand and local generation (Black et al. 2010). An effective way to reduce the mismatch is to reduce energy needs, a strategy which also provides advantages in terms of economic benefits (low energy buildings are significantly less prone to risks connected to volatility of costs/prices of conventional and renewable energy during their lifetime) and benefits associated to higher thermal comfort and user satisfaction (Hermelink 2013).

10.1.2.4 Balance Type/Energy Flows

The choice of a balance boundary and of which of the different energy end uses is to be included in the balance calculation has a major influence on the ZEB balance. For example, the inclusion or not of electrical plug loads or of central services can make a difference of 100 % in energy that must be generated to create the balance. In operation, to check that the so called energy import (delivered energy) and energy export (green line in Fig. 10.3) target is met requires sophisticated sub-metering. The difference between those two values is "self-consumption" (on the x-axis in Fig. 10.3), which represents the part of on-site generated energy which is instantaneously consumed in the building. It lowers the needed energy import and it is not fed into the grid.

During design analysis, energy use predictions for the balance calculation require three separate aspects of the energy flows to be simulated: the energy needed to maintain comfortable temperatures; the energy needed for appliances, hot water use and so on in the buildings; and the energy generated on site. Simulation

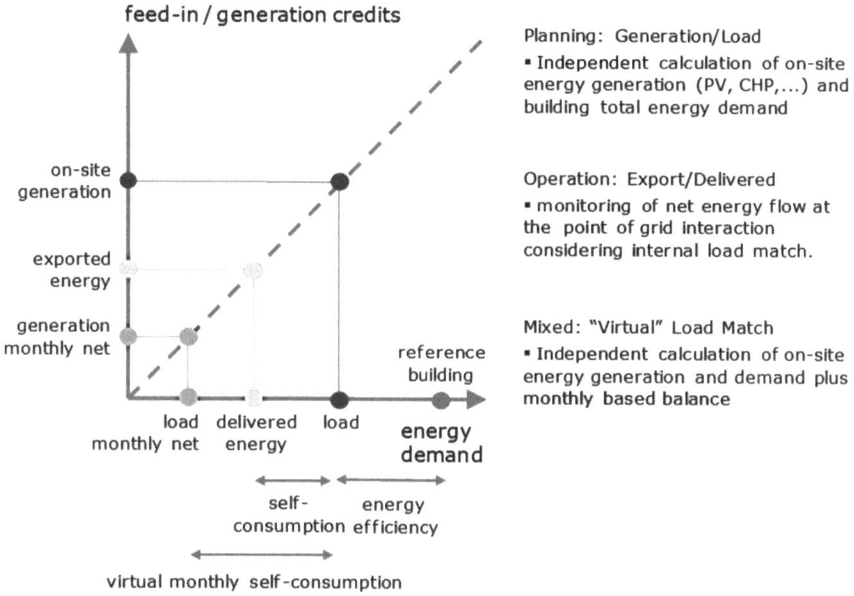

Fig. 10.3 Graphical representation of the three types of balance: import/export balance between weighted exported and delivered energy, load/generation balance between weighted generation and load, and monthly net balance between weighted monthly net values of generation and load (Voss et al. 2012)

of the first and the last of these energy flows is dependent on 'typical' climate data for the location. The variation from year to year of the climate from this typical value is rarely studied. Energy end uses are even more unpredictable being based upon predicting user behaviour. Data on end use patterns for appliances, hot water use, and so on with sufficient time resolution is required and is often unavailable. Because they are easier to predict with a degree of confidence, calculated on-site generation and energy demand are often balanced (load/generation balance, red line in Fig. 10.3), during the planning phase. These quantities do not cross the building system boundary. This approach at design time can often lead to subsequent pronouncements about a performance gap: the difference between the predicted performance and the actual performance (Voss et al. 2012). The standardized norms for people's behaviour and their use of equipment in a building used during design are proving to be about as accurate as the standardized norms for people's behaviour when driving a car that produce the manufacturers' fuel-efficiency comparisons. Put real people in the car (building) and the performance is very different.

In an effort to synthesize many of the issues covered in the previous sections, and to assess the advantages and disadvantages of different strategies and scenarios of NZEB definitions, an excel-based tool (NZEB evaluator tool) was developed by a group of experts from IEA SHC Task 40—ECBCS Annex 52 (International Energy Agency Solar Heating and Cooling Task 40 /ECBCS Annex 52 2008). The

Table 10.1 Net Zero Energy buildings definitions according to IEA

Definitions	Description
(IEA 2012)	**Net ZEB limited** Weighted energy use for heating, production of domestic hot water (DHW), cooling, ventilation, auxiliaries and built-in lighting (for non-residential buildings only) versus weighted energy supplied by on-site generation driven by on- or off-site sources. Static and symmetric primary energy factors are possible
	Net ZEB primary Weighted energy use for heating, DHW, cooling, ventilation, auxiliaries and lighting and every kind of plug loads (electrical car possibly included), versus weighted energy supplied by on-site generation driven by on- or off-site sources. Static and symmetric primary energy factors
	Net ZEB strategic Weighted energy use for heating, DHW, cooling, ventilation, auxiliaries, built-in lighting and every kind of plug loads versus weighted energy supplied by on- and off-site generation systems driven by on- or off-site sources. Weighting factors could be static and asymmetric, varying on the basis of the energy carrier, the technology used as energy supply system and its location
	Net ZEB emission Balance between building CO_2 equivalent emissions due to energy use for heating, DHW, cooling, ventilation, auxiliaries, built-in lighting, every kind of plug loads and the weighted energy supplied by on-site generation systems driven by on- or off-site sources. Static emission factors are used. They can be symmetric or asymmetric, depending on the energy carrier, technologies used as energy supply systems and their location

tool allows checking annual energy or emission balances as well as characterizing the load match and the grid interaction profile of a building by simplified indicators on the basis of four energy balance approaches (Table 10.1).

10.1.3 ZEH Best Practice/Examples

10.1.3.1 Plus Energy House

In the Plus Energy Houses, the supply energy systems installed in the building produce more energy than its owners actually need. In fact the energy balance is positive and exceed the performance of a ZEH or near ZEH, generating a surplus of clean energy (more often solar) into the house's power supply. Schlierberg Solar Settlement (Fig. 10.4) is a project developed by the architect Rolf Disch, a pioneer of solar buildings. The authors identified a full zero-fossil energy balance for the consumption of the 50 ground-based plus-energy terrace houses with heating and electricity on the demand side and solar electricity generation on the generation side.

This project had different goals, one of them being energy monitoring, a full balance of the consumption of the 50 ground based terrace houses with heating and electricity on the demand side and electricity generation on the generation side.

Fig. 10.4 Solarsiedlung am Schlierberg

Another goal was the special attention paid to the integrated urban planning. The orientation and housing density on the site were designed to take into account living quality, unobstructed solar radiation on the PV systems installed on the roof, sun shading for summer protection and sun exposure strategy for heating season (Heinze and Voss 2009).

The houses have a high energy performance as they were designed according to the passive house standards. Between the measures adopted: high insulation with a U-value for the building envelope of 0.28 W/m^2K, efficient ventilation system with heat recovery. Electricity saving appliances and appropriate user behaviour reduced the domestic energy consumption. Water savings tap fittings were installed.

Only a small remaining amount of energy is balanced by the PV yields installed on the roofs. This high energy efficiency on site reduces the consumption of the renewable energy and the requirements on transport and storage of energy in grids (Heinze and Voss 2009).

10.1.3.2 ÉcoTerra

A successful net zero-energy healthy house project is ÉcoTerra house, which won the Canadian federal government's EQuilibrium sustainable housing competition. The house was built in Eastman in the province of Quebec and it is currently open to the general public in order to sharpen the consumers' awareness of commercially available net zero-energy healthy housing today (Noguchi 2008). The house was constructed by making use of Alouette Homes' pre-engineered modular housing

Fig. 10.5 ÉcoTerra house: a during the modular construction and b after the completion

system (Fig. 10.5) that helped eliminate or reduce on-site construction nuisances, such as bad weather, theft and vandalism.

One of the key healthy housing design features applied to the ÉcoTerra house is the thermal comfort maintained by the application of high thermal insulation materials (e.g. U-value of nearly 0.1 W/m^{2}K in external walls) and the air-tight construction. These features are further combined with the continuous, balanced mechanical ventilation. In this house, the air-tightness is maintained at 1 air change per hour (or 1 ACH) at 50 Pa.

For energy generation ÉcoTerra integrated Building Integrated Photovoltaic Thermal (BIPV/T) systems. These BIPV/T systems have a great advantage compared with stand-alone photovoltaic (PV) arrays or solar thermal collectors because they generate both thermal and electrical energy simultaneously (Chen et al. 2007). Therefore, a 3 kW BIPV/T based on a system concept developed at Concordia University was installed in the ÉcoTerra house, having the capacity to produce approximately 12 kW of heat at 14 m^3/min of air flow according to the engineering team's previous experiment (Liao et al. 2007).

10.1.3.3 Casa Zero Energy

The Casa Zero Energy located in Udine (Italy) is not only a simple ZEB, it is also a prototype, which represents a demonstration project for proving that it is possible to build a house able to respect the environment in terms of impacts reduction and improvement of the inhabitants lifestyle (Fig. 10.6). Indeed, the aim of the project was not only the construction of a low-impact and low-energy building, but the construction of a building in which the role and the wellbeing of the users have a relevant importance. For such reasons, the design of the building was oriented to ensure the contact with the nature, guaranteeing a new and better living dimension for the people (Frattari 2013).

Fig. 10.6 Casa Zero Energy

The Casa Zero Energy is the prototype of a building that does not use energy produced from fossil sources, but that produces the needed energy with alternative energy systems. It was designed according to the criteria of bioclimatic architecture and integrated with passive systems for getting advantage from the climatic site characteristics for the heating in winter, and for the inner cooling and ventilation, in summer.

The Casa Zero Energy designed by Arch. Arnaldo Savorelli and Prof. Antonio Frattari, was developed in the years between 2007 and 2010 by the company Gruppo Polo—Le Ville Plus of Cassacco (UD). Some of the main features of the building design are: strong natural and bioclimatic characterization, similarity to passive house concept, the use of natural, renewable and recycled materials for the building construction, a new and innovative anti-seismic timber frame system, a new and innovative envelope to save energy, low energy consumption, the integration with energy systems using alternative and clean sources, the integration with an intelligent system (home automation) (Fig. 10.7) able to optimize the indoor comfort, the energy consumption and safety.

Two interesting aspects can be identified: the house is built with natural, renewable, recycled and recyclable materials. For this reason it is classifiable as a "natural building" and secondly the house is automated, through smart solutions that have been implemented for testing and verifying the utility and the effective use of scenarios for the flexible utilization of automated passive solar systems, quantifying the contribution of automated control for the lighting, shading and conditioning systems, experimenting the possibility to guarantee both safety and security

Fig. 10.7 Casa Zero Energy: **a** The sunspace closed in winter and **b** the sunspace opened in summer

to the house users. In the near future in the context of Smart City paradigm, the buildings need to be interactive/intelligent in order to satisfy the user need, the proper use of renewable energy and the interaction with the grid.

10.1.3.4 The Vermont House

ECOxIA is a French company in the green building industry, whose founders were very influenced by the experience of their Zero Energy Mass Custom Home (ZEMCH) Mission to Japan in 2007. It has been created with the corporate mission to democratize positive energy housing. The definition given back to 2010 was to build dwellings that are so sober in energy, it becomes easy, on a yearly basis, to produce more renewable energy on site than needed to operate the building. The challenge was not only to design a bioclimatic architecture with a very well insulated envelope but also a zero energy house. The company also wanted to develop a building solution that would be versatile (customization) and at an acceptable price premium (mass). The Smart Building Envelope (SBE) was the outcome of those specifications.

As a synthesis of best international practices, theory said that the SBE would have a low carbon footprint at the construction stage thanks to its all-wood structure and the off-site prefabrication; and at the use stage with a passive design (U-value below 0.125 W/m^2K in external walls and floors, and air-tightness below 0.6 air change per hour at 50 Pa). Comfort should be excellent with generous natural light (large window surface), good external sound proofing (design and air-tightness), an excellent indoor air-quality (high-end dual flow ventilation), and ideal indoor temperature and humidity levels in all seasons. Moreover, the architectural creativity should not be too much impaired by the concept of the SBE and the construction cost should remain under control, savings due to planning and off-site production counterbalancing, at least partially, the extra costs of features like triple-glazed windows, energy recovery ventilator or PV panels.

ECOxIA hence decided to build a prototype house to validate its SBE technology. However, more than an easy demo house, it had to be a guinea pig that would prove that the SBE can work in most situations. The house had hence to be single storey, with an elegant flat roof and rather small in size (about 100 m^2), which makes it more difficult from a cost and energy performance standpoint. Designed with American and French architects, the Vermont House was born.

The Vermont House (Fig. 10.8) was installed in Yerres, near Paris, France in November 2012.

The house it was prefabricated under 3D modules (Fig. 10.9a) at 200 km^2 from the location in Normandy with a high level of completeness (full envelope, partition walls, electricity, etc.). The three main modules and the elevated roof were installed on the low-impact foundation in one day. After finishing, the monitoring of the Vermont house in real life could start (Fig. 10.9b). It has been used as an office weekly by 3 employees and as a week-end house since April 2013.

The Smart Building Envelope is a true ZEMCH building solution. The life cycle analysis of the Vermont House measured 10 tons of carbon at the construction

Fig. 10.8 Vermont House by ECOxIA: positive energy Vermont House in Yerres, France

(a) (b)

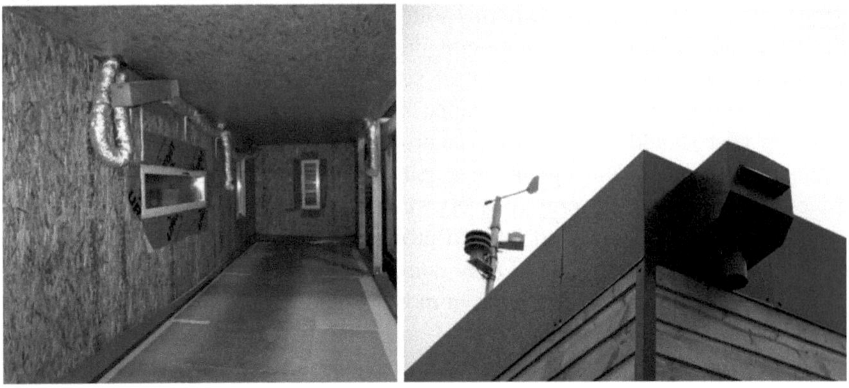

Fig. 10.9 Vermont House: **a** Modules under assembly in Normandy and **b** Weather mast on the roof

stage, and 6 tons of carbon sink. Its net building impact on the environment was 4 tons of carbon equivalent, which is very low.

From a usage standpoint, the purpose was to build a house at Net Zero Energy Cost. After 2 years of experience, the average total consumption for this all electric house is just under 4500 kWh per year, or 116 kWhPE/m^2. The 20 m^2 of PV panels installed on the roof (Fig. 10.10a), with a 3 kWp at a 0 % angle, produce about 2500 kWh per year. Consequently, the ECOxIA prototype yields an energy profit every year. Thanks to the advancement in PV and battery technologies, positive energy or even off-grid housing becomes possible.

From a mass customisation perspective, the Vermont house costed less than € 200,000 to build. Experts found it very cost effective, may be 50 % less than similar houses. However, it is still 20 % more than the new houses on the French market. More standard architecture and economy of scale would bring the premium to 10 %, which would lead to a very high cost performance ratio.

Unless you have a very high level of in-factory completeness and very flexible production units, 3D module prefabrication is not always panacea. As a feedback from the prototyping, the SBE is now delivered on site under flat packs (2D panels that comprise all the components to build passive). This means that the architecture is totally free, within the constraints of physics and bioclimatic principles (Fig. 10.10b).

The Vermont House is an interesting example of how zero energy cost project turned into a Zero Energy Mass Custom Home building solution.

10.1.3.5 LIVINGBOX

LIVINGBOX is a prototype designed by Prof. Antonio Frattari and developed at the Laboratory of Building Design of the University of Trento. It has been designed (Figs. 10.11 and 10.12) as a Near Zero Energy Building to minimize the impact of

(a) (b)

Fig. 10.10 Vermont House: **a** 3 kW PV panels on the elevated roof and **b** Windows facing South are the main radiators of the house

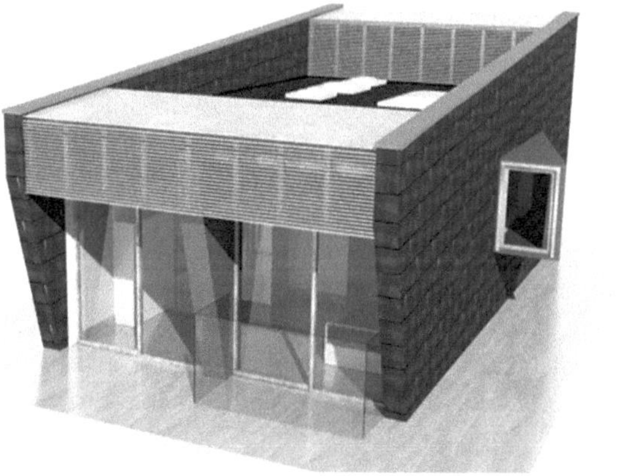

Fig. 10.11 LIVINGBOX 3D design

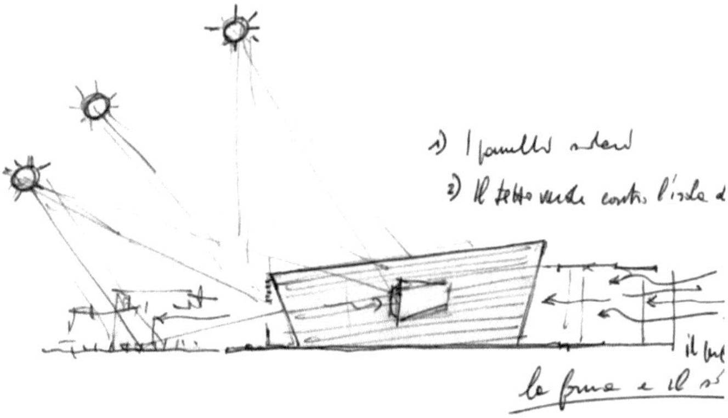

Fig. 10.12 Preparatory sketch for the bioclimatic design of the unit

the building on environmental matrixes: water, air, soil. To reach the target of NZEB it is equipped with systems for producing energy from renewable sources to minimize use of fossil energy. In addition, the unit is a low consumption building. A prototype at real scale has been built to verify the constructive feasibility.

The home is articulated in a large multi-purpose space that can be modified during the day and in a fixed kitchen-bath block. The unit can be expanded from 50 to 75 m^2 gross. In this last case, the kitchen and the bathroom block is bigger. The units can be put together to form terraced houses or block houses up to three storey. The living unit also can be organized functionally as hotel room. In this case the space is articulated in a room of about 15 m^2, in a multi-functional space (between the bedroom and the bathroom access, about 5 m^2), equipped with a wardrobe (length 2.40 m), and a bathroom, about 5 m^2. This unit can generate different types of hotels: blocks up to three storey or small settlements with the units spread over the territory. The living unit has been designed to minimize the impact of the building on environmental matrixes: water, air, soil. The used materials are natural, recyclable or recycled. It is characterized by extensive use of wood to limit CO_2 emission in the atmosphere and it was designed as a Near Zero Energy House.

It is built with load-bearing panels in massive laminated wood: Crosslam. The quantity of wood used embodies around 9.5 tons of CO_2 as a positive contribution to the greenhouse effect. Even the other used materials, mostly recycled, were chosen through a Life Cycle Assessment (LCA) including the aspects that minimize the CO_2 emitted in the atmosphere during the construction, maintenance and disposal. In addition, to further reduce the environment impact, have been studied architectonic and constructive solutions for maximising the free contributions offered from the surroundings in terms of solar and wind contributions to produce clean energy.

The design is based on bioclimatic solutions. A mini-greenhouse is on the south façade to improve winter heating and natural ventilation for summer cooling. LIVINGBOX is opened to the south, partly opened to the east with a screened window to minimize the negative sun in summer and it is totally closed, without any window, on the north front. The unit has been design to minimize the heat island effect through a green roof and, possibly, green walls. Roof and walls are ventilated to decrease use of insulating membranes and to ensure the natural breathability. The employed materials such as the paints are low-emissivity. For example, the water paint with acetic acid used for aging the larch of the interior fittings or of the battens wooden of the outer walls.

In addition, the low consumption of the living unit is enhanced by the envelope's stratigraphy that insures a transmittance value between 0.20–0.25 W/m^2K, and a thermal lag between 10–15 h. With these constructive solutions, the energy consumption is around 16 kW/m^2 for year. Particular attention has been paid to the prevention of thermal bridges realizing a continuous isolated envelope in rock wool. The thermal efficiency is approaching that of a "Passive Envelope" inspired by the "Passive Haus". To optimize the relationship between comfort and energy consumption LIVINGBOX is equipped with a modular home automation system. It allows managing with integrated modular packages the lighting and the heating, the daylighting, the shading and the natural or mechanical ventilation.

10.2 Criteria to Build or Refurbish to a ZEH Standard

As previously pointed out, there are different definitions and approaches to achieve Zero Energy House. The provisions of the Energy Performance of Building Directive (EPBD 2010) introduced in Article 9, "nearly Zero-Energy Buildings" (nZEB): "by 31 December 2020, all new buildings are nearly zero-energy buildings; and after 31 December 2018, new buildings occupied and owned by public authorities are nearly zero-energy buildings" will be the starting point of the project. Anyway, despite the EPBD Recast focuses on new buildings, the energy and CO_2 emissions associated to the existing buildings refurbishment towards nZEB is worth investigating because of the huge potentialities of energy savings.

Figure 10.13 shows a typical retrofit process contrasted with a deep retrofit process. On the right, in green, the deep retrofit process is shown. The blue steps on the right indicate the additional considerations that need to be taken in a net zero energy project process.

The process of taking an existing building to net zero energy is similar to that of a deep energy retrofit (RMI 2015) with some additional considerations. A deep energy retrofit involves a whole-building analysis process that delivers much larger energy cost savings, i.e. sometimes more than 50 % reduction, and fundamentally

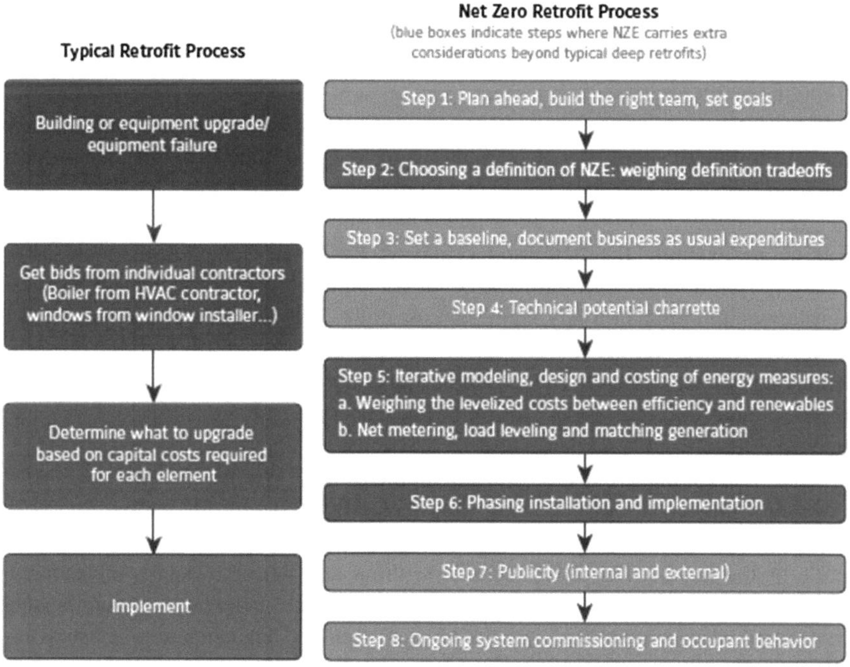

Fig. 10.13 Typical retrofit, deep retrofit and net zero retrofit process considerations (Carmichael and Managan 2013)

enhances the building value. The following analysis describes the process of completing a net zero retrofit (NZR), which was put together by the Institute for Building Efficiency—Johnson Controls, Inc. (Carmichael and Managan 2013).

Step 1 Plan Ahead, Build the Right Team, and Set goals. The need for a retrofit can be a sudden and not-so-subtle milestone in a building's life, often preceded by the degradation or failure of a key piece of equipment. When equipment fails, it is useful to analyse the life-cycle cost of replacing that equipment, along with complimentary retrofit measures. This can result in a better long-term outcome for the building.

Planning for equipment replacement and starting with the replacement of load reduction measures over time, such as upgrading windows, reducing plug loads, and improving lighting controls, enables better decision-making when equipment fails. Good planning can also help avoid common pitfalls, such as following rules of thumb on sizing and equipment selection based on obsolete data and assumptions that were relevant a generation ago.

Assembling the right execution team helps as well. The team should include designers and engineers who can design across systems and understand whole building benefits, as well as building maintenance personnel, and building operators and/or users more importantly. The right team will also be able to identify and mitigate risks that may arise during the project. The team should gather the energy use and cost data, upgrade history, equipment performance and life expectancy data, capital expenditure forecasts, lease structures, and major lease rollovers (for investor—owned buildings, mainly commercial buildings). Working together, the owner and design team can put forth goals that can include achieving net zero energy. The team should start planning early and keep all key stakeholders engaged throughout the process.

Step 2 Choose a Definition of NZE and Weigh Defined Trade-offs. All net zero energy buildings share a goal of maximising energy efficiency and then meeting remaining power needs with renewable energy. The key difference between types of net zero energy buildings is how and where the renewable energy is generated. Taking as example the definition proposed by NREL (Torcellini et al. 2006), (Table 10.2), all four definitions account for annual operations, even if there are surpluses and deficits on any single day or night. Net zero site energy is the most commonly used definition, and most in line with the spirit and intentions of achieving net zero. Each of the definitions has trade-offs regarding cost and tracking metrics, and different types of renewable energy can be used to meet each definition.

Step 3 Set a Baseline and Document Business-as-usual Expenditures. At the onset of a project, the project team should clearly document the energy use, costs, and how the building is performing today, and then lay out the anticipated future costs or the business-as-usual scenario without any net zero energy investments. Under business-as-usual, there will be costs involved in operations, maintenance, repair and replacement that should be documented. The business-as-usual case should include estimates for anticipated end-of-life capital investment needs in addition to anticipated future energy costs. Knowing future costs under the business-as-usual scenario is critical for comparison with the future costs of

Table 10.2 Summary of definitions and tradeoffs for Net Zero Energy (Torcellini et al. 2006)

Definition	Summary	Metric	Pros	Cons
Net zero site energy	Renewable energy must be generated on the building or site.	Site kBtu	Simple accounting low external fluctuations (i.e., not dependent on energy prices)	Annual energy bills may not be $0. Assumes electricity exported from the site can be used to offset natural gas needs on site. May emphasize an all—electric strategy if PV is the primary renewable energy system
Net zero source energy	Energy use is accounted for at the source, including the energy used for extraction, generation and distribution	Source kBtu	More accurate depiction of total environmental impact	Annual energy bills may not be $0 more complex accounting (acquiring site-to-source conversion multipliers, source energy technology changes)
Net zero cost energy	The amount the owner pays the utility for the energy is less than or equal to the amount of money the utility pays the building owner for the renewable energy the building exports to the grid	dollars	Energy costs are $0 simple accounting	
Net zero emissions energy	The building offsets all of the greenhouse gas emissions produced from the energy it uses through renewable energy production and carbon offsets (for up to 50 % of net energy consumption)	Co_2e	Uses greenhouse gas metric that aligns with carbon disclosure efforts and climate change	Annual energy bills may not be $0 challenging to track. Questions/concerns regarding carbon offsets

operating and maintaining a net zero energy building. The project team can then build a comprehensive and compelling business case in which investments in energy efficiency can reduce loads to the point where mechanical equipment can be downsized or eliminated, reducing capital and operating expenses.

Step 4 Organise a Technical Design Charrette. Every net zero energy project should start with a team charrette or brainstorming session to identify the technical potential for the building—the lowest possible energy use that could be provided by efficiency using available technology and best practices. This approach pushes both architects and engineers to focus on major, whole-systems improvements, fundamentally changing the design question from "We can't do this because ..." to "We could do this if..." it gets participants to think outside the box about options for maximizing the efficiency of each building system, about the types of on-site generation options that may be available, and about the ways different strategies interact to form an integrated design.

Step 5 Conduct Iterative Modelling, Design, and Costing of measures. An energy model of the building is critical to selecting a compatible bundle of energy measures. Since many retrofit projects occur over multiple years, if the model is set up early, it can be calibrated based on actual building energy meters and updated on an ongoing basis. Each individual energy measure, as well as different combinations of measures, can be modelled to see how they affect load throughout the day, season, and year. The energy model should also be used to analyse the cost-effectiveness of different measures because in that event, investments in complementary measures can be analysed together.

10.3 Design, Construction and Operation of New Zero Carbon Homes

The design and construction criteria of new Zero Energy/Carbon Homes (ZEH) are explained in this section. The UK construction practices, regulations and standards in general and the Code for Sustainable Homes (CSH) in particular are being explored to describe the design and construction stages as well as operational energy requirements of ZEH.[1]

10.3.1 Code for Sustainable Homes

Housing industry is one of the major sectors that should contribute towards the UK Government's objectives to reduce carbon emissions by 80 % by 2050 (DECC 2009). Domestic sector stands for around 29 % of all the CO_2 emissions of the UK 66 % of which is related to space heating, 17 % to hot water, 15 % to lighting and appliances, and 3 % to cooking (DECC 2011). In 2006, the UK government announced its ambition to make new homes carbon neutral by 2016 (EST 2009; DCLG 2008a) through gradual amendments in building regulations based on the Code for Sustainable Homes (CSH) standards (McManus et al. 2010; Osmani and

[1]Some parts of this chapter are adapted excerpts from 'Code for Sustainable Homes: opportunities or threats for offsite manufacturing and mass-customization' (Hashemi and Hadjri 2013).

Table 10.3 Gradual improvements to building regulations based on CSH standards (DCLG 2007a)

Year	2010	2013	2016
Energy/carbon improvements over building regulations part L (2006)	25 %	44 %	Zero carbon
Code for sustainable homes level	Code level 3	Code level 4	Code level 6

O'Reilly 2009). However, in July 2015, in an unexpected statement, the Conservative government announced that:

The government does not intend to proceed with the zero carbon Allowable Solutions carbon offsetting scheme, or the proposed 2016 increase in on-site energy efficiency standards (HM Treasury 2015). Although CSH has been scraped, its design approaches and strategies are still considered as one of the best example for achieving ZEH. This section therefore concentrates on CSH to explain design and construction stages of zero carbon homes defined by the Code for Sustainable Homes. Introduced in 2007 (DCLG 2009), CSH classified houses under six levels where Code Level 6 was the most sustainable and achieved zero carbon emissions (DCLG 2008b). The energy saving/improvement figures over the UK's Building Regulations, Approved Document L (2006) for Code Level 1 to Code Level 6 were estimated as 10, 18, 25, 44, 100 %, and finally, zero carbon for Code Level 6 (DCLG 2006). Code Level 6 was supposed to be implemented through the building regulations in 2016 when zero carbon homes would become mandatory (Table 10.3) (DCLG 2007a).

According to the Department for Communities and Local Government (DCLG 2007a) a house could be considered as zero carbon if it genuinely produces a net annual zero carbon for the consumed energy for heating, cooling, washing, cooking, lighting, ventilation, hot water and electric equipment. This house could be described as Code Level 6 in the Code for Sustainable Homes (EST 2009). Three requirements must be met for a home to be considered as a zero carbon home (Zero Carbon Hub 2013):

1. Complying with "Fabric Energy Efficiency Standard" (FEES) in terms of U-values, airtightness, etc. for the building envelope. The FEES is the maximum energy required for space heating and cooling;
2. Complying with the established "Carbon Compliance" limits (Table 10.4), established for zero carbon homes. Carbon Compliance is the maximum permitted CO_2 emissions from heating, cooling, hot water, lighting and ventilation;
3. Reducing any remaining carbon emissions to zero after considering items 1 and 2.

The third requirement can be met by intentional over-performance of the first and second requirements (by using, for example, photovoltaic panels, solar hot water, etc.) or can be achieved by investing in "Allowable Solutions" (Zero Carbon Hub 2013). CSH was originally very ambitious requiring all regulated (heating, cooling, hot water, ventilation, auxiliary services and lighting) and unregulated

Table 10.4 Fabric energy efficiency and carbon compliance requirements set by CSH standards (Zero Carbon Hub 2013)

Building type	Fabric energy efficiency standard (FEES) ($kWh/m^2/year$)	Carbon compliance ($kgCO_2/m^2/year$)
Detached house	46	10
Semi-detached house	46	11
End of terrace house	46	11
Mid terrace house	39	11
Apartment house	39	14

energies (home appliances) to be zero carbon (DCLG 2007a). The Allowable Solutions was proposed in 2008 to provide some flexibility due to the difficulties, such as high costs and feasibility issues, of delivering zero carbon homes based on entirely "on-site" strategies (Zero Carbon Hub 2011). The idea was that developers could pay to an Allowable Solution, which could be a small, medium or large offsite carbon saving project, to offset the remaining on-site carbon (Zero Carbon Hub 2011).

Design and Construction Approaches. Following are three design and construction approaches, which may lead to achieve a zero carbon home (Zero Carbon Hub 2013):

1. Balanced
2. Extreme Fabric
3. Extreme Low Carbon Strategies

The Balanced approach can be achieved by meeting the FEES requirements and considering reasonable onsite low carbon technologies. The remaining carbon emissions (e.g. 11 kg $CO_2/m^2/year$) will be eliminated by the Allowable Solution. In the Extreme Fabric approach, a very high performance fabric, significantly in excess of FEES, is considered to achieve a high standard building envelop comparable to Passivhaus (e.g. U-value 0.1–0.15 $W/m^2/K$; Air Permeability 1 $m^3/h/m^2$ @50 Pa; Thermal Bridge (y-value) 0.02 $W/m^2/K$; Space heating/cooling demand 25–30 $kWh/m^2/year$). Very little or no on-site low carbon energy technologies is used in this method. Similar to Balanced approach, Extreme Fabric approach fully complies with the requirements once the remaining carbon emissions are reduced to zero using Allowable Solutions. In the third approach, Extreme Low Carbon Strategies, on-site low carbon technologies along with high performance fabric (e.g. Passivhaus standards) are considered to considerably reduce the emissions beyond Carbon Compliance requirements for regulated emissions to zero (Code Level 5 in CSH). There is therefore no need for using Allowable Solutions to reduce emissions to zero (Zero Carbon Hub 2013). It should be noted that to achieve a fully on-site zero carbon home, both regulated (space heating, hot water, lighting and ventilation) and unregulated (e.g. appliances and cooking) energy/emissions should reduce to zero. This can be defined as Code Level 6 of CSH.

Code for Sustainable Homes is effectively a sustainability pointing/rating system in which buildings receive scores based on their design, construction, energy performance and implementation of sustainability strategies. There are minimum standard requirements for categories such as energy, CO_2 emissions, surface water, daily water consumption, waste and use of sustainable materials for roof, walls, floors, windows and doors. There are also some categories with no minimum standards such as pollution, health and wellbeing, management and ecology. A home must comply with the minimum standards as well as gain additional points on other categories in order to achieve a code level. For example, 60.1 and 64.9 additional points are required to achieve Code Level 5 and Code Level 6 respectively. Implementation of sustainable design strategies, such as Lifetime Homes (Goodman 2011) (up to 4 points), appropriate daylighting (up to 4 points), outside private space accessible by disabled people (1 point), cycle storage (1.2 points) and home office (1.2 points), are some of the design strategies which would count towards achieving higher points and levels of standard (DCLG 2006).

Barriers and Drivers for Delivering ZEH. The key drivers for delivering ZEH are legislations and regulations. Limited knowledge and skills in the construction industry (Heffernan et al. 2012) as well as considerable extra costs (DCLG 2011) are also the major barriers towards delivering ZEH. Achieving high standards for ZEH is not only difficult but is also expensive. Dwellings built based on Code Level 6 in 2016 could have been up to 50 % more expensive compared to 2010 regulations. According to the Department for Communities and Local Government, the total costs for achieving zero carbon homes is around £34 billion with an economic return of around up to £22 billion (DCLG 2007b). It has been estimated that Code Level 3 and Level 4 households can, respectively, save around £25–105 and £25–146 per annum. Code Level 6 households may save up to £359 per annum based on the entirely on-site solutions (DCLG 2007a). This is while another study by DCLG in 2011 indicates that the extra costs for a three-bed semi-detached house may vary between £16,407 and £29.326 for Code Level 5 and between £31.127 and £36,191 for Code Level 6 (Heinze and Voss 2009). A major portion of costs for achieving zero carbon homes is related to energy efficiency, which is achieved through fabric improvements (insulation, airtightness). For example, up to 79 % of the "extra over costs" for a semi-detached three bed house is related to improvements on the energy efficiency while it accounts only for around 36 % of the weight for the allocated points towards zero carbon homes (DCLG 2011).

Moreover, although zero carbon homes can be achieved by traditional methods of construction (DCLG 2013), considering uncertainties in the quality and construction period of traditional methods it is becoming more and more difficult and expensive to meet the requirements using traditional methods of construction (Miles and Whitehouse 2013). Traditional methods and practices have increasingly become less productive while their costs have increased significantly (Buildoffsite 2012). It has been suggested that offsite/prefabricated methods of construction can help to achieve zero carbon homes (DCLG 2007a) thanks to their higher quality

(Burwood and Jess 2005) compared to traditional methods of construction. However, a major barrier towards broader application of these methods is their extra immediate costs (Miles and Whitehouse 2013) compared to traditional methods of construction. In fact, cost is the key factor, which should be considered to increase the share of ZEH in the construction industry.

10.3.2 Examples

A permanent exhibition, known as Building Research Establishment (BRE) Innovation Park, has been established since 2005 by the UK's Building Research Establishment in Watford, near London, with the purpose of introducing Modern Methods of Construction, zero carbon homes and other innovative technologies (BRE 2013). The following are some examples of houses built in the Innovation Park. Kingspan Lighthouse is the first home certified to Code Level 6 in the UK and Barratt Green House is the first zero carbon home built by a major UK house-builder. The BRE Innovation Park provides builders with an opportunity to test and showcase their construction technologies and capabilities. It is also a great opportunity for designers, experts, and the public to see the innovations and emerging technologies and approached towards achieving a sustainable construction industry (BRE 2013).

Kingspan Lighthouse. Kingspan Lighthouse was constructed in 2007 in the Innovation Park. The Lighthouse is a CSH Code Level 6, two and a half storey, two-bedroom detached house with an area of 93 m^2 (Fig. 10.14). The annual heating cost for the house (including water and space) is about £30, which means around 94 % saving on fuel costs. The energy bills of a similar house with the same size and shape built based on the 2006 Building Regulations would cost around £500 (Kingspan 2009). Kingspan's TEK Building System, which is an offsite SIP (Structural Insulated Panel) system, has been used in the Lighthouse. Heat-losses through the building envelop, compared to a standard house, have decreased to around one third thanks to the very low U-value of the walls, roof and floor (0.11 W/m^2K) along with the airtightness of less than 1 $m^3/hr/m^2$@50pa. Triple glazed windows (0.7 W/m^2K), low energy lighting, photovoltaic (4.7 kW, 46 m^2), wood pellet boiler (10 kW), rainwater harvesting, and the 88 % heat recovery mechanical ventilation, in addition to A++ rated white goods (Kingspan 2009) make Kingspan's Lighthouse considerably energy efficient. Moreover, proper use of thermal mass, passive ventilation, and solar shading helps to reduce energy consumption as well as maintain the indoor air quality. The rather low average daylight factor of 1.5 to 2 % is however an area where improvements could have been made to make Lighthouse even more environmentally friendly. As the first house certified to Code Level 6 standards, Kingspan's Lighthouse was a proper example for the UK's housebuilders and manufacturers during its rather short life from 2007 to 2012. The house was dismounted adopting a sustainable "zero-waste-to-landfill" approach in 2012 (BRE 2013).

Fig. 10.14 Kingspan Lighthouse

Barratt Green House. The Barratt Green House is a three-storey, three-bedroom family home built to Code Level 6 Standards (Barratt 2013) (Fig. 10.15). Barratt Green House was constructed in 2008 in the Innovation Park (BRE 2013). It is the first home built by a major UK housebuilder, which meets the requirements to achieve zero carbon emission. The Barratt house is constructed from wall with aircrete masonry blocks with thin-joint mortar, concrete floor slabs, Structurally Insulated Panels (SIP) roof and low U-value triple glazing. Similar to the Kingspan's Lighthouse, Barratt Green House can achieve very low energy bills thanks to its high levels of insulations (180 mm = 0.11 W/m^2K), airtightness (1 $m^3/hr/m^2$@50 Pa), use of PV panels, rainwater harvesting, solar shades, heat recover mechanical ventilation, and highly efficient appliances. Application of triple glazed windows with a low U-value of 0.68 W/m^2K has also helped to achieve a good glazing to floor area ratio of 25 % providing sufficient natural lighting while maintaining low heat-losses through the window (Barratt Developments PLC).

Fig. 10.15 Barratt Green House

10.4 Application of the ZEH Construction Criteria to the Refurbishment Phases of the Existing Buildings

The reduction of buildings' environmental impacts is an internationally agreed agenda. Energy conservation and carbon emission reduction drove the quest to design, build and operate high performing and zero energy buildings in order to meet national carbon reduction target agendas. However, existing buildings will continue to offset any new construction savings if not addressed. In fact, existing buildings finds it increasingly difficult to compete with the high performing new construction. Hence, retrofitting of existing building stocks offers significant opportunities for reducing global energy consumption and greenhouse gas emissions.

Contextual data from developed and developing countries all comfort the need to consider the retrofitting of the existing building stock, with the residential sector leading the way as a dominant building type. In Europe, the residential sector represents 75 % of the built environment and in many of these countries the housing stock is over 50 years old. The UK, for example has the oldest stock with around 27 million existing housing units, while only around 120,000 new housing units are build every year. In many developing countries the residential building stock is often newer, but has been developed under the pressure of fast mass production, lacking any building energy efficiency measures, resulting on uncomfortable living homes and energy intensive building operation. Establishing common global standards for high performance retrofitting is difficult as building conditions vary greatly from one country to another. However, initiatives to address existing building retrofitting are fast developing. Examples of retrofitting initiatives, cases, standards and tools as well as challenges and key drivers are addressed in this section.

10.4.1 ZEH Retrofitting Standards and Tools

Producing Net Zero Energy buildings by the target 2030 date (Architecture 2030 2015) drove numerous initiatives including revisiting building codes and standards, development of building rating systems, design and development of design and assessment tools aiming at producing or retrofitting buildings. This section presents some of the efforts, relevant related standards and tools applicable to energy efficient and Zero energy buildings' retrofitting.

The recently published ANSI/ASHRAE/IES Standard 100-2015 addressing the Energy Efficiency in Existing Buildings, derives from an extensive revision of the 2006 version, is intended to be a code-enforceable standard for adoption by local authorities (ASHRAE 2015). The standard provides comprehensive and detailed descriptions of the processes and procedures for the retrofit of existing residential and commercial buildings in order to achieve greater measured energy efficiency. It provides the minimum requirements for energy efficient design and operation of existing buildings. Included in the revised standards are criteria for energy use surveys, auditing and requirements related to implementation and verification. Additionally, life-cycle cost analysis procedures as well as identification of potential energy conservation measures are included in appendices. The standard structure is based on setting a single upper limit on site energy use (kBtu/sqft/year) for each of 48 commercial and institutional building types and 5 residential building types in 17 climate zones. Energy use calculation includes all fuel, steam, hot water, chilled water and electricity crossing the building site boundary, net of energy exported from the building (including excess production of electricity and thermal energy).

ASHRAE's Vision 2020 on the other hand is an ad hoc committee that sets to develop guidance and strategy for the development of energy-related products, research in renewable energy systems, and the sequencing of the various identified activities that will produce net zero energy usage for all types of facilities by 2020. It targets a building community, including those who design, build and operate buildings, that will create or retrofit market-viable net zero energy buildings by the year 2030 (ASHRAE 2020 Vision 2008).

Energy efficient home retrofitting has spurred numerous guideline programs, aiming to assess viability and guide the retrofitting process. In this category, the European Retrofit Advisor developed an online decision tool (E2Rebuild) to assess the viability of retrofitting strategies. The vision of E2ReBuild is to transform the retrofitting construction sector from the current craft and resource based construction towards an innovative, high-tech, energy efficient industrialised sector (E2Rebuild 2015). In the US, the Regreen program was developed through a partnership of the American Society of Interior Designers and the U.S. Green Building Council (USGBC) (Regreen 2008). It guides a green home retrofit and addresses the major elements of a green retrofit, including site consideration, energy and atmosphere, material and resources and indoor environmental quality. It provides product selection, building system integration and technologies into integrated green strategies (Regreen 2008). Similarly, the E-Retrofit-Kit or the passive house retrofit Concept kit is a European initiative to provide a tool kit that guides an energy efficient home retrofit. Social housing companies in 14 mainly northern European countries were given the chance to benefit from a tool kit designed to help them carry out retrofitting in such a way as to considerably reduce primary energy consumption (by up to 120 kWh/m^2 a year). The tool kit includes best practices, "Passivhaus" standards and a methodology that guides the retrofitting process (E-Retrofit-Kit 2015a, b).

10.4.2 Process, Barriers and Drivers to ZEH Retrofitting

The challenges of a ZEH retrofitting are similar in nature to new ZEH construction. Challenges lie both within the decision-making process as well as within the retrofitting process itself. Lack of knowledge of the decision makers for retrofitting in general and limited awareness of owners are the initial barrier. Additional obstacles are of financial (investment costs) and social nature (building architectural and cultural value, structure of ownership). Further, the existing building structure and technical system may limit retrofitting potential. Direct actions that will benefit the existing housing retrofitting include review of policies, codes and standards processes, innovative financial scenarios and business models, awareness campaigns, training and industrial collaboration to standardize the process (Sutherland 1991; Dowson et al. 2012; Yudelson 2010; Building Technologies Program 2010; ZenN 2013; Anderson and Roberts 2008).

The key drivers to greening existing buildings, beyond concerns over energy use and carbon emission, are anticipated to lay in the following (Yudelson 2010)

- Return on investment
- Building occupants, tenants and stakeholders pressure
- Responsible property investing and future market competitiveness
- Corporate sustainability associated with leadership position
- Concern over energy prices and future volatility

Deep energy retrofit includes measures that not only increase energy efficiency but also improve other performance factors such as indoor air quality, thermal and

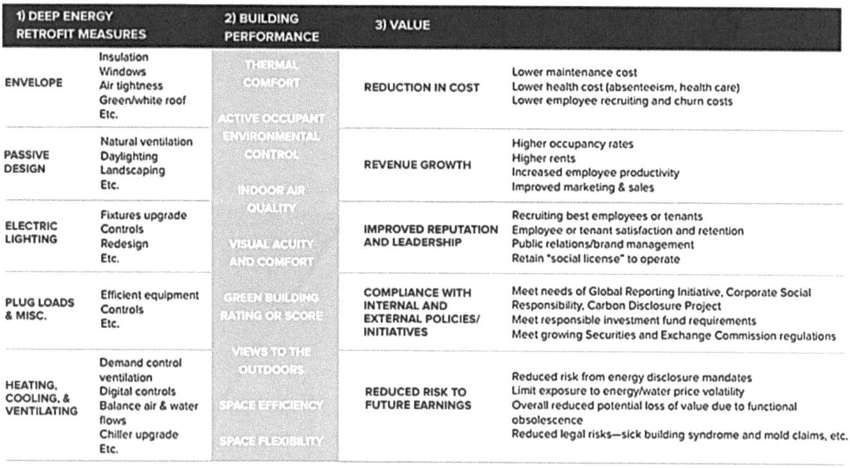

Fig. 10.16 Deep energy retrofitting measures, building performance and overall impacts (RetroFit Depot 2012)

visual comfort, which in turn create a benefit such as improved occupant health, organizational reputation and property value (Fig. 10.16).

10.4.3 Examples; Zero Energy Mass Housing Retrofitting

High performing residential retrofit cases are readily available and documented (Baeli 2014). Technologies to retrofit a house to net-zero energy consumption are now within the realm of possibility, but while it is possible, large-scale housing retrofits remain limited. Examples of ZEH individual retrofitting cases do exist, although not as numerous as new construction. Mission Zero House in Ann Arbor, MI, is one example of a Net Zero Energy Home retrofit, certified under the Living Building Challenge (Living Building Challenge 2015). The project initiated by the owners is a historic rehabilitation with the dual challenge to upgrade the house to the highest energy efficiency standards achievable today while preserving its mandated historic heritage. The strategies included insulation and sealing the building through an upgrade of the single pane window glass to low-e storm windows, wall, basement and attic insulation and photovoltaic panels covering a whole side of the roof (Fig. 10.17).

The above is an example of a deep-green renovation that can be achieved in the existing housing stock. However, it remains of limited value when it comes to mass housing retrofitting, especially with the financial and social restrictions it usually carries.

In this regard, the Dutch government driven initiative referred to as is seen as an achievement venture of zero energy social housing retrofitting (Energiesprong 2015a). Energiesprong is considered as one successful result of the Dutch

Fig. 10.17 Mission Zero House—Ann Arbor, MI (Mattgrocoff 2010)

Government funded Platform 31 program, an innovative program that brings together different actors for out-of-the-box thinking and approaches to complex problems. The result is a renovation program that commits to refurbish existing buildings (mainly social housing) to net zero energy, within a week, at zero cost to the tenant, a 30-year builder's guarantee and no subsidies. So far, 800 units have been retrofitted in the Netherlands and 100,000 are planned to undergo the same. The program calls for mass demand for deep refurbishments and is based on aims to meet the following criteria:

1. Quality and energy performance guarantee; by implementing quality standards, manufacturing and delivery methods, inspection and verification that enable a long-term performance warranty to be offered. The net zero energy refurbishment package needs to come with a long year (i.e. 30) energy performance warranty on the house, backed usually by an insurer.
2. Affordability; assured performance, coupled with mass-customized industrialization and delivery process efficiencies, will reduce costs and allow for the refinancing of the upfront investment. This will be achieved through guaranteed

energy cost savings (energy performance contracting), generation of on-site renewables and real estate value improvements and will thus make the solution affordable, independent of public grant. The ability to finance an investment requires a business case and in this case, the net present value of the energy cost savings over the lifetime of the package sets the price target.

3. Desirability: low disruption, fast process, improved aesthetics and comfort levels, increased asset value of individual dwellings plus neighborhood renewal and social impact of mass implementation (hassle-free solution) (Energiesprong UK 2015b). Quick delivery and least occupants interference (one week in this case); The instalment of the packaged does not require more than one week and allow occupants to continue living in the house for the greater part of the works.

Success ingredients overcoming a set of traditional barriers have been jointly considered in the Energiesprong venture and resulted so far in a growing experience that is being extended to the United Kingdom and France. The main idea behind this program is a transition of the building sector from project to product. The refurbishment drives the construction to innovate and start developing and producing integrated solutions that are industrialized instead of project based. The second idea resides in the financial trade off, i.e. cost versus performance, while fostering new industry collaborations (Fig. 10.18).

The result is an improved energy performance of the existing housing with added real estate and aesthetic value (Fig. 10.19).

Fig. 10.18 Prefabricated elements designed for a fast renovation of social housing, Energiesprong or the Dutch experience (Energiesprong 2015a, b)

Fig. 10.19 Examples of mass housing retrofitting in the Netherlands (Energiesprong 2015a, b)

10.5 Conclusions

Zero Energy Homes (ZEHs) have been receiving an increased attention in recent years as a result of constant concerns for energy supply constraints, decreasing energy resources, increasing energy costs and rising negative impacts of greenhouse gases on global climate. Among different strategies for decreasing the energy consumption in the building sector, ZEHs have the promising potential to significantly reduce or eliminate the total energy use, as well as to increase the overall share of renewable energy use. There are many different possible combinations of passive design energy-efficiency measures, utility equipment and on-site energy generation technologies able to achieve the net zero energy performance; thus, there is no common standard approach to designing a zero energy building. Nonetheless, this chapter attempted to define ZEHs with parameters that help demonstrate the cases. It also extended to identifying criteria applied to the design, construction and operation of new and refurbished zero carbon homes. The examples presented in this chapter led to clarify the possibility of calibrating net zero operational energy performance in housing where a passive design approach that helps reduce the energy demands can be considered a base for the progression and active technologies such as PV systems are supplied with the aim to cover or exceed the remainder of energy use in most cases.

References

Aelenei L (2012) Design issues for Net Zero-Energy Buildings. Oral presentation at the ZEMCH 2012 Conference, Glasgow, 20–22 Aug 2012

Aelenei L, Lollini R, Gonçalves H, Aelenei D, Noguchi M, Donn M, Garde F (2011) Passive cooling approaches in Net-Zero Energy Solar Buildings: lessons learned from demonstration buildings. In: CISBAT2011 Conference, Lausanne, 2011

Aelenei L, Aelenei D, Gonçalves H, Lollini R, Musall E, Scognamiglio A, Cubi E, Noguchi M (2013) Design issues for Net Zero-Energy Buildings. Open House Int Zero Energy Mass Custom Home Res Paradigms 38:7–14

Aelenei L, Gonçalves, H, Aelenei D (2013) The nZEBs in the near Future Overview of definitions and guidelines towards existing plans for increasing nZEBs. Paper presented at the Portugal SB13, Guimarães, 30 Oct 2013–1 Nov 2013

Anderson R, Roberts D (2008) Maximizing residential energy savings: net zero energy home technology pathways. Technical report NREL/TP 550-44547. National Renewable Energy laboratory, Golden

Architecture 2030 (2015) The 2030 challenge. http://www.architecture2030.org/2030_challenge/the_2030_challenge. Accessed 23 Nov 2015

ASHRAE (2015) Standard 100-2015, energy efficiency in existing buildings (ANSI Approved/IES Co-sponsored). I.P edition. The American Society of Heating, Refrigerating and Air-Conditioning Engineers, Atlanta, GA

ASHRAE 2020 Vision (2008) Producing Net Zero Energy Buildings. https://www.ashrae.org/file%20library/doclib/public/20080226_ashraevision2020.pdf. Accessed 28 Nov 2015

Ayoub J (2009) Factsheet towards Net Zero Energy Solar Buildings. Task40/Annex 52, Solar Heating and Cooling, Quebec

Baeli M (2014) Residential retrofit: twenty case studies. Edition RIBA Publishing, London

Barratt (2013) The Barratt House. Barratt Developments PLC. http://www.bre.co.uk/filelibrary/Innovation_Park/Barratts_Green_HouseBrochure.pdf. Accessed 29 Nov 2015

Black C, Ooteghem KV, Boake TM (2010) Carbon neutral steel building systems research project—a case study investigating the relationship of operational energy and embodied energy in achieving a holistic carbon neutral retail building. In: Proceedings of american solar energy society, national solar conference 2010, Phoenix, 2010

BRE (2013) Building research establishment. www.bre.co.uk/innovationpark. Accessed 02 Aug 2013

Building Technologies Program (2010) Summary of gaps and barriers for implementing residential building energy efficiency strategies. Building America, Building Technologies Program, U.S. Department of Energy, http://apps1.eere.energy.gov/buildings/publications/pdfs/building_america/49162.pdf. Accessed 29 Nov 2015

Buildoffsite (2012) Buildoffsite Review 2012: The business case for offsite. Buildoffsite, London

Burwood S, Jess P (2005) Modern methods of construction: evolution or revolution. BURA Steering and Development Forum, London

Carmichael C, Managan K (2013) Reinventing existing buildings 8 steps to net zero energy. Institute for Building Efficiency, Johnson Controls Inc, Milwaukee

Chen Y, Athienitis AK, Galal KE, Poissant Y (2007) Design and simulation for a solar house with building integrated photovoltaic-thermal system and thermal storage. ISES Solar World Congress, Beijing, vol 1. Springer, Heidelberg, p 327, 18–21 Sept 2007

DCLG (2006) Code for sustainable homes: a step-change in sustainable home building practice. Department for Communities and Local Government, London

DCLG (2007a) Building a greener future: policy statement. Department for Communities and Local Government, London

DCLG (2007b) Final regulatory impact, assessment building a greener future. Department for Communities and Local Government, London

DCLG (2008a) Definition of zero carbon homes and non-domestic buildings. Department for Communities and Local Government, London

DCLG (2008b) The code for sustainable homes setting the Standard in Sustainability for New Homes. Department for Communities and Local Government, London
DCLG (2009) Code for sustainable homes, impact assessment. Department for Communities and Local Government, London
DCLG (2011) Cost of building to the code for sustainable homes updated cost review. Department for Communities and Local Government, London
DCLG (2013) Annual Report and Accounts 2013–14. Department for Communities and Local Government
DECC (2009) Climate act impact 2008 assessment. Department of Energy and Climate Change, London
DECC (2011) Energy consumption in the UK: 2013. Department for Energy and Climate Change, London
Dowson M, Poole A, Harrison D, Susman G (2012) Domestic UK retrofit challenge: drivers, barriers and incentives leading into the Green Deal. Energy Policy 50:294–305
Energiesprong (2015a) Transition zero http://energiesprong.nl/wp-content/uploads/2014/06/Transition_zero.pdf. Accessed 28 Nov 2015
Energiesprong UK (2015b) Transition zero. http://www.energiesprong.eu/. Accessed 28 Nov 2015
EPBD (2010) The Directive 2010/31/EU of the European Parliament on the Energy Performance of Buildings Directive (EPBD recast). Official Journal of the European Union, 53
E-Retrofit-Kit (2015a) European retrofit advisor. http://www.e2rebuild.eu/en/events/workplan/pd/retrofitadvisor/Sidor/default.aspx. Accessed 28 Nov 2015
E-Retrofit-Kit (2015b) Tool-kit for "Passive House Retrofit"—Intelligent Energy Europe. http://retrofit.energieinstitut.at/. Accessed 28 Nov 2015
EST (2009) Low carbon futures: zero carbon case studies: CE310. Energy Saving Trust, London
Frattari A (2013) CasaZeroEnergy: an Italian prototype of zero energy building. Open House Int 2013(38):277–286
Goodman C (2011) Lifetime homes design guide. HIS BRE Press for Habinteg Housing Association, London UK
Hashemi A, Hadjri K (2013) Code for Sustainable Homes: opportunities or threats for offsite manufacturing and mass-customization? In: Proceedings of ZEMCH 2013 international conference: the visibility of zero-energy housing, University of Miami, Miami, USA, 30 October 2013–1 November 2013
Heffernan E, Pan W, Liang X (2012) Delivering zero carbon homes in the UK. In: Smith SD (ed) Proceedings 28th annual ARCOM conference, Edinburgh, Association of Researchers in Construction Management, pp 1445–1454, 3–5 Sept 2012
Heinze M, Voss K (2009) Goal: ZERO ENERGY BUILDING Exemplary experience based on the solar state solarsiedlung Freiburg am Schielrberg, Germany. J Green Build 4(4):93–100
Hermelink A (2013) Towards nearly zero-energy buildings: definition of common principles under the EPBD. Ecofys, London
HM Treasury (2015) Fixing the foundations: creating a more prosperous nation. Presented to Parliament by the Chancellor of the Exchequer by Command of Her Majesty, London
International Energy Agency Solar Heating and Cooling Task 40 / ECBCS Annex 52 (2012) Towards Net Zero Energy Solar Buildings (NZEBs). http://task40.iea-shc.org/publications. Accessed 28 Nov 2015
Kingspan (2009) Level 6 net-zero carbon house, fact file, lighthouse, Kingspan. www.kingspanlighthouse.com/pdf/lighthouse.pdf. Accessed 01 Aug 2013
Liao L, Athienitis AK, Candanedo L, Park KW, Poissant Y, Collins M (2007) Numerical and experimental study of heat transfer in a BIPV-thermal system. J SolEnergy Eng 129:423–430
Living Building Challenge (2015) Mission Zero House, http://living-future.org/case-study/missionzero. Accessed 28 Nov 2015
Marszal AJ, Heiselberg P, Bourelle JS, Musall E, Voss K, Sartori I, Napolitano A (2011) Zero energy building—a review of definitions and calculation methodologies. Energy Building 43:971–979

Mattgrocoff (2010) Ann Arbor, MI—Mission Zero House—Own work. Licensed under CC BY-SA 3.0 via Wikimedia Commons. https://commons.wikimedia.org/wiki/File:America%27s_Oldest_Net_Zero_Energy_Home_color_corrected_Grocoff.jpg. Accessed 28 Nov 2015

McManus A, Gaterell MR, Coates LE (2010) The potential for the Code for Sustainable Homes to deliver genuine 'sustainable energy' in the UK social housing sector. Energy Policy 38 (4):2013–2019

Miles J, Whitehouse N (2013) Offsite housing review. Construction Industry Council, London

Noguchi M (2008) Net Zero-energy home design strategies learned from Canadian experience. Open House Int 33(3):88–95

Osmani M, O'Reilly A (2009) Feasibility of zero carbon homes in England by 2016: a house builder's perspective. Build Environ 44(9):1917–1924

Regreen (2008) Residential remodeling program. http://www.regreenprogram.org. Accessed 28 Nov 2015

Retrofit Depot (2012) Guide for building the case for deep energy retrofits. RMI—Rocky Mountain Institute, Boulder

RMI (2015) Energy retrofits. RMI—Rocky Mountain Institute, Boulder

Sartori I, Napolitano A, Voss K (2012) Net Zero Energy Buildings: a consistent definition framework. Energy Build 48:220–232

Sutherland RJ (1991) Market barriers to energy-efficiency investments. Energy J 12(3):15–34

Torcellini P, Pless S, Deru M (2006) Zero Energy Buildings: a critical look at the definition. Paper presented at the ACEEE Summer Study Pacific Grove, National Renewable Energy Laboratory D. Crawley U.S. Department of Energy, California, 14–18 Aug 2006

Voss K, Musall E, Lichtmeß M (2011) From low energy to Net Zero-Energy Buildings: status and perspectives. J Green Build 6:46–57

Voss K, Sartori I, Lollini R (2012) Nearly-zero, Net zero and plus energy buildings—how definitions & regulations affect the solutions. REHVA HVAC J 06:23–27

Yudelson J (2010) Greening existing buildings, McGraw-Hill's Greensource Series. McGraw-Hill Education, London

ZenN (2013) Common barriers and challenges in current nZEB practice in Europe. ZenN—Nearly Zero energy Neighborhoods, IVL Swedish Environmental Research Institute, Stockholm

Zero Carbon Hub (2011) Allowable solutions for tomorrow's new home: towards a workable framework. Zero Carbon Hub, London

Zero Carbon Hub (2013) Zero carbon strategies for tomorrow's new homes. Zero Carbon Hub, London

Chapter 11
Building Performance and Simulation

Hasim Altan, Roberto Padovani and Arman Hashemi

Abstract This chapter emphasises on reasons behind building simulation and performance analysis, and provide a background on both advantages and disadvantages of using building performance and simulation for environmental analysis. Some of the learning outcomes anticipated as follow: thorough understanding of basic concepts of building performance simulation, develop the ability to apply these concepts in real life situations; understanding of the use of BPS and the need to use the appropriate software to answer a specific enquiry; understanding the relationships between the required outcome of a BPS exercise and the input details necessary to produce a valid answer. The idea behind this chapter is to provide a general knowledge of, and a basic ability to understand the use of building performance simulation, and to optimise the performance of different components used in buildings. Integration of renewable energy resources will also be addressed, e.g. solar and wind. The integrative effect of these components will also be studied.

H. Altan (✉)
Department of Architectural Engineering, University of Sharjah (UOS), Sharjah, UAE
e-mail: haltan@sharjah.ac.ae; hasimaltan@gmail.com

H. Altan
Welsh School of Architecture (WSA), Cardiff University, Cardiff, UK

R. Padovani
R&P Green Solutions, Docklands, Australia
e-mail: roberto.padovani@rpgreensol.com

A. Hashemi
University of Brighton, Brighton, UK
e-mail: a.hashemi@brighton.ac.uk

© Springer International Publishing Switzerland 2016

M. Noguchi (ed.), *ZEMCH: Toward the Delivery of Zero Energy Mass Custom Homes*, Springer Tracts in Civil Engineering,
DOI 10.1007/978-3-319-31967-4_11

11.1 Introduction

Building Performance Simulation (BPS), also called 'building simulation', 'building energy modelling and simulation', is the use of software to predict the energy performance of a building. In addition to energy performance of buildings, BPS can be used for environmental performance evaluation that may investigate daylighting, thermal performance and so on.

In the last two decades, "computer simulation is more pervasive today…" where "…Engineering and Science involves the use of computer modelling simulation to solve mathematical formulations of physical models of engineered and natural systems", which become "…tools for building performance evaluation (Glotzer et al. 2009; Augenbroe et al. 2004).

According to the definition by Oden et al. (2006), simulation "… refers to the application of computational models to the study and prediction of physical events or the behaviour of engineered systems." However, advances in computer simulation offer rich possibilities that essential to ensure the building can satisfy the needs of residents.

In principles of simulation, Oden et al. (2006) portrayed that the state of simulation built in "Advances in mathematical modelling, in computational algorithms, and the speed of computers and in the science and technology of data-intensive computing have increased the ... productivity of quality design and become vital in predictive of building behaviour to achieve the optimum performance as proposed by their designers". Moreover, "... unprecedented improvements in building design have brought ... the field of computer simulation to the threshold of a new era".

A typical simulation model (also referred to as 'energy' or 'thermal' model) will have inputs for climate; envelope; internal gains from lighting, equipment, and occupants; heating, cooling, and ventilation systems; schedules of occupants, equipment, and lighting. Computer simulation models will output building energy and environmental performance predictions in typical end-use categories: heating, cooling, lighting, fan, plug, process. In addition to energy units, most software includes utility rates input, and can predict energy costs. Energy-savings measures can be also calculated using a wide variety of bespoke software applications that can carry out BPS. The following are the potential applications of BPS:

- Building design: Many modern commercial or residential building codes require minimum energy performance. Therefore, energy or thermal modeling can be used to demonstrate compliance, or predict energy consumption of proposed developments through BPS.
- Life cycle cost analysis (LCCA): Comparing different building design alternatives to determine which is the least cost, including capital costs and (at a minimum) energy costs can be also determined through the use of BPS.
- Energy retrofit analysis: In conjunction with an energy audit, or deep energy retrofit, an energy or thermal model can be used to predict savings associated to proposed energy cost measures (often called ECMs).

In a large simulation exercise, the progress in technology has allowed simulation to become more extensively available, with building energy simulation tools such as DesignBuilder, IES VE, ESP-r, Ecotect, TAS, EnergyPlus and TRNSYS that now widely used. These software packages have a wide range of capabilities and can help designers compare various design options and lead them to more optimal and energy saving designs, as well as helping engineers define the energy saving potentials and evaluate the energy saving effects. However, the verification of any simulation model is necessary and important to give confidence in the accuracy and reliability of software.

In order to achieve a green building design, it is essential to consider all environmental factors that affect the performance of the building. Therefore, BPS has offered a greater support in building design, operation, analysis, commissioning and assessment of buildings in the last two decades. With BPS, the outcome results needs several parameters pertaining to building design and usage to be identified for improving the environmental performance along with recommendations for energy conservation measures.

11.2 Advantages and Disadvantages

11.2.1 Conceptual Design Versus Detailed Design

The use of Building Performance Simulation can assist the development of a residential project from the early concept stages all the way through to detailed design phase and post occupancy tuning. The use of Building Performance Simulation in the early concept stages have a positive influence in the architectural and services design and minimises the computational and development costs. In addition, the following advantages results from an upfront analysis:

- One of the main advantages of Building Performance Simulation is the ability to test different solutions, equipment efficiencies, system integration early in the development phases to achieve low emission targets and potential savings on the energy use of the building. The generation of simulated data assists the development of a business case and cost analysis like Return on Investment (ROI) and Net Present Value (NPV), which are crucial indexes used by investors to make decisions. The early integration of architectural design and building service solutions promote synergy and reduces development time, which can lead to a building that can have a successful post-occupancy performance.
- Enable optimisation of building façade and solar passive design in the early stages of the architectural project when the design is still fluid and glazing, massing parameters can be altered, and options assessed without affecting the development timing of the project and associated systems. The use of Building Performance Simulation assists the architects in determining the building massing and fenestration ratio that maximises visual and thermal comfort and

minimises solar gains through the façade, which is a crucial phase to design an energy efficient and comfortable building. Indoor Environmental Quality is also assessed at this stage to capture the requirements of indoor thermal and visual comfort levels.

An example of an architectural feature that can be assessed early on and that has the potential to positively impact the thermal and visual performance of a residential building is the use of *light shelves*. *Light Shelves* can significantly reduce direct solar gains through the external glazing, increase natural illumination into the interior spaces and reduce glare. Figure 11.1 shows an example of a complex façade shading analysis output from the building software simulation IES Virtual Environment.

- The use of building performance simulation enables the early integration of HVAC (Heating, Ventilation and Air-Conditioning) concepts into the architectural design allowing for a holistic approach. The selection of the most suitable system is therefore based on predicted performance and cost analysis rather than the use of "rule of thumb". The use of energy simulation for the cost benefit analysis of air conditioning systems is particular important when designing low cost homes in developing regions where active conditioning and mechanical ventilation is not commonly used to provide indoor comfort.
- Building Performance Simulation is also a powerful tool to estimate building potential for environmental rating schemes. Whole Building Performance simulation provides estimated data required for certification of Environmental Certification rating schemes such as LEED, BREEAM, Green Star and PassivHaus. The main analysis required for these certification systems are greenhouse gas emissions, thermal comfort and visual comfort. Energy efficiency and greenhouse gas emissions targets are particularly important and heavily weighted against other credits with Passivhaus being the more strict system with prescriptive values for cooling and heating loads and minimum levels of building fabric insulation.

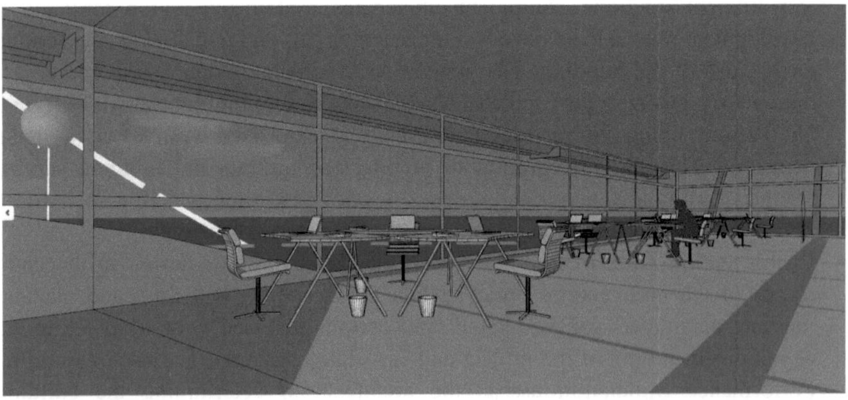

Fig. 11.1 Building facade analysis

On the tail end of the building design development, the use of Building Performance Simulation is very important to size air-conditioning systems and numerically validate energy efficiency and thermal comfort targets. Recent developments in building simulation algorithms enabled the sizing of HVAC components and systems more accurately since it can take into consideration the building envelope and complex façade systems in addition to more sophisticated analysis such as air stratification and stack up effect in Displacement Ventilations and Under Floor Air Distribution systems. Such refinement in the simulation software algorithm reduced the level of oversizing of systems and eliminated issues with thermal comfort like cold draughts. The accuracy of Building Performance Simulation when applied at detailed design stage is particularly relevant for certification schemes and design compliance to government energy efficiency rules such as the Section J of the National Construction Code of Australia and the California Title 24's Building Energy Efficiency Standards.

However, the reliance of designers and engineers on building simulation software also have its disadvantages as highlighted below:

- Building Performance Simulation is not a tool used to predict the actual energy consumption and indoor environmental quality indexes. A simulation model is built around assumptions that add many uncertainties to the results. Among those uncertainties are occupant behaviour, material properties, equipment performance and climate data.
- Incorrect modelling can lead to misleading results.
- Building simulation may take the understanding and questioning of building physics away from engineers and architects, when simulation is performed blindly without understanding of building physics and heat transfer phenomena and equations.
- Over complex models can lead to high computational costs and not necessarily meaningful results.

11.2.2 Time

The synergy and optimisation of solutions developed early in the concept and schematic design stages allow for significant reductions in design development timing. Testing and simulation of multiple alternative solutions enables a holistic simultaneous engineering and design approach leading to better integration of systems. There is however, a perception or misconceived idea that building simulation can add time and cost to the building project. Energy models in particular can be very detailed and take up to 3–4 weeks to be completed depending on the complexity of the project. However, model development can run in parallel with services and architectural design development with results and recommendations being simultaneously fed to the design team throughout the process. It is of crucial importance therefore that the simulation model is tailored to the clients' needs to avoid developing a model that is more complex and accurate than the required output, which could jeopardised the project development timing.

The level of details that is entered into the model has to be suited to the stage of the project development and the required outputs. It is important to note a Building Performance Simulation is based on a simplified and idealised version of the building that does not and cannot fully represent all of the intricacies of the building and its operation. As a result, the simulation model results only represent an interpretation of the potential performance of the building. This approach allows a decision making process that minimises computational time and provides timely feedback and design recommendations back to the design team which may include architects, engineers, project managers and building companies.

11.2.3 Accuracy/Reliability

The accuracy of computer simulations are limited by the margin of error of the modelling inputs. Most of the modelling input parameters comprise of idealized assumptions around the building occupant behaviours. Occupant behaviour such as people movements within a building, occupant density diversity, manual control of air conditioning, operable windows and lighting systems and equipment use can have a significant impact on the annual energy consumption and thermal comfort estimation.

In addition to occupant behaviour, the following modelling parameters have a significant impact on the accuracy of the simulation:

- Air conditioning operational conditions and efficiencies taken from testing specifications can differ from the simplified mathematical models used in the building energy simulations.
- The thermo-physical properties of the building envelope materials and all intricacies of the construction design are simplified in the energy simulation model due to software limitation and computational time restrictions. In addition, technical specifications of material constructions are based on a specific set of testing conditions but as installed product will have a deviation from the nominal properties due to manufacturing tolerances.
- The climatic weather data used in building energy simulations are a set of historical measured hourly data for a specific location. This historical data could represent data collected by a weather station over a period over 20 years or more. The uncertainty of the weather data increased by significant change in climatic conditions predicted by climate change scientists over the next few decades is another major source of error in the estimation of energy consumption and load calculations.

The only way to increase the accuracy and reliability of a Building Performance Simulation model is to correlate model results and input parameters to actual building operational data. In this case, the correlation would have to be done over a significant sample size to increase the reliability of the results. This method would also drive more resources, which in most cases would make the building simulation not financially viable especially in smaller residential projects.

11.2.4 Costs

An important process in the project development of a residential or commercial building is the economic and financial analysis. Building Simulation can have a significantly impact in lowering the lifecycle costs and building thermal simulation is especially useful for evaluating energy conservation initiatives in buildings. Multiple design alternatives can be run to determine an optimised solution with the lowest life-cycle cost that satisfies all the design criteria from architects and engineers. The parameters that affect the lifecycle costs are the initial costs of land acquisition and construction costs, operation and maintenance costs, capital investment, useful life of equipment, replacement costs, residual value of resale or disposal costs, loan interest charges, discount rate, energy price and energy price projections cost.

In order to be able to run a fair comparison between all design alternatives, cash flows must be calculated using a life-cycle cost that is made time-equivalent or in other words different life-cycle costs must be comparable when cost components are incurred in different times of the useful life of the building. In order to make the cost components of the life-cycle cost calculation time-equivalent, the life-cycle cost method uses the concept of net present values by discounting them to a common point in time. The interest rate used for discounting is a rate that reflects an investor's opportunity cost of money over time and over the performance of its portfolio of investments.

In addition to assessing lifecycle costs of multiple design alternatives, the use of the Building Performance Simulation can also cut time wasted in integration of the engineering services and architectural design throughout the different phases of the project. The assessment of different solutions at the early stages of the project allow for proper size allocation of systems within the available architectural spaces allocated for engineering services such as air conditioning ducts, indoor air-conditioning units and lighting fixtures.

The added costs of engaging an engineering consultant to perform the building simulation can be justified based on thought process of the life-cycle cost assessment. The costs associated with this service should however be based on the complexity of the analysis and budget allocated to the project. Smaller residential projects in particular would not have funds available for the same depth of analysis performed for larger commercial projects since the upfront investment cost play a more important role in the decision making process of the home owner.

11.3 Types of Environmental Analyses in Buildings

11.3.1 Energy Performance

Energy performance simulation is the most important analysis of a building design and is intrinsically dependent on indoor environmental quality design parameters. Climate change impact on the environment and human resources are driving governments to tighten the policies on carbon emissions. An example of the carbon policies is the target set by the European Union of 30 % carbon emission reduction by 2050.

Energy efficiency reduce the environmental impact associated with the production and consumption of fossil fuels, which directly and indirectly emit pollutants to the atmosphere such as carbon dioxide (CO_2). Coal, natural gas and oil are major sources of fuel used to generate energy consumed in the built environment. The process to extract and process the fossil fuels and the combustion process required to produce heat or electricity have a great environmental impact such as atmosphere and water pollution, soil erosion and contamination and emissions of gases with global warming potential. In addition, there is overwhelming scientific evidence that the processes associated with the energy generation from fossil fuels are driving climatic change and causing extreme natural events such as extreme heat waves, bushfires, severe storms and increase in ocean temperature.

The use of energy models to predict the energy consumption and cost of energy use enables the assessment of energy efficient systems and solar passive design. Energy models of multiple design alternative allows for the assessment of synergy and integration of solutions which leads to an optimised model that is much less based on rules of thumb and more based on building physics calculations. The following workflow is required to generate an energy model.

- Define Site, Location and climate. This step requires the selection of a weather file with hourly observation data for a specific weather station.
- Creation of building geometry and thermal zoning. This process is still a working in progress with most of the energy simulators designing all the intricacies of the building geometry directly on the simulation software. Building Information Modelling (BIM) technology is rapidly evolving and project teams are successfully transitioning the 3D architecture model to energy models with material properties assigned to the building envelope.
- Model of adjacent buildings, site obstructions and shading structures blocking direct sun radiation. Building orientation must be set at this step according to the architectural site layout. Solar shading calculations can performed at this stage and results carried over to the dynamic thermal simulation.
- Assignment of internal gains and operation profiles. Internal heat gains and electrical power consumption are associated with the use of receptacle power for office equipment, residential appliances, lighting power and any other equipment associated with occupant use. The sensible and latent heat gains associated with occupant metabolic rates must also be assigned to the model at this stage.

- Assignment of building envelope properties. Thermo-physical properties of building fabric materials are assigned to the energy model: external and internal walls, roof, ground floor, suspended floor, windows, doors and roof lights. The most important parameters considered in energy models are the U value, which represents the overall Heat Transfer Coefficient, Solar Heat Gain Coefficient (SHGC) of fenestration elements, material thickness, density and solar reflectivity.
- Define model of air tightness. Air tightness defines how well sealed the building is from external air conditions. Therefore, energy model should reflect the estimated air infiltration through the façade, which typically results from gaps in the window frame and doors. Sealing of the window framing and doors and the use of air locks in the entrance lobby area of buildings can significantly reduce air infiltration. The ideal scenario is to perform a pressure test in the building to have a precise value of the air tightness of the building or otherwise a conservative infiltration value based on the façade type and architectural specifications. Air infiltration have significant effects on the heating and cooling loads of the building and its model parameters should be representative of the actual building design.
- Develop Domestic Hot Water Systems. Domestic hot water consumption is estimated based on the number of occupants in the building and hot water fixtures such as showers and hot water taps. The model of the domestic hot water systems may include gas or electrical heaters, solar panels, circulating pumps and storage tanks. Model parameters include heating system efficiency, pump operating power, storage tank volume and heat losses associated with water storage and distribution.
- Develop HVAC model. Detailed model of air conditioning and ventilation systems is in most cases the most complex side of energy simulations but crucial to determine the potential of energy efficient measures with accuracy. The operation of all mechanical equipment such as chillers, air-handling units, fans, chilled beams, cooling towers and pumps have to be modelled to accurately mimic the operation and controls of the systems. Detailed plant equipment is modelled with part load efficiencies operating according to the testing data of manufacturers and thermal loads calculated according to the hourly simulation.
- Detailed models allow the optimisation of HVAC operation and the minimisation of energy consumption and carbon emissions. In addition, the operation of mixed mode ventilation buildings can be estimated with an integration of a HVAC model with windows operation. Design alternatives of air conditioning and mechanical ventilation system can be evaluated against a wide range of design criteria such as carbon dioxide equivalent, annual energy consumption, fresh air rates and thermal comfort levels measured in the Predicted Mean Vote (PMV) scale.
- Add any renewable energy source. Renewable energy sources such as photovoltaic panels, solar hot water systems, wind turbines and geothermal systems can be added directly to the energy model depending on the software capabilities or alternatively can be calculated using specific software dedicated for the

Fig. 11.2 Example of energy model and shading analysis

calculation of the renewable energy type. The addition of renewable energy to the building design is mandatory for zero carbon developments and accurate energy generation estimates are important to offset energy consumed by the building (Fig. 11.2).

11.3.2 Indoor Air Quality (IAQ)

Several indoor air quality criteria can be assessed using specific software simulation tools. The key design criteria used in most projects and green building certification tools are thermal comfort, level of indoor air contaminants, air change effectiveness and age of air. Good indoor air quality levels is a critical element of a healthy building and specific simulation enables design alternatives to be tested and optimised.

11.3.3 Thermal Comfort

Thermal comfort calculations are derived directly from the energy model and assesses the indoor air and radiant temperature, relative humidity and air velocity in specific occupant metabolic rates and clothing. Mechanically ventilated spaces are in most cases assessed according to the standard ISO 7730 that measures the variable Predicted Mean Vote (PMV).

Predicted Mean Vote (PMV) refers to a thermal scale that runs from Cold (−3) to Hot (+3), originally developed by Fanger and later adopted as an ISO standard. The original data was collected by subjecting a large number of people to different conditions within a climate chamber and having them select a position on the scale that best described their comfort sensation.

Most high-end energy simulation software packages have a mathematical model of the relationship between all the environmental and physiological factors derived from the original thermal comfort data. The result relates the size thermal comfort factors to each other through heat balance principles and produces the following sensation scale. Predicted Mean Vote sensation scale:

Value	Sensation
−3	Cold
−2	Cool
−1	Slightly cool
0	Neutral
1	Slightly warm
2	Warm
3	Hot

The recommended acceptable PMV range for thermal comfort from the standard ASHRAE 55 is between −0.5 and +0.5 for an interior ventilated space. This is the most accepted methodology by green building rating tools such as LEED and Green Star.

Naturally ventilated spaces on the other hand can have a wider temperature range following an adaptive comfort scale. Adaptive comfort theory is well explained in the ASHRAE standard 55-2010 and in essence shifts the acceptable temperature band within the occupied space according to the ambient conditions. The thermal comfort acceptability limits extracted from ASHRAE 55-2010 is shown on the graph below (Fig. 11.3).

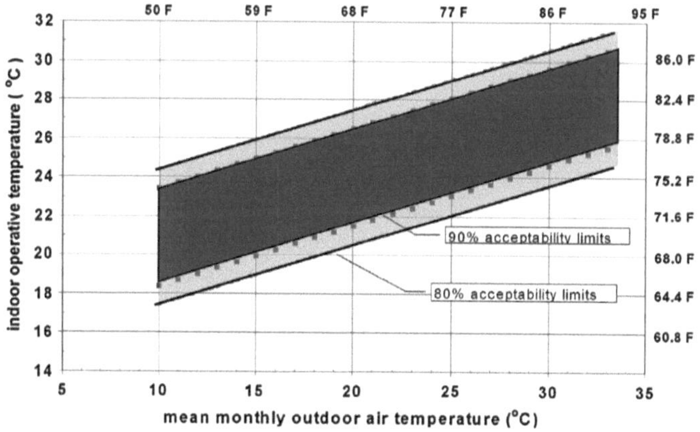

Fig. 11.3 Thermal comfort acceptability limits by ASHRAE 55-2010 (ASHRAE 2010)

11.3.4 CO_2 Concentration

CO_2 concentration levels is one of the indicators of the air quality within indoor spaces. Building occupants release CO_2 during respiration and increased levels of CO_2 create building spaces that are unhealthy. Insufficient renovation of air is one of the factors associated with Sick Building Syndrome, which is a condition that building occupants feel constantly tired with symptoms ranging from headache to chest pains and gastrointestinal problems. Outdoor air supply is therefore a critical parameter to be considered in the simulation of indoor air quality.

Outdoor air can be supplied mechanically by air-conditioning ducts or can be naturally ventilated via operable windows. Detailed energy models have the ability to simulate demand control ventilation systems where minimum supply of fresh air is determined based on the sensed concentration levels of CO_2 in parts per million (ppm). Carbon Dioxide supplied to the occupied zone by respiration vary according to the metabolic rate of the building occupants. Bulk-air movement analysis can be performed for early concept designs where accuracy of airflow patterns when accuracy is not as important as in the latest stages of the development. A Computational Fluid Dynamics (CFD) analysis may be required for a more accurate calculation of natural and mechanical ventilation systems. CFD models can predict CO_2 or CO concentration levels throughout the space at a finite element cell level as opposed to a dynamic thermal simulation that calculates fully mixed air within a thermal zone (Fig. 11.4).

11.3.5 Lighting (Visual Comfort—Daylighting and Glare Analysis)

Good levels of natural light can be beneficial to the wellbeing of the building occupants since it provides a connection to the outdoor environment. Natural light harvesting can also be directly used to reduce energy use and sensible heat loads associated with internal artificial lighting. Good levels of natural light enables a lighting control system to dimmer down in order to maintain the design illuminance level.

Daylight modelling is run using building simulation software packages that run static or dynamic simulations. Static simulations provide a snapshot of the daylight performance at a particular point in time. Static simulations can be suited for the façade analysis in the early stages of the project when the daylight factor (DF) can be assessed against the design criteria. The dynamic simulation or Daylight Autonomy (DA) provides the illuminance levels across different spaces throughout the year according to the location of the project and the climatic data. Daylight Autonomy is a more detailed analysis suitable to the later stages of the design development where the calculations test the illuminance levels to the minimum design target for a percentage of the year according to the criteria set for the project.

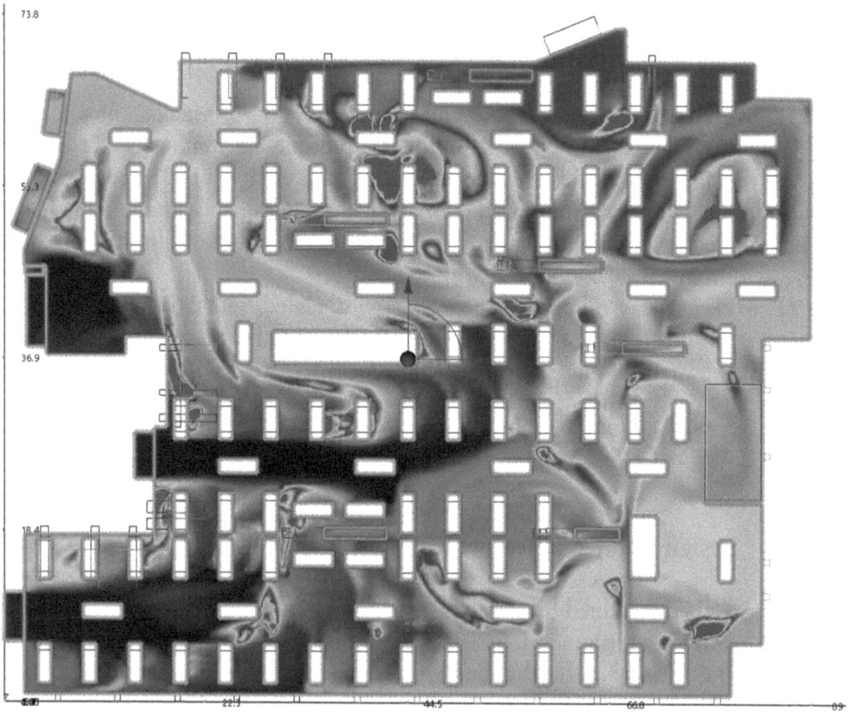

Fig. 11.4 CO_2 concentration result of a car park ventilation at peak hour

A daylight model consists of a three dimensional geometric model with material properties assigned to all internal and external surfaces that calculates the distribution of visible radiation in illuminated spaces. Material properties assigned to daylight models are Solar Reflectivity (R) and material emissivity for opaque materials and Visible Light Transmittance (VLT) and Visible Light Reflectance (VLR) for glazing elements.

Several sky model definitions can be used in a daylight calculation and each model has a specific use depending on the type of analysis. The sky model represents the luminance distribution on an obstructed sky hemisphere according to the site location and latitude.

- **CIE sunny clear day**, the sky distribution corresponds to the standard CIE clear sky condition with additional direct illumination from the sun included. Expect very bright patches due to direct illumination from the sun with relatively dark areas where direct sunlight does not fall.
- **CIE clear day**, the standard CIE clear sky distribution *without* direct illumination from the sun.
- **CIE sunny intermediate day**, sky with intermediate conditions between the overcast and clear skies (see graph below) with *direct illuminance from the sun included.*

- **CIE intermediate day**, the CIE standard intermediate sky with illuminance distribution conditions between the overcast and clear skies (see graph below). *Direct illuminance from the sun is not included.*
- **CIE overcast day**, The CIE Standard Overcast Sky, originally known as the Moon and Spencer Sky, was devised to represent the luminance distribution observed for overcast skies. Adopted as a standard by the CIE in 1955, this description is the one most frequently used for illuminance modelling. In this model, the sky brightness increases gradually with altitude from the horizon to the zenith, but it does not vary with azimuth.
- **CIE overcast day** described above but with scaled with a standard sky illuminance at the zenith of 10,000 lx such that daylight factors can be calculated simply as working plane illuminance values divided by 100. This option is frequently used for daylight factor calculations.
- **Uniform cloudy sky** where the sky illumination is completely uniform.

11.3.5.1 Design Builder Reference

The sky models are defined by standards and mathematical equations developed by the non-profit organisation CIE (Commission Internationale d'Eclairage or International Commission on Illumination). Sunny sky definitions are usually used for glare analysis where it is required to check high luminance contrasts and excessive luminance values in the field of view caused by direct light incident on surfaces. Overcast and uniform skies on the other hand are used to calculate the Daylight Factor (DF), which is the ratio of measured internal illuminance on a surface to a theoretical maximum external illuminance on an unobstructed plane. Daylight Factor is a true representation of the façade performance since it is independent from location and maximum external illuminance.

Daylight sensors can be used to measure illuminance at a particular point on an internal surface and results used on an energy model to dimmer down lights according to the design lighting levels.

Results for daylight and glare analysis can be presented on a specific working plane or in a specific 3D view such as Perspective, Hemispherical Fish-eye and Angular Fish-eye (Fig. 11.5).

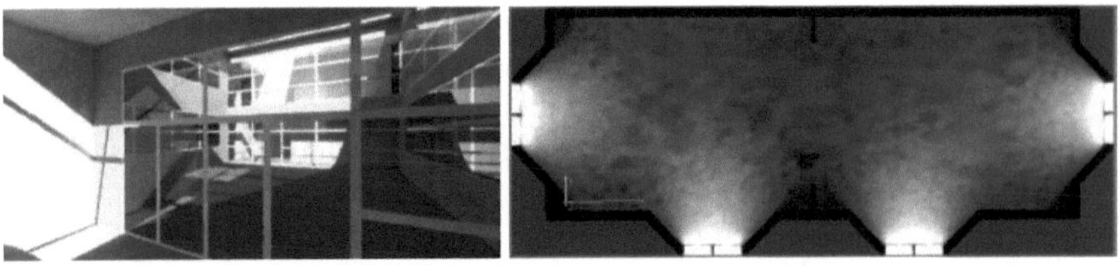

Fig. 11.5 Luminance Image and illuminance image on working plane

11.3.6 Ventilation

The mechanical or natural ventilation system of a building play an important role in providing optimum levels of thermal comfort. Poor ventilation design can lead to cold draught or insufficient ventilation to maintain temperature set point in areas regularly occupied. Operable windows in a natural ventilation system also need to be properly designed to ensure sufficient renovation of air and avoid excessive air velocities unevenly distributed across the ventilated space.

Air Change Effectiveness relates to how efficiently the supply air is distributed through an occupied space. The Age of Air is the average amount of time that has elapsed since a sample of air molecules at a specific location entered the building.

Air change effectiveness compares the age of the air in the occupied portions of the building to the age of air of a perfect mixing ventilation system. If air change effectiveness equals to 1 ach then the ventilation air within a space is perfectly mixed and the outdoor air flow rate to the ventilated space is the same as the minimum design requirement.

Performance simulations in the area of air change effectiveness are required to be performed by Computational Fluid Dynamics analysis. The configuration of the CFD model is performed with the proper setup of a finite element mesh and specific boundary conditions to mimic the mechanical and natural ventilation system.

11.4 Model Development

11.4.1 How to Get the Right Advice to Project Stakeholders

Building Performance Simulation is a powerful tool to provide design guidance in the early stages of a project when building form and glazing to wall area ratio can be assessed against environmental criteria such as building energy efficiency, natural ventilation, thermal comfort, daylight penetration and glare discomfort.

The building energy model in particular can be used to assess alternative solutions to maximise the potential for natural ventilation and minimize heating and cooling loads. This assessment is done through massing and shading analysis, simplified computational fluid dynamics of external and internal airflows and design sensitivity analysis of glazing area and thermal properties (U value and SHGC).

The energy model in the Concept Design stage should be done on a shoebox type of model to quickly assess the key parameters of a zero carbon design. The output of the initial model can be then split in different energy uses with a subsequent Energy Use Intensity calculation. It is imperative that the building simulation outputs at this early stage of the project development are benchmarked against performance metrics: cooling loads (W/m^2), heating loads (W/m^2), Lighting energy use (kWh/m^2/year), illuminance levels (lux), plug loads (kWh/m^2/year) and building total energy demand (kWh/m^2/year).

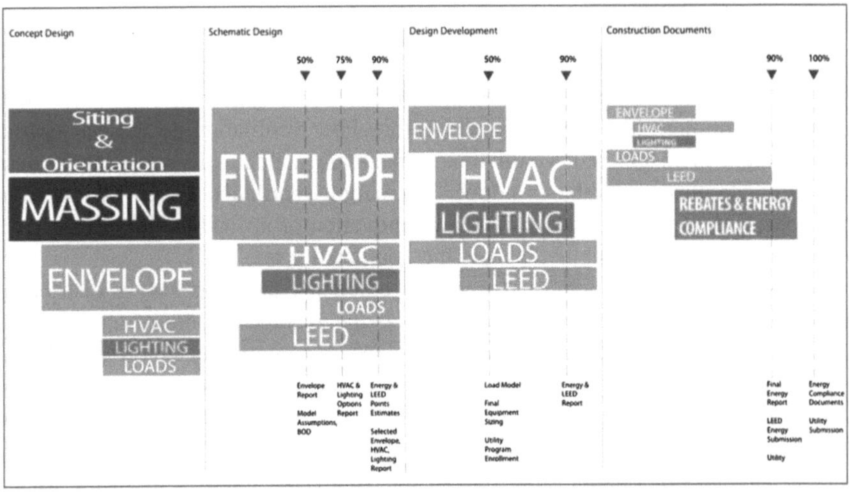

Fig. 11.6 Building performance simulation stages (Integral Group)

After the massing and natural ventilation analysis of the early concept design, the project moves to a schematic design where the services and building envelope start to be developed. At this point, a huge focus must be placed on the analysis of the building envelope such as insulation levels, airtightness, thermal bridging and thermal mass. The reduction of the Energy Use Intensity (EUI) of the building through the optimisation of the building form and envelope will allow for the specification of a smaller mechanical system to condition the building spaces. This strategy will shift a portion the investment costs from the mechanical and electrical services to the architecture project without incurring in an increase of the project overall costs (Fig. 11.6).

During the Schematic Design the building simulation or energy model can also evaluate and compare alternative HVAC design solutions in order to select the system that presents the potential for best synergy with other services, architectural design, local microclimate and ecosystem. System efficiency and life cycle costs can be compared between multiple design alternative solutions so that the design team can make an educated decision based on numerical data rather than rule of thumb or engineers personal experience.

The final step of the simulation process involves a detailed model that will inform stakeholders of the compliance of the building performance with local building regulations and codes and environmental rating systems such as LEED, BREEAM and Green Star. The detailed model is also very beneficial to analyse operation of natural ventilation, lighting and HVAC control systems so that these systems work synergistically and at its optimised efficiency. After the minimisation of the Energy Use Intensity of the building design and services, the zero energy target is achieved through the supply of onsite renewable energy, in most cases from Photovoltaic panels (Fig. 11.7).

Fig. 11.7 Bullitt Centre, Seattle: example of net zero energy building with integrated PV roof (Bullitt Centre 2015)

11.4.2 Holistic Building Analysis and Project Integration

Building performance analysis starts from the very beginning of the project in the development of concept ideas and architectural design. Freethinking charrette processes are an excellent form of encouraging the generation of solutions and imagining the place and its integration with the existing community and local ecosystem. During intensive charrette processes, stakeholders bring their experiences together in a collaborative way in order to develop a design intent document and performance targets for the building.

This process allows the Building Performance simulation to be an integral part of the building design process. The integrative process shifts the engineering costs upfront to the early stages of the project allowing for synergies and solutions that are numerically validated by models that account for building form, envelope thermal properties, services integration and occupant behaviour.

BIM (Building Information Modelling) is a tool that is being adopted by design teams in the recent years to speed up the development process and facilitate the communication between the architecture team, the building services team and the building simulation professional. BIM models allow a faster integration of design parameters into the building simulation models which in return provides a faster response to the project team in terms of simulation outputs to support the decision making process of design solutions.

11.4.3 Concept Design—Evaluation of Alternative Solutions

The concept design stage of a project is where the building simulation has the most significant impact into the solutions that will enable the development of a net zero building. At this stage of the project several analysis can be performed to evaluate solar gains, building envelope area, transmission losses, daylight penetration, thermal comfort and natural ventilation. The following parameters can be assessed over multiple alternative solutions (Fig. 11.8):

(i) Glazing to Wall area ratio
(j) Shading analysis
(k) Building form
(l) Building orientation
(m) Natural ventilation analysis
(n) Solar exposure for Photovoltaics

11.4.3.1 Detailed Design—Accurate Analysis for Final Design and Rating Submissions

11.4.4 Model Assumptions and Error Propagation

Building simulation models can provide an estimate of the building's performance. This estimate is based on a necessarily simplified and idealised version of the

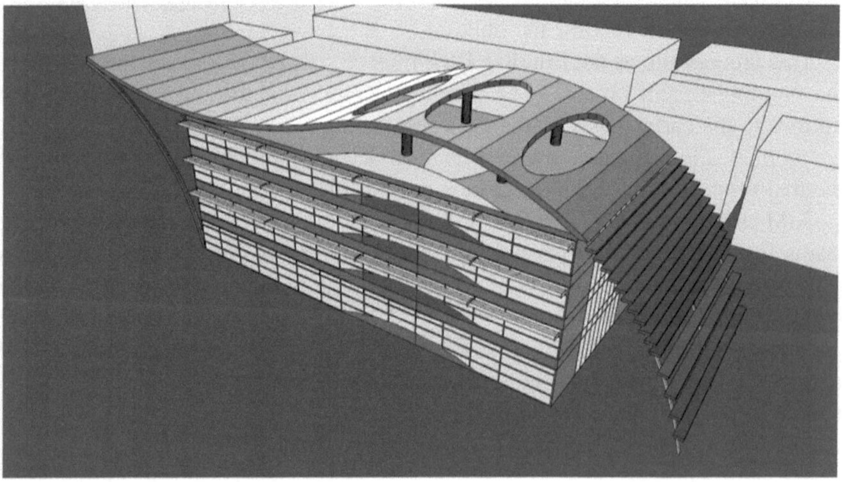

Fig. 11.8 Solar exposure analysis (IES 2015)

building that does not necessarily represent all of the intricacies of the building and its operation. As a result, the simulation results only represent an interpretation of the potential performance of the building.

Assumptions built into the models can propagate errors due to the unpredictability of occupant behaviour and building operation. Building Performance models are built with ideal parameters such as typical schedules for occupancy, lighting, equipment and HVAC operation. For example, if the energy performance of a building is validated by an energy model with air conditioning set point between 20 and 24 °C and the user of the building sets the thermostat at 19 °C in summer the energy consumption will be significantly higher than the estimated by the energy model.

11.5 Weather Data

Thermal simulation is adopted to predict building performance over a period of time (Cook and Short 2009). Interactions between the building and external environment highly affect the results of simulation (Belcher and Powell 2005). For this reason, all dynamic thermal simulation packages require weather data inputs, which are usually text format files (Jentsch et al. 2008). The weather data provide information on different climatic parameters such as temperature, solar radiation, wind speed, direction etc. on hourly basis (Degelman 2003).

Historical meteorological data are used to generate weather data for different locations. Several organisations including the U.S. National Renewable Energy Laboratory (NREL) the National Climatic Data Center (NCDC) and Sandia National Laboratory (SNL) have developed generic weather data for different locations. This weather data can be used either directly or by some modifications to study building performances under different climatic conditions (Jentsch et al. 2008; Crawley 1998).

Weather data can be categorized in four types of typical, extreme, actual and future weather data. Different types of weather data and their application are explained below.

11.5.1 Typical Weather Data

Typical weather data, such as test reference year (TRY) and typical meteorological year (TMY), are usually used to investigate annual energy consumption (Gazela and Mathioulakis 2001) and feasibility of application of renewable energy systems (Chow et al. 2006; Bhandari et al. 2012). According to Gazela and Mathioulakis (2001), typical weather data is a single year data, which includes 8760 different values. Reported values in typical weather data are representative of different meteorological variables including ambient temperature, solar radiation, relative humidity and wind velocity. Typical weather data is usually generated based on 20–30 years of actual measurements. Data for each month is based on statistical analysis which is believed to be representative of typical climatic condition of that month (Levermore and Doylend 2002).

11.5.2 Extreme Weather Data

Extreme weather data, such as design summer year (DSY), is usually generated based on extreme actual weather condition by considering the worst-case scenarios. Typical weather data represents typical weather conditions (excluding extreme weather conditions) and therefore typical weather data can predict building performance for typical weather conditions only. This is while, a comprehensive design should cover extreme conditions as well as typical conditions. For this reason a new series of weather data, known as extreme weather data, is used to be representative of extreme weather conditions. Extreme weather data may be used to size the HVAC systems and predict building performance during extreme climatic conditions. DSY may also be used to test the risk of overheating in naturally ventilated buildings (Adelard et al. 2000; Holmes and Hacker 2007). However, it should be noted that the selection of the extreme weather conditions should be reasonable/justified as it could lead to overestimation of demand and consequently designing oversized systems (Li et al. 2003), leading to higher costs.

11.5.3 Actual Weather Data

Actual weather data is based on the actual measurements of climatic parameters for a particular location at a particular time and is usually used to validate computer models. Actual weather data is typically used in retrofitted industry where actual performance of the building is recorded to be compared with the results of computer simulations. For example, the real and predicted energy consumption values for particular periods may be compared using actual weather data (Bhandari et al. 2012). It should be noted that due to the cost and time related issues measuring actual weather data for individual buildings is less practical and therefore designers need to rely on available data sources provided by different organisations. The accuracy of outputs is therefore highly affected by the quality and reliability of available weather data (Bhandari et al. 2012).

11.5.4 *Future Weather Data*

Results of several studies reveal that climate change is imminent due to extensive carbon emissions. Therefore a range of weather data, which is believed to be representative of future climatic conditions, are developed. Future weather data is used to predict the effects of future climate on the building performance (Guan 2009). Different methods including the extrapolating statistical, imposed offset, stochastic weather model and global climate models can be used to generate future weather data (Guan 2009). In these methods, different scenarios of carbon emission are assumed and effects of these scenarios on climate is predicted. The parameters in the current weather data are then adapted to consider these scenarios. The Chartered Institution of Building Services Engineers (CIBSE) of the UK, for instance, provides future weather data for four different carbon emission scenarios including low, medium-low, medium-high and high (Du et al. 2011).

11.5.5 *Physical Measurements*

Physical measurements are particularly important when evaluating the performance of existing/in-use buildings. Physical measurement may include building surveys; gas and electricity meter readings; interviews and questionnaire surveys; data recorded by sensors and data loggers; air pressure tests and infrared imaging. Figure 11.9 illustrates the typical sequence of physical measurements, which may then be fed into computer simulation.

11.5.5.1 Building Surveys and Meter Readings

Building surveys are carried out to collect information on the physical aspects of a building such as construction methods and materials, plans, facades, opening types/sizes, shadings, buildings orientation etc. Computer models are then developed based on the building surveys and are used for simulations.

Regular (e.g. monthly) gas and electricity meter readings may also be carried out for a certain period of time (e.g. one year) to evaluate the energy consumption of case study buildings. The recordings are then used to evaluate the energy performance of buildings by, for example, comparing the energy consumption with the benchmarks for different building types. The recordings may also be used to validate and/or compare with the results of computer simulations.

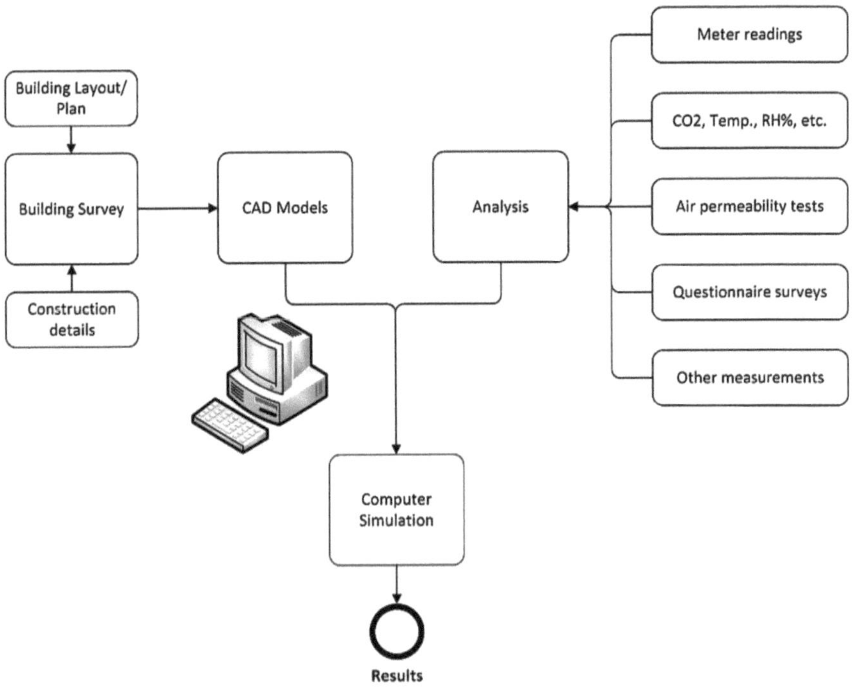

Fig. 11.9 Physical measurements

Fig. 11.10 Equipment used for air permeability tests

11.5.6 Questionnaire Surveys

Questionnaire surveys and interviews are usually carried out to investigate occupancy patters and behaviours in buildings. Depending on the building type, questionnaire surveys may cover subjects such as the number of occupants; occupancy times; opening/closing doors, windows, vents and shadings as well as sleeping, cooking, bathing, and washing patters. This information will then be used to create occupancy schedules for computer models.

Questionnaire surveys may also be considered to evaluate building performance before and after design interventions. For instance, a questionnaire survey may be carried out to evaluate Indoor Air Quality and thermal comfort of occupants before and after a major refurbishment aiming to improve comfort in a building. The results may then be compared with computer simulations for comparison and/or validation. Existing questionnaires may be used/adapted to carry out the surveys. Examples include Building Use Studies (BUS) survey (Bordass et al. 2000); Indoor Environmental Quality (IEQ) survey from the Centre of the Built Environment at the University of California (Huizenga et al. 2002); and a questionnaire developed by the Building Research Establishment (BRE) for evaluating sick building syndrome (Raw 1995). Different subjects such as the temperature, air movement and quality, visual comfort, noise levels and control over the environment may be investigated to evaluate occupants' comfort in a building. Pilot tests should always be considered to identify any possible issues prior to conducting a major questionnaire survey.

11.5.7 Air Pressure Tests

Air pressure tests are carried out to measure Air Permeability (AP) of buildings. AP may also be referred to as air leakage and airtightness. Airtightness can greatly affect indoor air quality and energy performance of buildings (Hashemi and Khatami 2015) and is therefore an important impute for accurate energy and IAQ simulations in buildings. Air permeability rates of buildings are measured by conducting air pressure tests in which escaping air through the building envelope under a certain pressure is measured. This is done by (A) attaching fan(s) to the building to (B) pressurizing/depressurising the building, and (C) measuring the drop in pressure (Fig. 11.10). The number and sizes of fans may vary depending on the size of the tested building. All external openings including windows and doors are kept closed and internal doors are kept open during the test. The gap around the doors and trickle vent as well as any other opening such as the extractor fans and cooker hood are sealed to accurately measure the air permeability rates. The test is repeated several times and the results are averaged over the total envelope area of the building taking into account external environmental conditions such as wind speed, temperature, and barometric pressure. AP, in principal, is a function of inside and outside pressure differences and required flow rate of the fan(s) to produce that pressure difference. CIBSE TM 23 explains the processes and standards for air pressure tests in domestic and non-domestic buildings (CIBSE 2000).

AP is measured/expressed in cubic meters per hour per meter square of building envelop under a certain pressure, for example, 50 Pa (e.g. 5 m^3/h m^2 @ 50 Pa). Building regulations may set a minimum standard for the AP. For instance, a limiting air permeability rate of 10 m^3/h m^2 @ 50 Pa is required by the UK building regulations as the minimum standard for new buildings (HM Government 2013). The following tables and figures have been extracted from an actual air pressure test carried out by EPS Group, UK on a two-bed flat (Tables 11.1 and 11.2).

Summary of the results. The dwelling's air leakage rate was checked by means of a depressurisation test, which incorporated the all of the building envelope. The result attained during this test was 6.997 m^3/(h m^2) at an imposed pressure differential of 50 Pa, when the aforementioned temporary sealing was in place (Fig. 11.11).

11.5.8 Sensors and Data Loggers

Sensors and data loggers are used to measure and record various indicators such as temperature, relative humidity (RH%), CO_2 (Fig. 11.12), and Volatile Organic Compounds (VOCs). Correct installation of sensors and data loggers is key to accurate and reliable recordings. Sensors and data loggers should be placed in an appropriate location based on their specifications and manufacturers' guidelines.

Table 11.1 Dwelling information

Dwelling type		Ground floor flat	
Net floor area, AF:	52 m^2	Envelope area, A_E	172.9 m^2
Volume, A_E	114.4 m^3	Year built	1920s, renovated 2012
Test date	2014-01-21	Test time	11.15
Dwelling type	Ground floor flat	Test method	B (Building envelope)
Client	A H	Test engineer	A W

Table 11.2 Results of air permeability tests

	Results	Uncertainty (%)
Air flow at 50 Pa, V [m^3/h] 50	1200	±0.8
Permeability at 50 Pa, Q [m^3/h m^2] 50	6.997	±1.0
Equivalent leakage area at 50 Pa [m^2]	0.06040	±0.8
Air changes, n50	10.55	±1.3

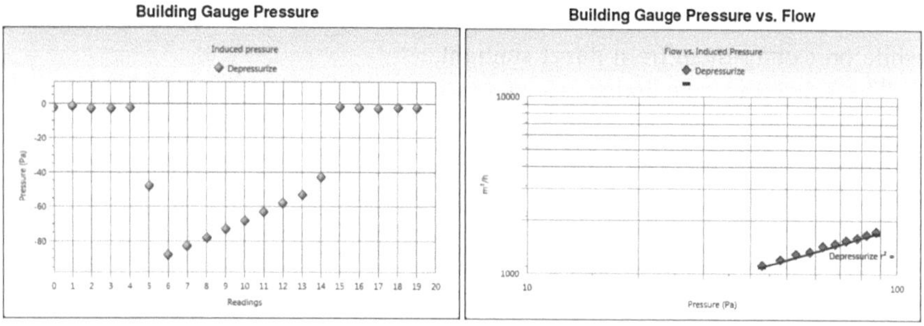

Fig. 11.11 Test data

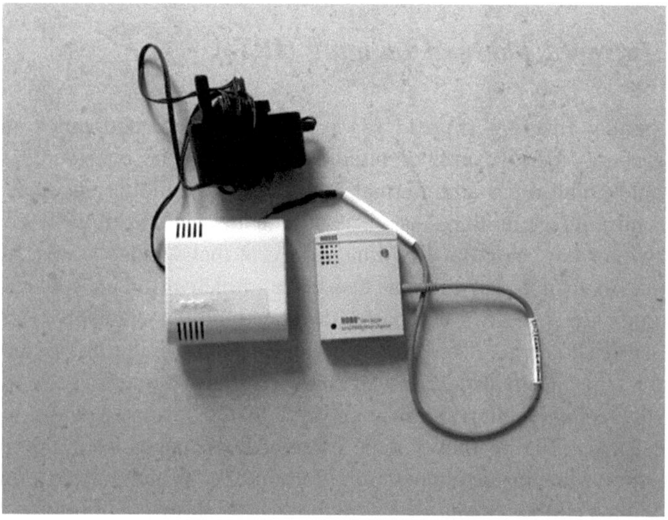

Fig. 11.12 HOBO U12/12 data logger (*right*) and CO_2 sensor (*left*) assembly

They should not be exposed to harsh environments and/or be installed in locations where inaccurate measurements are probable. They should also be away from direct solar radiation and heat sources (unless intentionally installed in such areas) to provide realistic and accurate measurements. Installation on external walls and near openings should be avoided as much as possible. Generally, sensors should be installed in locations where recorded data is representative of the overall conditions within an environment. BSRIA (1998) guidelines, for instance, recommends internal sensors to be installed at a height of 1.0 m above the finished floor level to represent the height of a seated occupant. External temperature sensors and data loggers should also be installed on the north side of the buildings and be protected from direct solar radiation (BSRIA 2009). Special plastic boxes with several holes have been developed for external sensors to allow air movement around the sensors while protecting them from direct sunlight.

The suitability and limitations of sensors and data loggers should always be considered prior to installation. For instance, CO_2 sensors can usually measure/record CO_2 levels up to 2000 ppm. Some sensors may measure CO_2 levels up to 4000–5000 ppm and higher although they are usually more expensive and/or less accurate. Similar limitations apply to other sensors such as light sensors. Generally, for commercially available off-the-shelf sensors and data loggers, the higher measurement capabilities are achieve with compensation for lower accuracy. Sensors and data loggers should be calibrated regularly and tested prior to installation on site to achieve accurate and reliable results.

11.5.9 Infrared Thermal Imaging (IRT)

Infrared thermal imaging (Fig. 11.13) is an appropriate and quick method for evaluating energy performance of buildings. IRT may be considered to identify weak thermal points/areas and thermal bridges in the building envelope. Areas of heat losses through the buildings can instantly be identified using IRT. For instance, in a cold winter day, external IRT images reveal that windows have higher temperature due to their lower thermal resistance, which result in considerable heat losses through the windows compared to other building elements (Hashemi and Gage 2014). It should be noted that in a heated building during winter when the pictures are taken from inside, low thermal resistant elements such as window, appear to have lower temperature compared to the insulated walls with higher thermal resistance. Figure shows some external IRT images from a double-glazed UPVC window when inside temperature is around 20 °C and outside temperature is around −5 °C. Yellow, red and white coloured areas show cold bridging around the frame as well as warm air escaping through the trickle vents on the top.

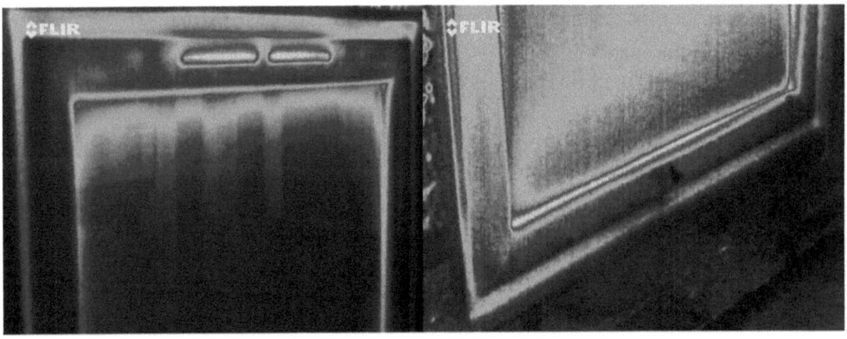

Fig. 11.13 Infrared imaging of a *double-glazed* UPVC window

11.6 Conclusions

The chapter provided a general knowledge of, and a basic ability to understand the use of building performance and simulation, and to optimise the performance of different components used in buildings as part of the building design. It also touched on both advantages and disadvantages of using building performance and simulation for environmental analysis at different stages of the building design. Moreover, additional information on related input sources, such as weather data, physical measurements, etc. and other methods have been discussed and explained to support building performance and simulation.

References

Adelard L, Boyer H, Garde F, Gatina JC (2000) A detailed weather data generator for building simulations. Energy Build 31(1):75–88

ASHRAE (2010) Standard 55-2010—thermal environmental conditions for human occupancy (ANSI Approved)

Augenbroe G, Malkawi A, de Wilde P (2004) A workbench for structured design analysis dialogues. J Archit Plann Res

Belcher SE, Powell DS (2005) Constructing design weather data for future climates. Build Serv Eng Res Technol 26(1):49–61

Bhandari M, Shrestha S, New J (2012) Evaluation of weather datasets for building energy simulation. Energy Build 49(2012):109–118

Bordass W, Ford B, Schiano R (2000) Specification of automatic vent opening devices for natural ventilation. Partners in technology DETR project report

BSRIA (1998) Building control and indoor environmental quality—a best practice guide. Technical Note TN 9/98. BSRIA—The Building Services Research and Information Association, Berkshire

BSRIA (2009) Illustrated guide to ventilation. BSRIA building applications guide BG 2/2009. BSRIA—The Building Services Research and Information Association, Berkshire

Bullitt Centre (2015) http://www.bullittcenter.org/

Chow T, Chan ALS, Fong KF, Lin Z (2006) Some perceptions on typical weather year—from the observations of Hong Kong and Macau. Sol Energy 80(4):459–467

CIBSE (2000) Testing buildings for air leakage, CIBSE technical memoranda TM23: 2000. The Chartered Institution of Building Services Engineers, London

Crawley DB (1998) Which weather data should you use for energy simulations of commercial buildings? ASHRAE Trans 104(Part 2)

Degelman L (2003) Simulation and uncertainty—weather predictions. In: Augenbroe MA (ed) Advanced building simulation. Spon Press, Abingdon

Du H, Edge J, Underwood C (2011) Modelling the impacts of new UK future weather data on a school building. In: Proceedings of building simulation 2011: 12th conference of International Building Performance Simulation Association, Sydney, 14–16 Nov 2011

Gazela M, Mathioulakis E (2001) A new method for typical weather data selection to evaluate long-term performance of solar energy systems. Solar Energy 70(4):339–348

Glotzer SC, Kim S, Cummings P, Deshmukh A, Head-Gordon M, Karniadakis G, Petzold L, Sagui C, Shinozuka M (2009) International assessment of research and development in simulation-based engineering and science, Available at: http://www.wtec.org/sbes/SBES-GlobalFinalReport.pdf (Accessed: 20 November 2015)

Guan L (2009) Preparation of future weather data to study the impact of climate change on buildings. Build Environ 44:793–800

Hashemi A, Gage S (2014) Technical issues that affect the use of retrofit panel thermal shutters in commercial buildings. Build Serv Eng Res Technol 35(1):6–22. doi:10.1177/0143624412462906

Hashemi A, Khatami N (2015) The effects of air permeability, background ventilation and lifestyle on energy performance, indoor air quality and risk of condensation in domestic buildings. Sustainability 7(4):4022–4034. doi:10.3390/su7044022

HM Government (2013) The building regulations 2010 (ed 2013), Approved Document L1 A, Conservation of Fuel and Power in New Dwellings. HM Government, London, UK

Holmes MJ, Hacker JN (2007) Climate change, thermal comfort and energy: meeting the design challenges of the 21st century. Energy Build 39:802–814

Huizenga C, Laeser K, Arens E (2002) A web-based occupant satisfaction survey for benchmarking building quality. In: Indoor Air. Monterey, CA

Integrated Environmental Solutions (IES) (2015) http://www.iesve.com/

Jentsch MG, Bahaj AS, Patrick AB (2008) Climate change future proofing of buildings—generation and assessment of building simulation weather files. Energy Build 40:2148–2168

Levermore GL, Doylend N (2002) North American and European hourly-based weather data and methods for HVAC building energy analyses and design by simulation. ASHRAE Trans 108 (2):1053–1062

Li DHW, Wong SL, Lam JC (2003) Climatic effects on cooling load determination in subtropical regions. Energy Convers Manag 44(11):1831–1843

Oden TJ, Belytschko T, Fish J, Hughes TJR, Johnson C, Keyes D, Yip SA (2006) A report of the National Science Foundation Blue Ribbon Panel on simulation-based engineering science: revolutionizing engineering science through simulation. Available at: http://www.nsf.gov/pubs/reports/sbes_final_report.pdf. Accessed 20 Nov 2015

Raw GJ (1995) A questionnaire for studies of sick building syndrome: a report to the royal society of health advisory group on sick building syndrome. Report 287. Building Research Establishment (BRE), Watford

Chapter 12
ZEMCH Business Operation in Japan

Masa Noguchi, Ercilia Hirota, Jun-Tae Kim, Carlos Formoso, Haşim Altan, Pamela Bell, Antonio Frattari and Laurent Riscala

Abstract Today, Japanese housing manufacturers mass-customise their highly industrialised houses with the aim to deliver quality affordable homes in which design quality, or housing performance, can be defined by buyers/users in consideration of their economic constraints. Moreover, in response to the drastic hike of energy prices and global warming issues, the prefabricators today tend to commercialise net zero energy homes. Their high-quality, zero energy mass custom homes (ZEMCH) that are also reasonably priced enjoy a good reputation. This chapter will identify the essential elements of Japanese housing manufacturers'

M. Noguchi (✉)
The University of Melbourne, Melbourne, Australia
e-mail: masa.noguchi@unimelb.edu.au

E. Hirota
State University of Londrina, Londrina, Brazil
e-mail: ercilia@uel.br

J.-T. Kim
Kongju National University, Gongju, South Korea
e-mail: jtkim@kongju.ac.kr

C. Formoso
Federal University of Rio Grande Do Sul, Porto Alegre, Brazil
e-mail: formoso@ufrgs.br

H. Altan
The British University in Dubai, Dubai, UAE
e-mail: hasim.altan@buid.ae

P. Bell
PrefabNZ, Wellington, New Zealand
e-mail: pam@prefabnz.com

A. Frattari
University of Trento, Trento, Italy
e-mail: antonio.frattari@unitn.it

L. Riscala
ECOxIA, Yerres, France
e-mail: laurent.riscala@ecoxia.com

© Springer International Publishing Switzerland 2016

M. Noguchi (ed.), *ZEMCH: Toward the Delivery of Zero Energy Mass Custom Homes*, Springer Tracts in Civil Engineering,
DOI 10.1007/978-3-319-31967-4_12

business operation being applied for the delivery of socially, economically and environmentally sustainable homes that satisfy the market needs and demands.

12.1 Introduction

There is a tendency for house-builders to rarely adopt or buy new products and services that to some extent interrupt their business operation that is often based on routine. However, the routine has been streamlined to a point at which home-builders achieve great efficiency of their activities. Working drawings are extremely simplified because craftsmen know how to carry out their task, based on traditional construction techniques, and housing specifications are also brief, because the designers know that craftsmen know how to fit materials together. In general, homebuilders tend to restrict their information search to the programmed decision, since the search for information for non-programmed decisions take too much time, money and effort (Noguchi 2008). Therefore, to open the gate for conventional homebuilders and housing manufacturers at global scale to consider the future production of their own net zero energy sustainable houses that help contribute to alleviating climate change issues, the 'Zero-energy Mass Custom Home Mission to Japan' (termed 'ZEMCH Mission to Japan' today) was initiated in 2006. The housing manufacturers practise a quality-oriented production approach that reflects their high *cost-performance* marketing strategy (Noguchi 2003). Today, they have been at the forefront of mass-customising net zero energy, carbon dioxide (CO_2) emission, and/or utility cost housing that satisfies the wants and needs of individuals as well as society.

Learning from the ZEMCH Mission to Japan experience, this chapter will identify the essential features of Japanese housing manufacturers' design, production and marketing approaches being applied for the delivery of their zero energy mass custom home (or ZEMCH) that secures social, economic, and environmental housing sustainability in local contexts.

12.2 ZEMCH Mission to Japan

In 2005, CANMET Energy Technology Centre (called 'CanmetENERGY' today)—Varennes, Natural Resources Canada, supported the organisation of 'Japan Solar Photovoltaic Manufactured Housing Technical Mission 2006.' The knowledge transfer educational event was held from 20th to 23rd February, 2006, with the aim to provide Canadian homebuilders, housing manufacturers, building component suppliers, architects, academics and government officers with opportunities to explore not only the state-of-the-art production facilities of the then four leading net zero energy cost housing manufacturers in Japan, i.e., Sekisui Chemical Co. Ltd., Misawa Homes Co. Ltd., PanaHome Corp., and SANYO Homes Corp., but also a sales centre that displays a variety of Japanese manufacturers' housing prototypes (Richard and Noguchi 2006). Based on the past experience, from 3rd to 5th September, 2007, the educational event was reorganised by MEARU, Mackintosh School of Architecture, The Glasgow School of Art, in collaboration with the Centre for the Built Environment in the United Kingdom. Then it was renamed to 'Zero-carbon Mass Custom Home Mission to Japan' and added a visit to Sekisui House Ltd. which had been committed to the development of a net zero CO_2 emission house towards the future exhibition during the G8 Hokkaido Toyako Summit that was expected to be held in the following year. In 2008, the title was again changed to 'Zero-energy Mass Custom Home Mission to Japan' and was held from 10th to 12th September. In 2010, the mission was organised as one of the 'Renewable Energy 2010' international conference's related events held from 23rd to 25th June. In addition to the company visits, a post-mission strategic 'Zero-energy Housing' workshop was also executed within the conference programme in which 20 delegates from Australia, Brazil, Canada, Italy, Korea, Portugal, Sweden, UK and USA were invited as panellists to report the mission outcomes and discuss issues around the delivery of net zero energy/carbon mass custom homes today. The workshop helped establish a forum of interactive discussion on the design, production and marketing approaches and solidify diverse individual expertise and experience into universal knowledge. After 2 years of absence, the ZEMCH Mission to Japan was resumed in 2013, held from 14th to 17th May. Since 2014, the Faculty of Architecture, Building and Planning, University of Melbourne, has begun offering the administrative support for the organisation and operation of ZEMCH Mission to Japan. In 2014 and 2015, it was run from 22nd to 25th July and from 20th to 24th July, respectively. In addition to the facility visits, the 2015 Mission encompassed a technical seminar with the support of Kogakuin University located in Tokyo, while the visits also included the University of Tokyo's demonstration smart house and Japan's first net zero CO_2 emission prefabricated housing prototype built by Sekisui House Ltd.

12.3 Evolution of Japan's Prefabricated Housing Industry

The prefabricated housing industry in Japan has successfully overcome the inferior image of industrialized houses prevalent in the industry's early stages when the mass production of housing was the key factor in creating monotonous, boxy units, which the public subsequently regarded as being of low quality. More importantly, Japanese housing manufacturers have married their production methods to specially developed marketing, or communication, techniques to skilfully satisfy local housing needs and demands. In the past, manufacturers encountered many difficulties including strict building codes, building officials, local unions, banks and, in particular, consumers' prejudices that limited the popularity of industrialised housing. Manufacturers have seemingly overcome these difficulties by developing their own quality-oriented production approach that focuses solely on responding to consumers' requirements,

Fig. 12.1 Midget House: the origin of mass produced prefabricated housing in Japan (Daiwa House Industry 2015)

desires and expectations for housing *quality*, while housing *affordability* has been less of a consideration.

In 1955, Daiwa House Industry Co. Ltd. (1999) pioneered Japan's first prefabricated housing called "Pipe House" and later the then Japan lightweight Iron Construction Association qualified the company to pave a way for the mass production of Japan's practical prefabricated housing unit called "Midget House" in 1959 (Fig. 12.1).

This commercialisation success spurred both the government and the private enterprise on to focus mainly on the productivity enhancement of prefabricated homes. In consequence, during the 1960s and 1970s, due to a simple mass production approach to the industrialised housing delivery, the public came to associate prefabricated houses with low quality. In 1963, in order to improve the public image of prefabricated homes, the Japan Prefabricated Construction Suppliers and Manufacturers Association (JPA) was established by the Ministry of Construction and the Ministry of International Trade and Industry. In 1976, the government proposed a nation-wide competition called "House 55" project to encourage housing manufacturers to improve the quality of their products and to demonstrate to the public that industrialised housing need not be of low-quality (Sackette 1986). Since then, Japanese housing manufacturers have been contributing to the continuous improvement of prefabricated housing quality in light of the design and performance, while adhering to higher standards than those prescribed in building regulations. Today, the manufacturers cater to the market demand for value-added quality homes.

In Japan, a total of 892,261 houses were newly built in 2014 and amongst them, 140,501 homes, or 15.7 %, were estimated to be prefabricated (JPA 2015). The prefabricated housing rate published is inclusive of all homes that were built using

Fig. 12.2 Modular housing unit in-factory assembly (Sekisui Chemical Co. Ltd.)

post and beam, panelised, modular housing systems. In Japan, a modular system is normally called a "unit system" as a modular house consists of a number of three dimensional modules, or units, fabricated in a factory prior to the on-site assembly (Noguchi 2011). The factory-built units include not only structural frames but also interior and exterior finishing materials and service facilities; therefore, a living room, dining room, kitchen, bathroom, entrance, cloakroom, staircase, and storage are all installed in the factory assembly sections (Fig. 12.2).

The modular housing system substantiates over 80 % of in-factory completeness of homebuilding components. After in-factory assembly, the prefabricated modules are shipped, on truck beds, to the building site. Accordingly, the size of each unit tends to be standardised based on the loading capacity of freight trucks, as well as the local traffic regulations applied to the transportation. At the building site, the sections are stacked to form a multi-storey modular home (Fig. 12.3).

Fig. 12.3 Modular housing on-site assembly (Misawa Homes Co. Ltd.)

Today, Japanese housing manufacturers still mass-market homes, yet the delivery is based on their quality-oriented production approach that reflects their high "cost-performance" marketing strategy (Noguchi 2003). The manufacturers tend to use the money saved from lowering production costs through mass production to equip homes with more standard features; thus, the selling price of their new housing is not reduced, whilst the housing quality is drastically upgraded. This approach may help distinguish their industrialised houses from conventional ones as to the housing quality and performance. Today, the manufacturers have been successful in commercialising their industrialised houses that are often equipped with a variety of technical features including renewable energy technologies such as PV power generating systems (Noguchi and Collins 2008). Due to the Japanese Residential PV dissemination programme offered between 1994 and 2003, the number of domestic PV installations in Japan drastically increased from 539 to 52,863 houses.

The following sections will identify the essential features of Japanese housing manufacturers' design, production and marketing approaches being applied for the delivery of commercially, culturally and environmentally sustainable prefabricated homes in the local contexts.

12.4 Quality-Oriented Production

Japanese housing manufacturers have normally acquired ISO 9000 and 14,000 accreditations that certify the quality control of their products, as well as the companies themselves. They set up higher standards than those prescribed in ordinary building regulations, maintaining uniform product quality by strict control over their products. In particular, most Japanese manufacturers establish their own quality standards, in order to improve structural resistance, durability and amenities. The manufacturers adapt resource-saving strategies. Toyota Home, which produces steel-frame modular homes in Japan, advocates that the durability of housing should contain structural durability, as well as design durability and flexibility (Noguchi 2003). Misawa Homes, which also produces steel-frame modular homes, emphasises the need for the production of houses that can be used by several generations of the owner family over 100 years. Surprisingly, the lifespan of housing in Japan is regarded as much shorter than that in other advanced nations (Matsumura and Tanabe 1996). In 1993, there were in total 45,940,000 houses, and of the figure, only 2,150,000 were pre-war houses. In Japan, the legal lifespan of wood-frame housing is said to be 30 years. However, 10 % of wood-frame houses statistically vanish within 18 years after the home is newly built and almost half of the houses are destroyed within 33 years. In consideration of the short durability of their older housing stock, Japanese housing manufacturers tend to demonstrate to the public that unconventional factory-built homes are structurally, environmentally and economically sustainable.

Japanese housing manufacturers have been producing better products with higher levels of safety and amenities that make the home not only a mere shelter but also a living environment where room temperature, indoor air-quality and soundproofing are well controlled. Due to high rates of seismic activity in Japan, earthquake-resistant housing is extremely important. For example, the Great Kanto earthquake of 1923 destroyed houses by exerting a horizontal force of approximately nine tons and 142,807 people lost their lives (Utsu 1995; Mishima 2000). Most housing manufacturers now establish their own standards for structural resistance based on the Great Kanto earthquake, producing homes that can resist 28.7 tons (1000 gal) of horizontal force. These higher standards set by prefabricated housing manufacturers have already paid significant dividends. On January 17, 1995, an earthquake of magnitude of 7.2 on the Richter scale battered Kobe, Japan's sixth largest city (population 1.6 million). The 20-second quake, which deprived over 5500 lives, destroyed a number of houses, buildings, bridges, port facilities and other urban infrastructure. According to the statistics provided by the City of Kobe (2003), 79,283 houses, or 15.1 %, of the 524,733 houses built in the city were completely destroyed by the Kobe quake. Furthermore, 74,234 houses, or 23.7 %, of the 313,085 houses built in six disaster-districts (Higashi nada, Nada, Chuo, Hyogo, Nagata, and Suma) were fatally damaged. Within the contexts, 58 % of the houses destroyed in the six wards were estimated to be those built in 1945 or before, 49 % between 1946 and 1965, 23 % between 1966 and 1975, and 6 % in 1976 or later. On the other hand, a number of the housing manufacturers proudly reported that their prefabricated homes withstood the Kobe quake without fatal collapse and their innovations physically protected the user lives and property (Daiwa House 1999; Misawa Homes 1998; National House 2001; Sekisui Chemical 2000).

Air-tightness is considered as an effective mean to reduce operating energy and cost for cooling and heating rooms since it reduces the ventilation heat loss of building envelopes. Indeed, most manufacturers have become increasingly aware of the benefits of energy efficiency that affects the operating cost. Misawa Homes, for instance, demonstrates that high insulation properties of modular homes are more cost-effective, producing homes whose air-leakage is 67 % less than that of conventional ones. As a result, annual heating and cooling costs can be reduced nearly by 32 % (Noguchi 2003). Ventilation is necessary to achieve a healthy living environment. When the house lacks ventilation, uncomfortable indoor air conditions, caused by humidity, condensation and the accumulation of gaseous toxic substances (i.e. volatile organic compounds) may cause illness. Toyota Home is one of the pioneers, who developed artificial ventilation systems in housing, introducing light cleaning ventilating fans. This fan consists of ultraviolet catalytic filters that decompose gaseous toxic substances including nitrogen oxide and formaldehyde. Noise is another concern that may reduce the levels of environmental comfort in housing. Using soundproofing techniques to alleviate sound associated with air vibrations and floor impacts, most housing manufacturers sets the higher standard to reduce the noise transfer inside and outside their houses than the Basic Environment Law regulated by the Ministry of the Environment—i.e. the maximum acceptable level of noise that can enter a house is 55 dB during the daytime and 45 dB at night in the residential area (JPA 2015).

In principle, their quality-oriented production aims to drastically improve housing quality in response to the wants and needs of individual homebuyers, as well as society. Today, the manufacturers are practicing unique communication approaches that aim to convey the value of high cost-performance housing to their potential clients. The following section describes the housing manufacturers' user-oriented communication approaches being applied for marketing prefabricated homes, which are usually more expensive than their site-built counterparts.

12.5 User-Oriented Communication

In 2014, the Japan Prefabricated Construction Suppliers and Manufacturers Association distributed questionnaires to 1000 prefabricated housing owners with the aim to identify the buyers' reasons for selecting prefabs and housing manufacturers in Japan (JPA 2015). Based on the results from 523 responses, the association unveiled the tendency of consumer preferences (Table 12.1).

The results indicate that the reliability of the company was the utmost reason for the selection that potential buyers concern. In fact, 64 % of the purchasers replied to this survey selected this buying factor. The second most significant factor that affected the buyers' preference was the product quality and performance. Ranking next to them was the sales staff's explanation. These results somewhat suggest that homebuyers in Japan tend to consider housing quality as the top priority and the consumers venture to purchase an expensive prefabricated house, if convinced of the superiority.

12.5.1 Cost-Performance Marketing Strategy

Japanese housing manufacturers, indeed, produce quality products and offer extensive customer services. Despite raised costs associated with the implementation of

Table 12.1 Buyer reasons for selecting their prefab company in Japan (After JPA 2015)

Selection factors given (523 respondents)	Year 2014 (%)
Reliability of the company	64
Superior product quality and performance	56
Convinced by the sales persons' explanation	54
Good proposal that reflects expectations	23
Good post-purchase services	15
Good external appearance and design	14
Recommendation from acquaintance	12
Enabled to confirm the actual products	10

quality-oriented production, the manufacturers have still retained a positive reputation for their innovations through their effective communication. According to a housing survey conducted in 1997 by the Government Housing Loan Corporation in Japan, the average initial cost of a conventional home was estimated at 175,404 JPY per square meter and a prefabricated home was at 190,033 JPY per square meter (Noguchi 2003). These results indicate that the price of industrialised housing in Japan is about 8 % more expensive than that of conventional site-built housing. On the other hand, the manufacturers emphasise that they have been producing better quality homes for about the same price as conventional ones.

In short, the housing manufacturers' way to commercialise expensive innovative products reflects a "cost-performance" marketing strategy, which is applied widely by other industries as well—e.g. the automobile industry. Although today's automobiles can be produced with lower production costs than those in the past, their selling price does not seem to be affected dramatically by higher productivity. Generally, consumers may still regard new automobile models as expensive. The list of items now offered as standard features in new cars, such as air conditioning, a stereo set, airbags, remote-control keys, power steering, power windows and adjustable mirrors, were offered only as expensive options in older models. Clearly, the quality of newer models is much higher than that of older models. The same is true for the housing industry in Japan. The quality-oriented production contributes towards the delivery of high cost-performance housing in which high-tech modern conveniences that are installed as optional features in conventional homes are now available as standard equipment.

Consumer engagement may accord with buyer motivation and the level of client involvement may be increased by such factors as the cost, consumers' interest, perceived risk in buying a product or service and the social visibility. Prior to entering into a contract with the client, Japanese housing manufacturers offer first-rate customer services and extensive information to potential homebuyers so as to allow them to learn the distinguishing characteristics of high cost-performance housing through their effective communication made particularly during the marketing process.

12.5.2 Advertisement

In Japan, the manufacturers tend to invest heavily in advertising, even though the costs are extremely high. For instance, a single page advertisement in a monthly housing catalogue in a standard format costs approximately $7800 (U.S. $). It is not unusual for a housing manufacturer to spend more than 3 % of its annual income on advertising. The purpose of advertising is obviously to draw the consumer's attention to products. In general, the earliest stage of client participation in purchasing an industrialised home is to enhance the consumer's motivation to gain knowledge of the products and services available from advertisements. Japanese housing manufacturers adopt various modes of advertisement to enhance the consumers' trust in their products. These include local media and newspaper advertising, inserts in monthly or quarterly housing pamphlets, mail solicitation sent to rental dwellers, and presentations at large companies or to social gatherings. Referrals from previous customers are another important source for new clients.

12.5.3 Brand Name Effect

In principle, advertising attempts to keep a brand name in the consumer's mind. Whether or not the advertisement can provide a convincing demonstration or an attractive display, repeating the brand name to the consumer places the brand at an advantage over an unknown brand, when the time comes for consumers to make their buying decision. Customers may be attentive to brand names. Most manufacturers in Japan intend to make the company well known, enhancing their brand awareness mainly through TV commercials, which reach a wide audience. The company's brand name can be rapidly spread through such mass media; instantly, a favourable image of the company, itself, and its products can be achieved. Consumers believe that a company that invests heavily in advertising has the ability to produce superior products because poor products destroy a reputation of the company. Seemingly, consumers tend to select brand name products, even when these products are somewhat more expensive. In principle, *reliability* is one of the key elements that influence consumers' motivation to select a specific company and the product they choose to buy and Japan's prefabricated housing market has no exception (JPA 2015).

Most large-scale housing manufacturers exploit print- and electronic-media to advertise the company's name and products: such advertisements include pictures with enthusiastic descriptions of their strengths. Furthermore, the manufacturers also use another unique method of advertising: housing exhibition. The exhibition is carried out in the form of 'housing parks' and 'housing information centres' where full-sized model houses and domestic components are exhibited to the general public. Potential homebuyers who are interested in the products can easily gain access to such exhibition venues in order to familiarise themselves with the companies and their products in question. Such a visit gives industrialised housing manufacturers an opportunity to establish personal contact with the potential homebuyers. Through the housing exhibition, the prefabricators may be able to gain credibility by allowing potential consumers to examine actual, full-sized products.

12.5.4 Housing Display Parks

Housing parks play an important role in local and regional advertising, generating remarkable sales of their products (Table 12.2). Housing park is a unique way of advertising and marketing homes and it offers a collection of display homes built by

Table 12.2 Sales of prefabricated homes per model home displayed in 1998 (Noguchi and Friedman 2002)

Company name	Number of operational show homes (units)	Annual prefabricated home sales per show house (units)
Sekisui Chemical Co., Ltd.	711	25
National House Industrial Co., Ltd.	380	26
Resco House Co. Ltd.	17	23

a variety of housing companies. In general, the exhibition parks are located in commercial centres readily accessible by making use of public transport. In the park, each company build and exhibit their housing models for a certain period of time. The scale of these housing parks varies in size, depending on the location; however, 20 to 40 housing companies commonly locate their model homes in a park so as to allow potential homebuyers to become familiar with their products. Housing parks are maintained by the companies, who will rent a vacant site and operate the show homes. Housing companies pay operation fees including land and management fees to the companies. Such a fee was estimated at approximately $12,000 (US $) per month in the case of a housing park in Sendai, in which Asahi Broadcasting Company, the largest of such operators, had maintained (Noguchi and Friedman 2002).

Housing parks not only display model homes and enable people to examine the actual products, but also serve to educate potential homebuyers concerning the superior qualities of their products relative to others. In the model home, the salesman will not only explain the distinguishing characteristics of the company product, but also address the client's own requirements with rough sketches of the housing layout. Sales staff are the company's greatest assets in reaching the clients, and they also assist the clients in designing their customisable homes by making use of housing catalogues that contain various options and arrangements of standard housing components.

For instance, Sekisui Chemical, which is also known as Sekisui Heim, is one of the largest modular housing manufacturers in Japan. The company marketed 17,990 detached housing units in 1998 while possessing 711 model homes that are exhibited throughout the nation. In consideration of these figures, each model house statistically contributed to marketing 25 housing units per annum. National House Industrial (known as 'PanaHome' today) is another major prefabricated housing company in Japan and sold 9720 detached housing units from April 1998 to March 1999. The company displayed 380 model homes throughout the country. Accordingly, the sales ratio could be estimated at 27 units per model home. Resco House, one of seven industrialized housing manufacturers that only produce concrete panelised homes in Japan, displays 17 model houses nation-wide, and sold 387 detached houses in 1998. In spite of its relatively small size, Resco House had a sales ratio of 23 units per model house that year. According to these results, housing parks play a significant role in advertising products, and model houses generally

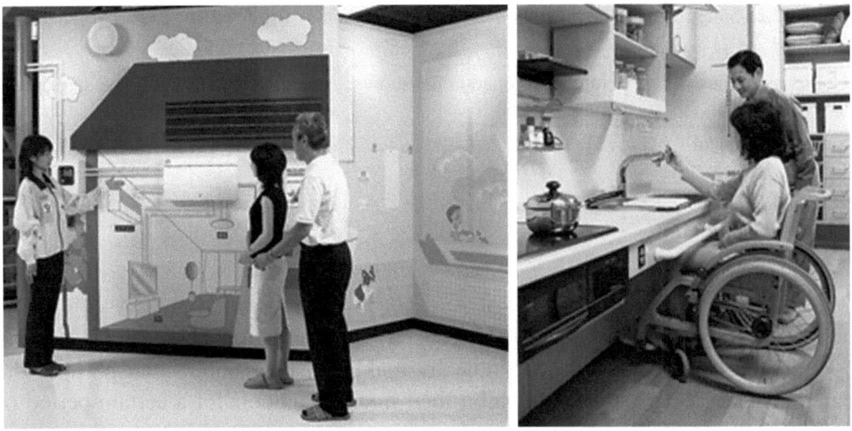

Fig. 12.4 Housing information centre for design experience (Sekisui House Ltd.)

translate into home purchases with a rate of roughly 23–27 units per model home in a year.

12.5.5 Housing Information Centres

Many manufacturers set up housing information centres in various parts of the country, offering technological information related to the housing materials, construction methods, renewable energy technologies, and amenities that people can try out (Fig. 12.4). The centre also functions as an exhibition and design consultation base, where experienced staff make concrete proposals concerning the external appearance and floor plans of their standardised, yet customisable home using advanced information and communication technologies (Fig. 12.5).

Fig. 12.5 Custom design demonstration at a housing information centre (Sekisui House Ltd.)

Moreover, clients can see and touch the material samples to confirm the superior qualities of the company's products and learn more about the company's proposals related to housing facilities. In consequence, such visual information, visits, and individual design consultations with housing experts may increase the clients' faith in the reliability of the company and its products, and thus increases the likelihood that the client will select the company, when purchasing a home.

12.6 Value Assurance

Price is associated with information-seeking, influencing a consumer's purchase decision. Nevertheless, the reluctance to purchase a product with a high initial cost can often be overcome, if the consumer's interest in the product is strong enough, if the degree of perceived risk in purchasing the product is low, and when the selling price is reasonable in comparison to the market value. The market survey of prefabricated housing conducted by JPA in 2008 indicates that 2 % of the homebuyers, who preferred to purchase a prefabricated home, considered the selling price as relatively low (Table 12.3).

However, the response appears to contradict the fact that the initial cost of industrialised housing is about 8 % more expensive than that of conventionally housing built based on the minimum quality standard. These homebuyers might be well convinced of both the superior quality and economic value of prefabricated homes through the manufacturers' systematic user-oriented communication.

Today, most housing manufacturers in Japan provide a twenty-year warranty and reassure their clients with free post-purchase maintenance and inspection services. Moreover, the companies remain in touch with their clients after sales by offering

Table 12.3 Buyers' motivation to purchase industrialised housing in Japan (Noguchi and Collins 2008)

Motivation factors selected	Year 2008
Reliability of the large-scale company	22 %
Superior product quality and performance	18
Earthquake resistance	17
Convinced by the sales persons' explanation	15
Good post-purchase services	2
Good design	4
Customisation according to needs and demands	5
Reliable safety measures	0
Short construction time	2
Wide variety of equipment selections	1
Confirmed by visiting housing exhibitions	1
Consideration of environmental issues	1
Recommendation from acquaintance	3
Relatively low price	2

periodic inspections on each product and sending the company's housing magazines and newsletters to their users with further information on maintenance and potential renovation. The offering of such long-term post-purchase maintenance services to clients, basically free of charge, is effective in assuring the homebuyers of product quality and value both before and after they purchase a home. The value assurance helps anchor and elevate the customer's trust in the reliability of the company, as well as the product—i.e. industrialised housing.

12.7 Mass-Customising Prefabs

Prefabrication is rooted in standardization of parts and components, or entire units; thus, the end-user products can be manufactured in the way of mass production where material and labour costs are likely reduced through economies of high volume work. Nevertheless, in consideration of today's depressing economic climate, growing demographic diversity, and global warming issues, house-builders may generally be becoming keener on pursuing not only housing cost reduction strategies, but also the social and environmental sustainability. On the other hand, today's consumers are no longer satisfied with monotonous products, even though the products themselves are reliable enough in terms of product quality. End-user products may need to be responsive to individual requirements, desires and expectations, particularly when they are relatively costly. In order to customise standardised prefabricated homes, Japanese housing manufacturers have been practicing "mass customization" over the last decades (Davis 1987; Noguchi and Friedman 2002).

In 1987, this paradoxical concept of mass customization was formally launched by Stanley Davis, who partially reviewed Misawa Homes' business operation. Today, Japanese prefabricated house-building industries are in the forefront of mass customizing industrialised homes. An important part of mass customisation is that the user directly determines the configurations from choices given as client input during the design stage. This could hardly be achieved without the standardisation of housing components for the structural, exterior and interior arrangements. The concept of component standardisation can be illustrated with Lego® building blocks. A number of simple, modularised blocks can be connected in a variety of ways, because of their interlocking tabs and holes. Likewise, Japanese manufacturers offer a variety of housing components to their clients and then encourage them to participate in combining the components to design their new home.

Housing components can be divided into three categories: structural, exterior and interior. The structural components are used to construct the housing models that determine the number and size of each room, while the interior and exterior components serve to coordinate both the decorative and the functional elements that customise housing.

12.7.1 Catalogue of Offerings

The catalogues play a variety of roles in advertising and educating clients. Moreover, at a design stage, the catalogues also function as design communication tools used between the manufacturers and the client. Through consultation with housing experts, the clients can choose the housing type and components from the catalogues to customise a home that fits their needs and demands. Clearly, catalogues can be regarded as synthesised information sources that integrate and simplify extensive data through an inductive process.

At the marketing and design stages, Japanese housing manufacturers usually provide their clients with four types of catalogues: general catalogue, commodity style catalogue, technology catalogue and design component selection catalogue. A general housing catalogue contains overarching information about the company's profile and products, while the commodity style catalogue offers more detailed information about the design concept and technology applied to the selected housing prototypes. A technology catalogue packages information about the distinguishing features of product quality. The manufacturers effectively use these three catalogues during the marketing process unlike a housing component selection catalogue, which is designed to assist the client in selecting the various standard housing components for housing customisation. The catalogue elaborates in detail on each component in terms of material, size, colour, texture, and functions. Such catalogues do not contain the price of each component, so that the client will choose each item according to the use—not the cost. However, the cost will be estimated according to user choices with the support of a computer system. The catalogues are visually attractive and well edited in a way that homebuyers can make choice of standard housing design components for customisation.

12.7.2 Digital Communication

The use of a computer-aided design (CAD) system is important in the creation, modification, analysis, or optimisation of a design, allowing the consumers to customize their choice in housing. Most housing manufacturers use a CAD system as a communication tool to offer flexibility in design. The benefits derived from applying a CAD system in the design stage include the short time elapsed between the receipt of a consumer order and the delivery of the proposal, the accuracy of material and cost estimates, and the standardisation of drafting and documentation. Ease of visualisation of a drawing boosts clients' comprehension of the layout of their new home. The interactive CAD system basically contributes to the creation of line drawings; however, advanced geometric modelling in three dimensions with shade and colour helps to display more information on the graphic screen. Another advantage of a CAD system is data communication. The purpose of data communication is to transfer data between a computer and its peripherals that can be defined as any input or output device. Some housing manufacturers send sales staff directly to their client's house and using laptop computers, proposed housing images are presented to homebuyers for their design review and further modification. In consequence, clients can mass-customise their house at home comfortably.

12.8 ZEMCH Manufacturers in Japan

Under the Kyoto Protocol, Japan was assigned the target of reducing its emissions of CO_2 by 6 % from the 1990 level between 2008 and 2012. Today, a growing number of PV systems are being installed for residential use in Japan (Noguchi 2011). The renewable energy technology helps ease the environmental impact as it produces energy for the operation of a house. In Japan, PV systems rated at 3 kW or less used to be common in the past (Ikki and Matsubara 2007). Today, the high-capacity systems at power levels of 5 kW are beginning to be installed in housing. The leading housing manufacturers have created a new market for residential PV systems, developing all-electric houses whose energy can be fully or partially supplied by the micro-power generation systems. In response to meet the societal demand for environmental sustainability in housing, Japanese manufacturers today commercialise low energy mass custom homes, which secure the operating energy cost and carbon dioxide emissions to be low or zero with the support of the installation of renewable energy technologies.

12.8.1 Sekisui Chemical Co. Ltd

Sekisui Chemical Co. Ltd. is also well known as 'Sekisui Heim' that derives from the name of Japan's first modular house that they built in 1971. Sekisui Chemical is aiming to develop business that focuses on the production of high performance and value-added housing products where their modular construction system technologies are brought into effect. In general, modular housing systems ensure a high level of in-factory completeness of housing components. The company claims that nearly 80 % of housing components used for building a house can be assembled at their factory. Today, Sekisui Heim possesses 8 factories nationwide, which have the capacity for producing wood- and/or steel-frame modular homes through their highly mechanised assembly lines. In April 2004, Sekisui Chemical launched a new net zero-energy-cost home, called "Parfait AE," and the housing model further expands the company's zero utility expense specifications. Added-features include thermal barrier-free systems applied for the foundation and floors, a passive ventilation heat-blocking system for the high heat dissipation skylight and heat-blocking screens that control the amount of sunlight. Moreover, in collaboration with Sumitomo Trust and Banking, Sekisui Chemical are jointly developed a new housing loan which helps homebuyers to purchase a house equipped with

Fig. 12.6 Modular housing rooftop PV Arrays (Sekisui Chemical Co. Ltd.)

high-capacity PV systems (Fig. 12.6). The company explains that "the higher the power generating capacity of photovoltaic generator, the lower the loan's interest rate" (Sekisui Chemical 2004). The rate was estimated at as low as 2.8 %, comparing favourably with long-term fixed-rate bank loans. The industry partnership with the bank might enhance the economic appeal of their PV solar homes. Given the cumulative sales of over 85,000 solar homes to date, the company is recognised widely as the largest modular manufacturer of net zero utility cost housing in Japan.

12.8.2 Misawa Homes Co. Ltd

Misawa Homes Co. Ltd. was founded in 1967, headquartered in Tokyo, Japan. Misawa Homes, together with its subsidiaries, engages in the development, construction, and sale of prefabricated mass custom homes and housing materials. Their house is also constructed based on 3-dimensional modules consisting of rigid steel frames. Yet, their homes are characterised by the use of large-sized non-structural 'precast autoclaved lightweight concrete' (PALC) panels that the company produces (Fig. 12.7). Misawa Homes started business operations upon acquisition of the housing industry's first certification from the then Ministry of Construction in 1962. The company spread a 'Home Core' design concept by building the demonstration homes in Abu Dhabi, UAE in 1968. Afterwards, Misawa Homes put 1-million JPY Home Core houses on the local market in 1969 and the concept was well exhibited at Expo '70 in Osaka. In 1980, the company commenced with research on the development of its own zero energy housing prototype and in 1998, it successfully developed Japan's first zero energy home.

Fig. 12.7 Precast autoclaved lightweight concrete panels (Misawa Homes Co. Ltd.)

12.8.3 PanaHome Corp

Since 1963, PanaHome, formerly called 'National House Industrial Co. Ltd.', has been operating as a housing company within Panasonic Group. Yet, the group's housing business itself actually dates back to the introduction of its first house prototype called 'Matsushita Type-One Housing Units' in 1961. Then, the Koto Plant (present Head Plant) also opened in Shiga Prefecture (National House Industrial 2001). Today, PanaHome produces steel-frame panels applied to their industrialised mass custom homes through highly automated production lines. In line with their state-of-the-art automation system, the company has also been at the forefront of solar community developments (Fig. 12.8).

The PanaHome-city Seishin Minami prefabricated housing estate development in Hyogo Prefecture dates back to 2004, comprising of over 100 PV mass custom homes. In 2009, the development was selected as one of 100 excellent regional projects that introduced new types of energy hosted by the Ministry of Economy, Trade and Industry, and the New Energy and Industrial Technology Development Organization. Afterwards, their proposal was also adopted in a scheme, which was led by the Ministry of Land, Infrastructure, Transport and Tourism, and entitled 'Initiative Model Project for Developing Long-Life Houses.'

12.8.4 SANYO Homes Corp

The forerunner of SANYO Homes Corp. is 'Kubota House Co. Ltd.' which was established in 1969 and won the first place for 'the 1st Pilothouse Technique Invention Competition' held in 1972 by the then Ministry of Construction in Japan.

Fig. 12.8 PanaHome City Seishin Minami housing estate in Hyogo Prefecture (PanaHome Corp.)

In 2002, the company was renamed to 'SANYO Homes' as it became owned by SANYO Electric Co. Ltd. Today, SANYO Homes emphasises that housing should be gentle to people and the earth, stemming from the company's business concept of an 'Eco & Safety' life that aims to enhance their housing performance as to local power generation, energy-efficiency, and earthquake resistance and emergency measures. In 2004, SANYO Homes received a 'Good Design' award with their housing prototype called 'Logia-Type E' which had a simple cubical exterior appearance designed based on a minimalism concept for space use efficiency. The company is thriving in their mass custom homes' architectural integration with SANYO HIT solar cells that have the world's highest energy conversion efficiency of 23 % (Fig. 12.9). The company also introduces solar water heating panels to their homes equipped with an air-source heat pump, which aims to control and secure the supply of domestic hot water according to weather conditions.

12.8.5 Sekisui House Ltd

In 1960, the company incorporated with the capital stock of 100 million JPY, stemming from the housing division of Sekisui Chemical Co. Ltd. Today, Sekisui House is recognised as the largest housing manufacturer in Japan given the cumulative sales of 2,001,722 homes estimated as of 31st January 2010. The company is committed to taking the initiative for the development of zero carbon emission housing based on the company's 'Action Plan 20' that aimed to introduce

Fig. 12.9 Building integrated PV (Sanyo Homes Corp.)

a variety of energy-efficient housing measures (Ishida 2008). In conjunction with the G8 Hokkaido Toyako Summit held from 7th to 9th July, 2008, Sekisui House announced their so-called 'Zero Emission House.' The house was a single-storey steel structure prefabricated house with a total floor area of approximately 200 m^2, featured particularly by a 14.5 kWp capacity photovoltaic power generation system, energy efficient lighting, a household fuel cell, and energy saving domestic appliances. Moreover, in order to assist clients in understanding the value of their products, Sekisui House opens the home amenities experience studio, called 'Nattoku Kobo,' to the general public. The studio provides a variety of information on housing functions and performances such as lifetime barrier free design, home security, earthquake and typhoon resistance, energy and space use efficiency, and local power generation. Today, the company markets a housing prototype called "Green First Hybrid" which is equipped with a building integrated roof tile PV panels and a fuel cell in addition to a lithium-ion battery, whose energy storage capacity ranges from 4.6 to 9.3 kWh. The house is also designed to import electricity from an electric car parked at the house, when needed—e.g. blackout. The company claims that the daily electricity use of a house can be covered by the electricity imported from an electric car equipped with a battery which has energy storage of 24 kWh.

In general, Japanese housing manufacturers emphasise distinguishing features of their high-cost and high-performance (i.e. high cost-performance) houses. As a result, a variety of renewable energy technologies including a PV rooftop system, a CO_2 refrigerant heat-pump water heater and a combined heat and power system today are introduced to the homes, installed as standard equipment rather than options. Moreover, to realise net zero-energy-cost housing operation, a house needs to be connected to a commercial grid where electricity can be exported and imported between the energy user and the supplier under a net metering programme. In addition, the power companies in Japan also offer to their users a variable time zone contract that affects the price of electricity based on the time of transaction. Accordingly, the buying price of electricity often becomes higher than the buying price through the time-related feed-in-tariff arrangement; for instance, the users are allowed to sell electricity generated by PV systems back to the power supplier at a higher price during the daytime and buy the power at a lower price during the night time. In consequence, the net zero-utility-cost housing drastically reduces the energy consumption and CO_2 emissions when it comes into operation (Richard and Noguchi 2006).

12.9 Conclusions

Japanese housing manufacturers are successful in mass-customising low to zero-energy homes today, whose energy cost and CO_2 emissions can be low or zero with the support of the net metering, feed-in-tariff and time zone contracts that the local power companies offer to their users. To initiate and maintain the moderate sales of zero energy/emission mass custom homes, the manufacturers implement a high cost-performance marketing strategy that aims to install some expensive amenities including renewable energy technologies today in their homes as standard features rather than optional ones. Based on this value-focused marketing strategy, Japanese housing manufacturers do not merely reduce their selling price, yet the product quality is continuously improved towards the constant delivery of sustainable mass custom homes that satisfy the wants and needs of individuals as well as society.

References

Chemical Sekisui (2000) Unit technology for conformable human life of future. Sekisui Chemical Co., Ltd., Tokyo

Chemical Sekisui (2004) Annual report 2004. Sekisui Chemical Co., Ltd., Tokyo

Daiwa House Industry (1999) Technology catalog. Daiwa Housing Industry Co., Ltd., Osaka

Daiwa House Industry (2015) Corporate History. http://www.daiwahouse.co.jp/English/history. Accessed 20 Nov 2015

Davis SM (1987) Future perfect. Addison-Wesley, New York

Homes Misawa (1998) Home club 8. Misawa Homes Co., Ltd., Tokyo

Ikki O, Matsubara K (2007) National survey report of PV power applications in Japan 2006. International Energy Agency, Paris

Ishida K (2008) The global warming prevention strategy for housing in japan. Open House Int 33 (3):38–47

Japan Prefabricated Construction and Suppliers and Manufacturers Association (JPA) (2015) News. http://www.kyomu.purekyo.or.jp/sites/default/files/2014sumaidukuriannketo%20news%20release.pdf. Accessed 12 Feb 2015

Matsumura S, Tanabe S (1996) Kinmirai jyutaku no gijyutsu ga wakaru hon. PHP Kenkyujyo, Tokyo

Mishima S (2000) Jyutaku: Nihon no kaisya. Jitsumu kyouiku, Tokyo

National House Industrial Co. Ltd. (2001) Sumai to kurashi. No.377 National House Industrial Co. Ltd., Osaka

Noguchi M (2003) The effect of the quality-oriented production approach on the delivery of prefabricated homes in Japan. J Housing Built Environ 18(4):353–364
Noguchi M (2008) Structuring a choice model for mass customisation. Int J Mass Customisation 2 (3/4):264–281
Noguchi M (2011) Enhancement of industry initiative through the zero-energy mass custom home mission to Japan experience towards commercialisation. Int J Mass Customisation 4 (1/2):106–121
Noguchi M, Collins D (2008) Commercialisation strategies for net zero-energy-cost housing in Japan. Open House Int 33(3):96–104
Noguchi M, Friedman A (2002) Manufacturer-user communication in industrialised housing in Japan. Open House Int 27(2):21–29
Richard RB, Noguchi M (2006) The Japan solar photovoltaic manufactured housing technical mission 2006. Nat Res Can, Varennes
Sackette JG (1986) Japan's manufactured housing capacity: a review of the industry and assessment of future impact on the U.S. market. Energy Design Resources, Saint Luis
Utsu T (1995) Daitoshi chokka no katsudansou ga ugoita. Asahi Graph 3794, Tokyo, pp 67–71